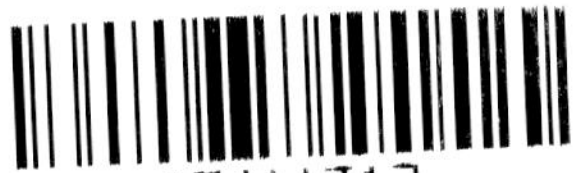
AF411317

Symmetry in Human Evolution, from Biology to Behaviours

Symmetry in Human Evolution, from Biology to Behaviours

Editor

Antoine Balzeau

MDPI • Basel • Beijing • Wuhan • Barcelona • Belgrade • Manchester • Tokyo • Cluj • Tianjin

Editor
Antoine Balzeau
UMR 7194 CNRS and MNHN
Musée de l'Homme
17 place du Trocadéro
75016 Paris, France

Editorial Office
MDPI
St. Alban-Anlage 66
4052 Basel, Switzerland

This is a reprint of articles from the Special Issue published online in the open access journal *Symmetry* (ISSN 2073-8994) (available at: https://www.mdpi.com/journal/symmetry/special_issues/Symmetry_Human_Evolution_Biology_Behaviours).

For citation purposes, cite each article independently as indicated on the article page online and as indicated below:

LastName, A.A.; LastName, B.B.; LastName, C.C. Article Title. *Journal Name* **Year**, *Volume Number*, Page Range.

ISBN 978-3-0365-5593-5 (Hbk)
ISBN 978-3-0365-5594-2 (PDF)

Contents

Editorial

Special Issue "Symmetry in Human Evolution, from Biology to Behaviours"

Antoine Balzeau [1,2]

[1] PaleoFED Team, UMR 7194, CNRS, Département Homme et Environnement, Muséum National d'Histoire Naturelle, Musée de l'Homme, 17, Place du Trocadéro, 75016 Paris, France; abalzeau@mnhn.fr

[2] Department of African Zoology, Royal Museum for Central Africa, 3080 Tervuren, Belgium

Citation: Balzeau, A. Special Issue "Symmetry in Human Evolution, from Biology to Behaviours". *Symmetry* 2022, 14, 1808. https://doi.org/10.3390/sym14091808

Received: 24 August 2022
Accepted: 26 August 2022
Published: 1 September 2022

Publisher's Note: MDPI stays neutral with regard to jurisdictional claims in published maps and institutional affiliations.

Our knowledge of human evolution has made particular progress over the last twenty years, thanks to the discovery of new fossils and the use of new methods and multidisciplinary approaches. This consideration may seem true for all periods since the beginnings of prehistoric sciences. Nevertheless, the accumulation of data and knowledge seems particularly notable lately. At present, this allows us to approach, with more ambition, complex subjects on the margins of research in paleoanthropology. In particular, studies of departure from symmetry, including variations in fluctuating or directional asymmetries, have contributed to the expansion of this knowledge in various fields of paleobiology and archaeology. This Special Issue brings together articles dealing with symmetry and human evolution. Eight papers are original studies, proposing either new tools to investigate bilateral variation or new results on the brain, skull or skeletons during human evolution. Finally, two papers are reviews of the state of the art of our knowledge of limb preferences in the animal kingdom, and of the potential (and future) interactions between paleoanthropology and the field of neurosciences.

Lin et al. [1] proposed an original approach while quantifying and visualizing the variation in endocast asymmetry in modern humans using diffeomorphic surface matching. This type of development is important in order to break free from the limits related to techniques that require 2D or 3D landmarks, but also because it is thus particularly suited to the particularities of work on the brain endocast. Their results are congruent with well-documented classical anatomical asymmetry of the human brain/endocast, proving the validity of such a new methodology.

Hurst et al. [2] addressed bilateral variation in dimensions of the occipital lobes in chimpanzees. This anatomical area has been the subject of much discussion about the supposed characteristics of different species during human evolution. This work on 83 specimens brings us information on the anatomical variation in the shape and proportions of the occipital lobes in our closest living relatives.

Zhang and Wu [3] investigated bilateral variation in morphometric data of the cerebellar lobes on virtual endocranial casts of a large sample of fossil hominin species, including *H. neanderthalensis* and *H. erectus*, and a comparison with *H. sapiens*. This anatomical area is known to have a specific morphology in our species compared to other hominins. It is interesting here to see information for a unique fossil sample and to observe differences in terms of shape and in bilateral variations among hominins.

Buzi et al. [4] performed a virtual reconstruction of an important fossil specimen, the Stenheim skull. This fossil is highly distorted, making the description of its anatomical traits difficult. The obtained retrodeformed model of Steinheim will be of interest for future studies of the craniofacial variation among Mid-Pleistocene hominins.

Melchionna et al. [5] proposed a new R tool that allows for automatic numerical quantification of fluctuating and directional asymmetry. Moreover, it produces a chart of the quantified bilateral variation directly on the analyzed 3D model. This graphical production gives immediate visual information on the intensity, topology and direction of departures from symmetry, being a useful tool for description and illustration.

Profico et al. [6] proposed a new tool added to an R package to assess the lateralization of the distribution of cortical bone along the entire diaphysis of long bones.

Zhao et al. [7] were interested in a regularly studied subject, the bilateral variations in the biomechanical properties of the humeri. However, they proposed a novel approach. Instead of conducting qualitative comparisons at a limited number of locations along the diaphysis, they included a comparison of biomechanical asymmetries quantified by morphometric mapping all along the length of the bone. This approach is informative and promising as it details more complex and subtle variation in the analyzed parameters in the different parts of the humeral diaphysis.

Bardo et al. [8] aimed to investigate the link between the form and the function of the human hand and also deal with the question of human laterality and dexterity. To do so, they measured grip strength in a very large sample of volunteers and tested the potential effects of age, sex, asymmetry (hand dominance and handedness), hand shape, occupation and practice of sports and musical instruments that involve the hand(s). This original study gives original detailed information and is particularly interesting from the perspective of trying to decipher the evolution of human behaviors and capacities.

Boulinguez-Ambroise et al. [9] detailed the state of the art of the knowledge on the limb preferences in animals in order to contextualize how recent research has revolutionized our perception of the specificities of manual laterality in hominids, and how we now study those aspects. New methods, such as functional neuroimaging, but also original developmental approaches are crucial today to propose a new vision of the mechanisms underlying human handedness.

Finally, Balzeau and Mangin [10] discussed the recent developments of their respective fields of research, namely, paleoanthropology and neurosciences. In the future, the contribution of neuroimaging will allow us to better define the relationship between the brain and its reflection on the internal cranial bone surface, the endocast, which is the only material available for fossil hominins, to approach the evolution of the human brain. Moreover, documenting the anatomy among past human species and including the variation over time within our own species are approaches that offer us a new perspective through which to appreciate what really characterizes the brain of humanity today.

Funding: A.B. received funding from the Agence National de la Recherche (ANR-20-CE27-0009).

Conflicts of Interest: The authors declare no conflict of interest.

References

1. Lin, S.; Zhao, Y.; Xing, S. Asymmetry of Endocast Surface Shape in Modern Humans Based on Diffeomorphic Surface Matching. *Symmetry* **2022**, *14*, 1459. [CrossRef]
2. Hurst, S.; Holloway, R.; Pearson, A.; Bocko, G. The Significance of Chimpanzee Occipital Asymmetry to Hominin Evolution. *Symmetry* **2021**, *13*, 1862. [CrossRef]
3. Zhang, Y.; Wu, X. Asymmetries of Cerebellar Lobe in the Genus *Homo*. *Symmetry* **2021**, *13*, 988. [CrossRef]
4. Zhao, Y.; Zhou, M.; Li, H.; He, J.; Wei, P.; Xing, S. Biomechanical Evaluation on the Bilateral Asymmetry of Complete Humeral Diaphysis in Chinese Archaeological Populations. *Symmetry* **2021**, *13*, 1843. [CrossRef]
5. Profico, A.; Zeppilli, C.; Micarelli, I.; Mondanaro, A.; Raia, P.; Marchi, D.; Manzi, G.; O'Higgins, P. Morphometric Maps of Bilateral Asymmetry in the Human Humerus: An Implementation in the R Package Morphomap. *Symmetry* **2021**, *13*, 1711. [CrossRef]
6. Melchionna, M.; Profico, A.; Buzi, C.; Castiglione, S.; Mondanaro, A.; Del Bove, A.; Sansalone, G.; Piras, P.; Raia, P. A New Integrated Tool to Calculate and Map Bilateral Asymmetry on Three-Dimensional Digital Models. *Symmetry* **2021**, *13*, 1644. [CrossRef]
7. Buzi, C.; Profico, A.; Di Vincenzo, F.; Harvati, K.; Melchionna, M.; Raia, P.; Manzi, G. Retrodeformation of the Steinheim Cranium: Insights into the Evolution of Neanderthals. *Symmetry* **2021**, *13*, 1611. [CrossRef]
8. Bardo, A.; Kivell, T.L.; Town, K.; Donati, G.; Ballieux, H.; Stamate, C.; Edginton, T.; Forrester, G.S. Get a Grip: Variation in Human Hand Grip Strength and Implications for Human Evolution. *Symmetry* **2021**, *13*, 1142. [CrossRef]
9. Boulinguez-Ambroise, G.; Aychet, J.; Pouydebat, E. Limb Preference in Animals: New Insights into the Evolution of Manual Laterality in Hominids. *Symmetry* **2022**, *14*, 96. [CrossRef]
10. Balzeau, A.; Mangin, J.-F. What Are the Synergies between Paleoanthropology and Brain Imaging? *Symmetry* **2021**, *13*, 1974. [CrossRef]

 *symmetry*

Review

What Are the Synergies between Paleoanthropology and Brain Imaging?

Antoine Balzeau [1,2,*] **and Jean-François Mangin** [3]

[1] PaleoFED Team, UMR 7194, CNRS, Département Homme et Environnement, Muséum National d'Histoire Naturelle, Musée de l'Homme, 17, Place du Trocadéro, 75016 Paris, France

[2] Department of African Zoology, Royal Museum for Central Africa, 3080 Tervuren, Belgium

[3] Baobab Research Unit, CEA, CNRS, Neurospin Department, Université Paris-Saclay, 91191 Gif-sur-Yvette, France; jean-francois.mangin@cea.fr

[*] Correspondence: abalzeau@mnhn.fr

Abstract: We are interested here in the central organ of our thoughts: the brain. Advances in neuroscience have made it possible to obtain increasing information on the anatomy of this organ, at ever-higher resolutions, with different imaging techniques, on ever-larger samples. At the same time, paleoanthropology has to deal with partial reflections on the shape of the brain, on fragmentary specimens and small samples in an attempt to approach the morphology of the brain of past human species. It undeniably emerges from the perspective we propose here that paleoanthropology has much to gain from interacting more with the field of neuroimaging. Improving our understanding of the morphology of the endocast necessarily involves studying the external surface of the brain and the link it maintains with the internal surface of the skull. The contribution of neuroimaging will allow us to better define the relationship between brain and endocast. Models of intra- and inter-species variability in brain morphology inferred from large neuroimaging databases will help make the most of the rare endocasts of extinct species. We also conclude that exchanges between these two disciplines will also be beneficial to our knowledge of the *Homo sapiens* brain. Documenting the anatomy among other human species and including the variation over time within our own species are approaches that offer us a new perspective through which to appreciate what really characterizes the brain of humanity today.

Keywords: brain-endocast correspondence; paleontology; interdisciplinarity; artificial intelligence

Citation: Balzeau, A.; Mangin, J.-F. What Are the Synergies between Paleoanthropology and Brain Imaging?. *Symmetry* **2021**, *13*, 1974. https://doi.org/10.3390/sym13101974

Academic Editor: Kazuhiko Sawada

Received: 23 September 2021
Accepted: 15 October 2021
Published: 19 October 2021

Publisher's Note: MDPI stays neutral with regard to jurisdictional claims in published maps and institutional affiliations.

1. Introduction

The brain is important to us as humans beings. Its anatomy contributes to the biological definition of our species, *Homo sapiens*, but is also important to discuss evolutionary patterns along the last 7 millions years of human prehistory. It is also the center of all our thoughts, the tool we even use to study it. It has long been considered unique in its functioning and in its morphology compared to all other living beings. Technical progress and the multiplication of diverse approaches means that we are learning more about the biology and the functioning of our brain. However, an approach combining neuroimaging and paleoanthropology opens up new perspectives, as it could help us to better understand the characteristics of the *Homo sapiens* brain by integrating its variability over time. Studying related human fossil species closely will also allow us to better characterize what makes our brain unique and the evolutionary development of these specificities. This perspective, in light of our knowledge of past human behavior, will also allow us to better appreciate the mysterious functioning of our brain.

Paleoanthropology seeks to understand the evolution of the human brain by studying the shape of skull fossils [1]. For this reason, the first historical milestone of interest for this paper is phrenology, a nineteenth century endeavor to link personality traits with the morphometry of the bumps of the scalp, building upon the hypothesis that the extent of these bumps is related to the extent of underlying brain convolutions [2]. Phrenology was fiercely criticized but very influential in its time. It was, however, rapidly considered as a pseudo-science and is not difficult to invalidate with modern imaging methods. For instance, it was shown recently that scalp curvature is not related to brain gyrification [3]. However, very few such studies have been carried out on the links between cortical morphology and the internal interface of the skull, which gives rise to the endocasts of paleoanthropology [4]. This is an important topic addressed in this study. It should be noted that despite the lack of scientific methodology behind the work of phrenologists, they were among the first to hypothesize the idea of "functional specialization" or "segregation", which is central to our current understanding of the brain's organization [5].

Towards the late 19th century, functional specialization was made more concrete thanks to the advent of clinical neuropsychology, based on the observation of the consequences of brain damage. For instance, this strategy was used by Paul Broca to show that different areas of the brain are responsible for articulation and the understanding of speech [6]. Clinical neuropsychology and paleoanthropology share a weakness, however: they have to make do with the samples that nature offers them and extrapolate the rest, even if the sample distribution is not optimal. In this paper, we discuss the possibility of improving the extrapolation performed in paleoanthropology by taking into account the models established in the world of modern neuroimaging regarding the intra-species variability of brain morphology.

At the beginning of the 20th century, the spatial heterogeneity of the microscopic organization of the cortex was highlighted in 2D brain sections observed under the microscope, giving rise to several major maps partitioning the cortex according to the distribution of cell types (cytoarchitectony [7]) or the myelination of cortical layers (myeloarchitectony [8]). Despite their importance for modeling the organization of the human cortex, these mappings are currently still inaccessible in vivo. They have only been achieved in 3D for about ten postmortem brains, each with its own idiosyncrasies [9]. In this sense, this particular field of neuroimaging shares with paleoanthropology the scarcity of samples from which a representative model of the brain of a species and its variability must be inferred.

During the last forty years, functional imaging has revolutionized brain research, allowing major advances in the understanding of brain regionalization and its anatomical characterization. Moreover, it is now possible to access huge databases of *Homo sapiens* brains combining morphological and functional imaging, but also maps of large axonal bundles whose evolution is probably key to the acquisition of certain abilities [10]. The study of the relationships between the inter-individual variability of morphological features and that of fiber bundles or functional areas could probably contribute to the interpretation of the differences observed between the endocasts of ancient species. The largest current database, UKbiobank, which will soon include 100,000 brain images but also an exhaustive map of the genome for each subject [11], and the progress of paleogenomics [12], now make it possible to study the impact of genes inherited from our ancestors on our brain structures [13,14]. There are now also very large databases on brain development [15,16], which will allow studies associating ontogeny and phylogeny. Finally, there is a major interest in the neuroimaging of non-human primates, which should also create synergies between neuroimaging and paleoanthropology [17–19].

Paleoanthropology and the Evolution of the Brain

In prehistoric sciences, the archaeological and paleontological record is scrutinized to explore directly several facets of past human populations. The available biological information obtained on fossil specimens is crucial to explore human variation and evolution but also to try to trace some relationships with past behaviors. Indeed, the anatomy of humans may provide some clues about this last aspect, though it is difficult to interpret [20–23]. The question of the available evidence related to brain anatomy for ancient humans is necessarily the first restriction and a crucial challenge for such studies. The debate about the potential interpretation of anatomical traits in terms of past functions is also important. In this context, the rich anatomo-functional correlations observed with modern neuroimaging can be inspiring.

There is a huge body of research, spanning over a century, about the anatomical asymmetries of the extant human brain and those traits are still widely studied for their functional, physiological and behavioral implications [24]. However, the comparison with fossil hominins is complex for various reasons. Moreover, the question of the date of appearance of particular anatomical traits, including brain asymmetries, in the hominid lineage is still widely debated [25–29]. Among the aspects considered at this interface, the combination of right frontal/left occipital protrusions, usually associated with the 'torque' pattern, has been studied on brain endocasts (the imprints left by the brain on the internal surface of the skull), from both recent humans and fossil hominins. The larger anterior/frontal and posterior/occipital projection (petalia) is coupled with another component, a larger lateral extension of the more projected hemisphere (lobar asymmetries). Globally, the most common pattern in humans is the combination of right frontal/left occipital protrusions, which is also associated with the well-known Yaklovian "torque" pattern of the human brain. Several other aspects of hominin brain evolution have been also investigated, such as the shape of the third frontal convolution, the development of the parietal lobes in fossil *H. sapiens*, or particular areas with supposed functional implications. The field of paleoneurology is now very active and more and more actors are concerned. Nevertheless, an important constraint on these approaches is that the link between the structure of the brain and the information available on the endocast is not yet fully understood, whereas the possible peculiarities of the different human species must be addressed by this proxy.

In addition to this pronounced interest for the brain anatomy of our predecessors, there has been a new focus on our own particularities. This is why the study of the observed specific anatomical traits and structural asymmetries of the brains of living humans is of major importance as they are considered as an anatomical substrate of functional asymmetries in *H. sapiens*. Indeed, a new field of research is emerging in which these data are considered in comparison with those of great apes and fossil hominins, to understand the structural basis of modern human cognition and to investigate potential interpretations of the brain anatomy of fossil hominins.

In this paper, we contextualize the most recent improvements in neuroanatomy in the context of past studies of the human brain and of the brain endocast of our predecessors. In addition to detailing the current knowledge in "paleoneurology", we explore how up-to-date methodologies from different fields may help in the future to explore in more details the anatomy of the brain of other human species and to improve our deductions about their past behaviors.

2. A Synthesis on Past and Living Brains

Evolving Methodologies in the Study of Human Brain Morphology

The rise of computational neuroanatomy over the past 30 years has had a tremendous impact on the study of brain morphology. Previous methods were often cumbersome to implement, due to the manual delineation of structures they involved, not very reproducible, and often biased, due to a two-dimensional approach to quantification. For example, a gyrification index calculated in 2D was biased by the orientation of the slices used or by the large thickness of these slices at the early stages of MRI. Furthermore, as in paleoanthropology, each study led to the design of a specific ad hoc methodology, leading to huge difficulties when trying to synthesize research results, as can be observed, for instance, in the study of the asymmetry of the planum temporale [30].

The substantial requirements of neuroimaging research have led to the design of robust and automatic methodologies for brain morphology analysis. In spite of an abundance of proposed methodologies, Darwinian-style pressure has selected a small number of software packages (SPM, Freesurfer, FSL) that are sufficiently simple to be used by more than a thousand research teams using MRI in one way or another. This de facto standardization of the analysis of brain anatomy has largely contributed to the success of the field and is linked to the emergence of a paradigm a la Kuhn that is difficult to escape without loss of credibility. The software is based on a powerful idea: "let's align brains with a template brain before comparing them".

Voxel-Based-Morphometry (SPM, FSL), born in the 1990s, encompasses methods that practice this alignment in 3D ("non-linear warping") [31]. They include approaches that work point-by-point but also ROI-by-ROI, with the ROI also being defined in the template space. VBM is a versatile technique that can be used for the cortex and for subcortical structures. The feature to be compared across subjects is a kind of grey or white matter density supposed to be a proxy for local tissue volume. A specific branch is dedicated to asymmetry studies, which usually involve the use of a specific symmetric template. The tools used in this area have generated much discussion [32,33]. The main issue lies in the fact that there is no clear ideal alignment across brains with varying morphologies (Figure 1), particularly with respect to the cortical folding that is supposed to be partially printed in endocasts [34].

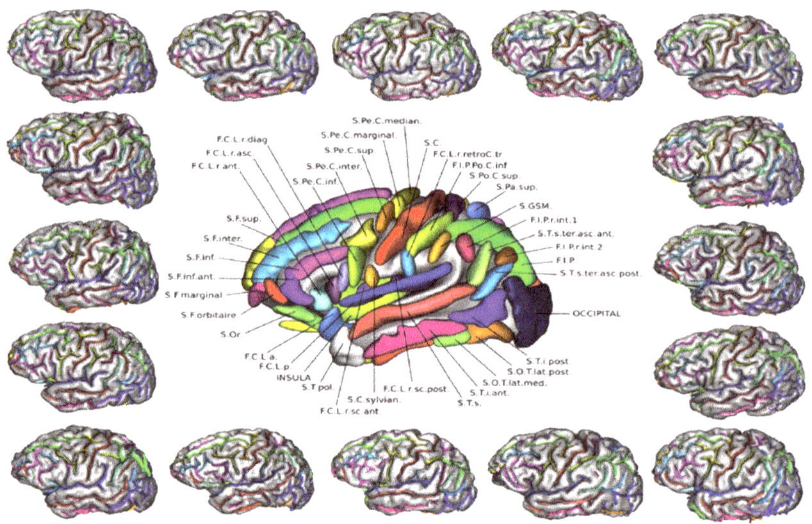

Figure 1. A nomenclature of cortical sulci applied to 16 different brains to illustrate the variability of the folding pattern.

The template used is usually an average brain in order to overcome the bias induced by the idiosyncrasies of specific brains. At the onset of VBM, this template was fuzzy because of the poor alignment of the folding patterns across the brains to be averaged; however, thanks to methodological advances, average brains are now very similar to actual brains but with regularized folding patterns (see Figure 2). The choice of the template, however, still raises questions: should it be adapted to the study population, should it be blurred to reflect variability, or should it resemble a real brain? Should it be symmetrical or asymmetrical? Should it be age-specific?

Surprisingly, geometric morphometrics, which is the mainstream strategy in paleoanthropology [35] has not been successful in neuroimaging. One could look for a technical explanation but this lack of interest is probably mainly linked to sociological phenomena. The rare use of geometric morphometrics in neuroimaging can be explained by the "winner takes all" phenomenon. The usual computational neuroanatomy methods are based on the concept of spatial normalization forged for functional imaging, the modality at the origin of the neuroimaging boom. There was probably no room for a radically different vision based on landmarks, all the more given that landmarks are difficult to define unambiguously in the human brain. The fate of geometric morphometrics in the world of neuroimaging is that of all methods that have sought to deviate from the paradigm of their field.

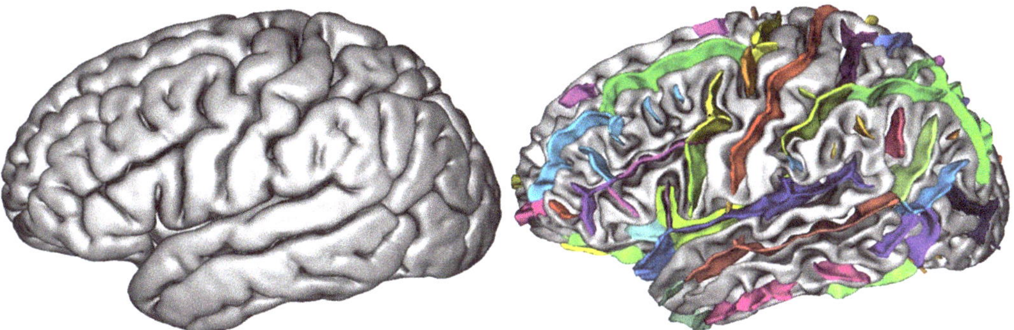

Figure 2. The ICBM 152 realistic template of the MNI (McGill University, Montreal), resulting from the averaging of 152 different brains, and its regularized sulci, with the nomenclature of Figure 1.

Surface-based morphometry (Freesurfer, CIVET) is very similar to VBM in spirit, but is dedicated to the cortical surface, which is inflated and mapped to a sphere before being aligned across subjects [36]. It was designed to simplify the alignment of large sulci and to quantify parameters with real anatomical meaning: the thickness of the cortex or the surface area of a convolution. Because this approach is more computationally complex, there are far fewer tools available than for VBM. It would be interesting to compare this surface-based strategy with methods that seek to align endocasts, i.e., surfaces with the trace of certain furrows. The major difference is that the neuroimaging approach unfolds the cortex, whereas the endocast approach can only manipulate the external part of the cortical surface. Morphometry of the shape of the cortical sulci (length, depth, etc.) can also be performed using brainVISA, whose output is illustrated in Figures 1 and 2 [37].

3. Virtual Anthropology and Paleoneurology

The use of imaging methodologies in paleoanthropological studies appeared to be of great benefit as early as the mid-1980s [38,39]. Among their first applications, the determination of endocranial volume aroused wide interest. Indeed, the resolution of the tomographic data was of the order of a millimeter, thus complicating the detailed study of fine character, but being well suited to overall quantifications of large structures. Fortunately, the technique has largely progressed, as has its application to the human fossil record. The term "virtual anthropology" has been proposed to name this emerging field [40]. Imaging facilities are now considered one of the classic techniques in the toolbox of paleoanthropologists (Figure 3). However, although they are very important, providing important possibilities, they also feature limitations.

Imaging data allows more robust studies. Fossils, of course, can only be studied by methodologies based on X-rays. MRI approaches are not applicable to our dry specimens, which are composed of highly mineralized and fossilized bones. It has recently been demonstrated that X-ray methodologies, when used at adapted settings for the classic study of fossils, have no influence on the preservation of the structure of the fossil and that they do not cause damage to the preservation of ancient DNA [41]. However, they have some effect on ESR dating [42]. These aspects have to be considered. Imaging methodologies play a crucial role in the preservation of our heritage. Moreover, thanks to this approach, the samples to be analyzed in the context of the study of human evolution may be much larger. From a methodological point of view, it is much easier to improve and test any protocol and methodologies may be more easily repeated. These aspects are particularly important as the original fossils are housed all over the (ancient) world. Nevertheless, progress is still expected in the way we share the imaging datasets. Among technical limitations are those related to the size of the datasets and the necessary informatics environment to manage the analyses. The resolution is now potentially very high, allowing very precise analyses. Fortunately, computers and software have also progressed. In addition, paleoanthropologists could rely on the massive computational infrastructures that are currently emerging to support neuroscience research. For example, the virtual models of endocasts scattered all over the world could be gathered on Ebrains (https://ebrains.eu/, accessed on 11 October 2021), the platform resulting from the European flagship Human Brain Project, and give rise to synergies with other communities.

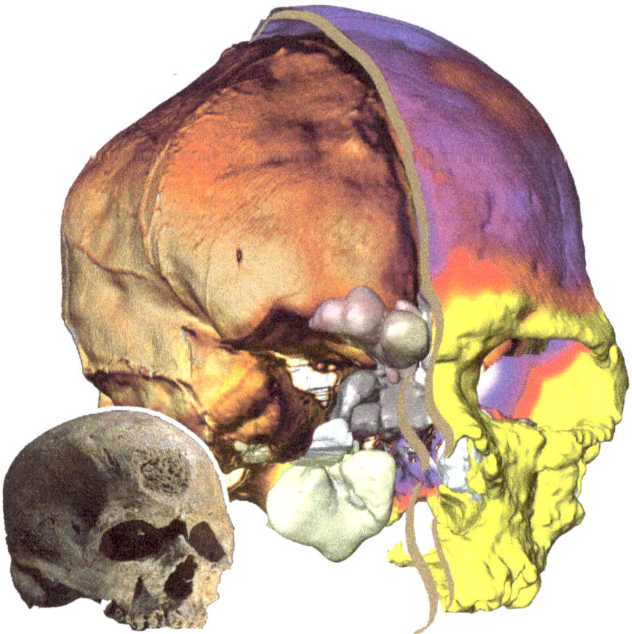

Figure 3. The original skull of Cro Magnon 1: 3D reconstruction of the endocranial cast (in orange) of the paranasal pneumatization and of the right half of the skull, on which are shown variations in bone thickness (thinner areas are in white and blue, intermediate areas in purple, and thicker areas are in red and yellow).

In fact, the main concern in "virtual anthropology" is probably an unexpected aspect. Virtual models may be reconstructed with mirror images, from templates obtained on comparative samples, or by estimation of the missing areas. As such, the new "virtual" fossils are not real reflections of the original specimens. It is, of course, particularly important to remove distortions related to post-mortem alterations, but it is crucial to keep a detailed record of all the modifications made to a model. For example, none of the *H. neanderthalensis* specimens analysed in a study of the evolution of the brain [43] preserve this anatomical area. The study is by itself interesting and important in a comparative perspective but raises some questions about the interpretation of the results that could be obtained beyond this particular context. The extreme and tautological case is when a reconstructed model is used as an essential milestone in a systematic approach.

3.1. Does the Endocast Reflects the Brain?

Paleoneurology is a fascinating topic, dealing with anatomical and biological aspects of past humans and, in addition, potential behavioral implications. The field is, of course, highly debated, for multiple reasons.

The main reason relates to the complex nature of the material that researchers analyze. Indeed, the soft tissues that constitute the brain never fossilize. Scientists only have to deal with the shallow imprints of the convolutions that the brain forms on the internal surface of the skull. This incomplete reflection of the brain is named the (brain) endocast. The brain presses on and leaves marks on the inner surface of the skull throughout a person's life. This was true for the humans who lived a few million years ago, but also for all of us. The phenomenon is particularly intense during the period of accelerated growth of the brain, and therefore of the cranial box which surrounds it, during the first years of life. The whole process is intertwined, so that the shape of the adult skull is reminiscent of the moment of peak brain development. The behavior of the skull can be described as that of a morphological black box, retaining information that later makes it possible to reconstitute its original contents. Therefore, when a fossil skull is discovered, its inner surface is molded, either physically or virtually, using imaging methods, to reconstruct its endocast. This model represents the preserved imprints of the external surface of the brain. However, the correspondence between these limited records of convolutional patterns and details of the surface of the brain remains to be demonstrated in modern humans. A few pioneer studies have considered this problem [4,44]. Moreover, it is necessary to develop new tools for the automatic and reliable determination of the endocranial sulci [45].

In the context of the PaleoBRAIN project, financed by the ANR, we are conducting a direct investigation of the correlation between the shape of the brain and that of the intracranial cast within a sample of modern humans using MRI (for Magnetic Resonance Imaging) acquisitions, including some with a specific sequence that allows the characterization of bone tissues. The comparison of morphometric data and anatomical traits between the brain and the endocast will be performed using state-of-the-art quantification methodologies. But our large dataset could probably also be used to refine the methodology dedicated to the sulcus detection in the endocast. Current methodologies use differential geometry to detect sulci as ravine or crest lines [4,44]. A key component in the design of such robust detectors is the amount of local smoothing performed before detection, which is usually tuned to the scale of the features to be detected. The T1-weighted MRI of our dataset can be used to define the ground truth using the sulci detected by the Brain VISA software. Subsequently, the optimal smoothing can be estimated using an inverse problem framework. Thanks to the large dataset, we can probably afford to include the estimation of regularized spatial variations of the optimal amount of smoothing, which may help to achieve a more consistent sulcus detection throughout the endocast. This could help to overcome some of the weaknesses observed in the superior part of the brain [1]. Once we have acquired a better understanding of the reliability of the endocast-based definition of the folding pattern within our own species, we will be able to use this model to address the shape of the brain/endocast in well-preserved fossil hominin specimens.

This project will also contribute to answering a key question about the evolution of the human brain. In many studies, the endocast is analyzed with distances characterized at maximal points of extension, maximal length or maximal width, or that correspond to intracranial points, such as endobregma or endolambda, for example [27,46], or with 3D methodologies that consider the surface as a whole [47,48]. These methodological approaches are justified by the complex nature of the material. Indeed, gyri and sulci are difficult to identify on the endocast (Figure 4). In this context, there is little information available about variations in the global size of the different lobes and their relationship with each other between hominin species.

In a previous study [28], we demonstrated clear differences in brain organization when considering the relative contribution of the different lobes to the surface of the complete endocast. Asian *H. erectus* specimens show a significantly smaller relative size of the parietal and temporal lobes than all other samples of the genus Homo. This field of research could benefit from the recent revival of interest in the study of the laws of allometry that govern the relative variations of the various cerebral structures, linked to the large databases of modern brain images [49].

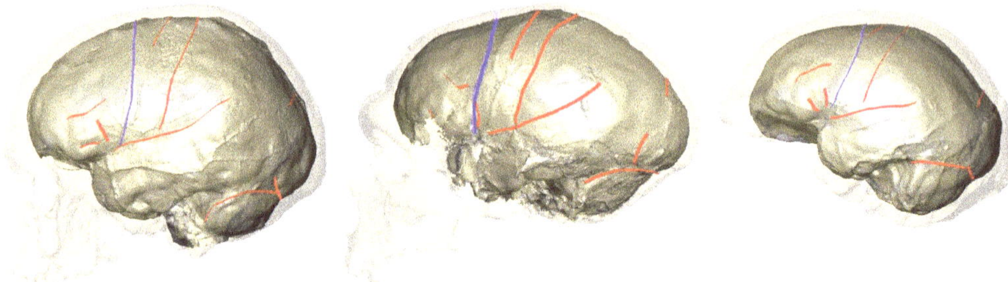

Figure 4. Comparison between the position of the main sulci of the endocranial surface (in red) and the shape and position of the skull, including the course of the coronal suture (in blue) in Cro-Magnon 1, an Upper Paleolithic *Homo sapiens*; La Chapelle-aux-Saints 1, an *Homo neanderthalensis*; and Sambungmacan 3, an *Homo erectus*.

Moreover, *H. neanderthalensis* and fossil *H. sapiens*, which have the largest endocranial volume of all hominins, show different brain structures (Figure 4). These results illustrate that differences existed in the structure of the brain in addition to the well-known variation in size during human evolution. An important contribution to this topic will be to improve our ability to determine the location of the sulci and gyri on fossil hominin endocasts. To do so, a better knowledge of the anatomy and characteristics of hominids is necessary [50,51], together with a better knowledge of the brain–endocast relationship in living humans. Finally, it is fundamental to obtain a more generalized and simplified access to high-resolution endocranial data for fossil specimens. Indeed, this material is so complex that multiple appreciation by the few researchers dealing with paleoneurological information would certainly enhance our capacity for anatomical determination. It would also certainly help to minimize potential conflicting interpretations, which are very frequent in this small field of research.

3.2. What Can Be Deduced about a Species' Folding Pattern from a Few Samples?

The very high intra-species variability of the cortical folding of Homo sapiens, illustrated by Figure 1, is a major difficulty for modern brain mapping. It should also warn us about the risk of over-interpretation inherent in the small number of samples available in paleoneurology. The idiosyncrasies of a specific brain are not necessarily representative of the folding pattern of its species. The amount of intra-species variability is species dependent. In great apes, it is less than in humans but still significant, especially in the frontal lobe. In baboons or macaques, it is almost non-existent. In species with a variable folding pattern, the match between the folds of an individual and its nomenclature can be difficult to establish and leads to confusion, especially when only an endocast is available [1,52]. In modern humans, the large sulci described in anatomical books are often split into pieces and reorganized into unusual folding patterns that are difficult to decipher (see Figure 5) [53]. Notably, these phenomena occur in the general population without developmental pathologies.

The mysteries hidden behind the variability of cortical folding have led to the emergence of a multidisciplinary community that aims to understand these variations and their meaning. It associates biologists, who focus on the developmental phenomena that are at the origin of cortical folding (spatially heterogeneous neurogenesis, spatially heterogeneous chronology of synaptic development, etc.) [54], and physicists, who model the mechanical phenomena that result from these growth heterogeneities [55]. This new community also includes anatomists, who study the links between folding and the organization of cortical areas and fiber bundles [56], and computer scientists, who geometrically model the variability observed in the general population, and the specificities of developmental pathologies [34,57]. In our opinion, the progress made by this community could contribute to a better exploitation of the scarce data observed in the endocasts of the folding of extinct species. A better understanding of the rules driving cortical folding dynamics would provide insight into the architectural changes at the origin of the changes observed across species in endocasts. Endocasts are used as a proxy of the folding pattern, but the folding pattern is only a proxy of architecture, which is even more difficult to reverse-engineer. Current efforts for cracking the code behind folding patterns could contribute, for instance, to the discussion around the third frontal convolution when comparing sapiens, great apes, and extinct hominids. Joint modelling of folding variability and of functional variability will help to understand which features of the folding pattern can be used as landmarks of key cytoarchitectonic areas (see Figure 6) [58].

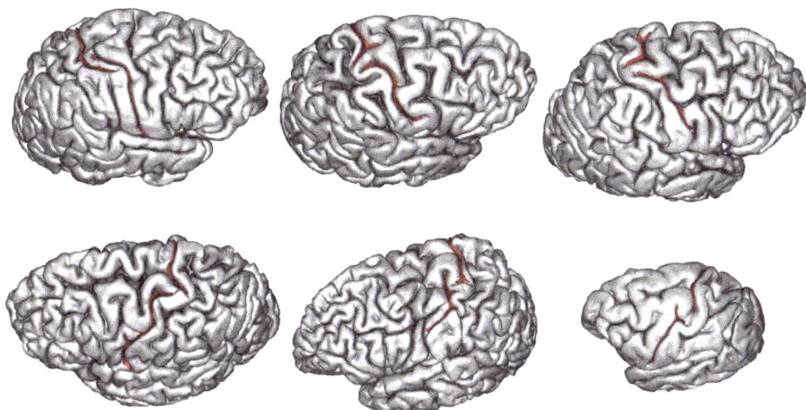

Figure 5. The hemispheres of five *Homo sapiens* and one chimp, with interruption of the central sulcus, which hosts sensorimotor areas (0.5% of occurrence). This kind of interruption is frequent in associative areas and leads to folding configurations that are difficult to decipher, which can be observed here in the frontal lobes.

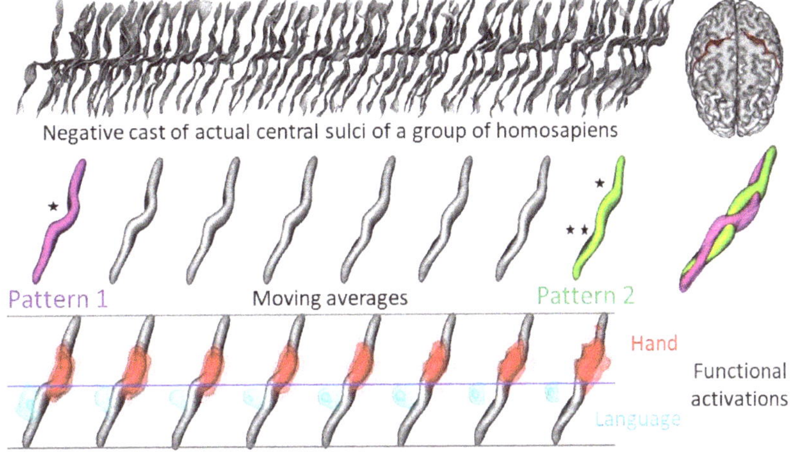

Figure 6. Machine learning can be used to model the variability of folding patterns. Here, the variability of the shape of the central sulcus is projected into a one-dimensional manifold representing the transition between a single knob and a double knob pattern. Functional mapping performed along this manifold shows that the two different folding patterns correspond to different localizations of functional areas along the central sulcus.

3.3. How Are Brain Asymmetries Quantified in the Hominin Fossil Record?

The number of brain structural asymmetries observable on endocranial casts and, consequently, in fossil specimens is limited due to several factors, which are of course related to the specificity of our material, which concerns only the external surface of the brain. By chance, features on the brain and endocast for which bilateral variation studies are possible are among the most consistent features available for cross-taxa studies on large samples. One important limiting point needs to be considered. Indeed, the difficulty in defining structural parameters and in establishing left-right homologies makes studies of brain asymmetries complex. Moreover, gross anatomical asymmetries of selected pairs of points may reflect combined asymmetries in brain subregions. The quantification of surface morphology, distance, or volume of discrete anatomical areas may not fully express real bilateral variation if their pattern of asymmetry is defined in reference to global anatomical brain areas. For example, previous works have proposed the quantification of the volume or regional surface areas of endocasts in hominin fossils [28,59,60].

Another limitation is that the methodologies employed in most previous studies of cerebral or endocranial asymmetries involved qualitative assessment or a simple index of bilateral traits and did not analyze departures from symmetry and different patterns of asymmetry (i.e., fluctuating and directional asymmetry, antisymmetry) in efficient and adapted ways [61]. It has indeed been shown that the brains of extant hominids demonstrated high levels of fluctuating asymmetry, allowing pronounced developmental plasticity and therefore making brains highly evolvable [62]. The quantification and analysis of the morphology—including the asymmetries—of the endocranial cavity need further development. Currently, the most advanced computational tools used in analysis of bilateral shape asymmetries rely on the standard framework of landmark-based morphometrics [63,64]. In this context, in addition to homologous landmarks between shapes for population studies, one must define homologous landmarks between the two sides of each shape under study. Analyses can then be carried out by using slight modifications of the linear distance-based [65] or superimposition [66–69] methods. However, we identified methodological problems underlying the theory and its application to the assessment of bilateral asymmetries [48,70,71]. Moreover, a limited set of landmarks is likely to be inadequate to capture the shape of intricate anatomical structures, or that of structures with few obvious salient features, such as brain endocasts. New methodological improvements are therefore necessary to better characterize and quantify bilateral asymmetries [72,73]. A specific methodology has been developed and tested on the endocast of the Cro-Magnon 1 fossil [28]. This approach is promising as it allows for an independent characterization of the asymmetries without referring to the potential global asymmetry of the object that is analyzed. New approaches based on machine learning could also be a source of inspiration. They allow us, for example, to establish the asymmetry of folding patterns without requiring the definition of homologous landmarks across subjects and hemispheres. For instance, the double-knob configuration of the central sulcus, depicted in Figure 6, is more frequent in the left hemisphere [74]. These new approaches could contribute to the old question of the language-related asymmetry of the third frontal convolution, which is difficult to tackle because of the large intraspecies variability of the related folding patterns [75].

3.4. The Complex Definition of Brain Features and of Their Application to the Fossil Record

A general problem concerns the lack of homogeneity in the definition of brain asymmetries and of the methods used to quantify them. For example, one of the most studied brain asymmetries on brain endocasts concern the petalias. LeMay [76,77] initially considered the antero-posterior projection of the frontal and occipital lobes, respectively. By contrast, later studies generalized the term 'petalias' to a wide range of anatomical traits. Some studies indeed referred to bilateral differences in the lateral extension of the posterior area of the frontal lobes [78], to other anatomical areas of the brain, and even to volumetric variations between hemispheres [79–83]. It is therefore difficult to compare data obtained on petalias if studies do not consider the same brain features. Nevertheless, it was largely accepted that this pattern of asymmetries appeared with early Homo [27,78,84] and is more common in right-handed individuals [77,78,85–89]. Based on an original methodology applied to the largest samples ever used, we demonstrated a shared specific pattern of protrusions of the frontal and occipital across all hominids, including extant African great apes, modern humans, and hominin fossils [21,73]. These asymmetries are a topic of debate in non-human primate brain studies [76,79,80,90–92] and paleoanthropology [25,26,78,84,93–95] because of their relationship with handedness and other specific aspects of human cognition. Similar results were obtained recently by an independent team [48]. *H. sapiens* appear to have more asymmetrical petalias than other extant great apes, but a shared pattern is observed, suggesting that a globally asymmetric brain is the ancestral condition. A recent study questioned this observation [96]. However, this is a good example of differences in the definition of the anatomical traits that are analyzed. These authors measured the bilateral variation in lateral extension of slices of the brain. This trait is not directly comparable to our analyses of the 3D position of the occipital poles [29] or to the 3D displacement between the left and right corresponding anatomical area. Another good illustration of the problem is Broca's area, whose extension is defined differently according to authors [97]. This functional area is impossible to characterise on brain endocasts. However, we conducted a comparative study on the size, shape, and position of the third frontal convolution in great apes, *H. sapiens*, and hominin fossils [29]. The neuroanatomical asymmetries as quantified in our work show a pattern that is different from what was previously accepted based on qualitative data. Our main finding was a shared pattern of asymmetry in Broca's area in all hominins and Pan paniscus, as well as an increase in the size of this area during human evolution. We also identified that Pan troglodytes and Pan paniscus have differences in their asymmetry patterns in the third frontal convolution. This topic is of great interest for future research. More generally, brain and endocranial studies have to rely on a clear definition of the anatomical features that are analyzed and an effort to use similar protocols will certainly enhance the reproducibility of our studies.

3.5. How to Grow a Hominin Brain?

The knowledge of ontogenetic patterns in fossil human species is scarce [98–101] and, to date, no information is available about the evolution of brain lateralization during growth and development. Both *H. neanderthalensis* and *H. sapiens* have enlarged brains compared with other hominins, but their respective organizations and morphologies are different, each of the two species having "grown" large brains through specific evolutionary processes. Much remains unknown about what these processes are, and how they are rooted in the hominin evolutionary tree. In the case of *H. neanderthalensis*, although some changes in gross cerebral morphology during childhood are documented, researchers have presented conflicting results concerning how their endocranial growth patterns relate to those of other primates. While the post-natal Neandertal ontogenetic trajectory is deemed closer to that of chimpanzees than to that of *H. sapiens* by some researchers, emphasizing a unique globularization phase in *H. sapiens* [101], others find that the mode of cerebral growth is largely similar in *H. neanderthalensis* and *H. sapiens*, emphasizing instead the characteristic morphologies of each species at birth, and refuting the idea of the derived nature of the post-natal cerebral growth trajectory in *H. sapiens* [102]. Nevertheless, these studies only consider the global shape of the internal surface of the skull. Additionally, available data addressing cerebral growth do not provide enough details, so that much of "how" the Neandertal brain grows remains unknown (e.g., do the contributions of the different lobes to total brain volume remain stable throughout infancy and childhood?).

We previously demonstrated that the two species have distinct brain organizations [103], but this important biological aspect has not yet been considered in the study of brain growth in *H. neanderthalensis*. The emergence of large databases on the brain development of sapiens, and to a lesser extent of extant non-human primates [104], could contribute to these debates.

3.6. Brain Endocast and Function

The question of the relationship between brain shape and function in hominins has been explored in previous studies [105]. According to their authors, they "show that Neanderthals had significantly larger visual systems than contemporary anatomically modern humans (indexed by orbital volume) and that when this, along with their greater body mass, is taken into account, Neanderthals have significantly smaller adjusted endocranial capacities than contemporary anatomically modern humans." For the authors, these results had implications for interpreting variations in brain organization in terms of social cognition. Indeed, larger visual systems would have implied smaller adjacent anatomical areas, including the parietal areas related to social skills. Their final conclusion was that the extinction of *H. neanderthalensis* was due to weaker social cognition compared to modern humans. This study suffered from methodological limitations. The main problem was that they were improperly interpreting data mostly derived from the research of one of the authors of this paper [103]. These authors considered that our data for the external extension of the occipital lobe were directly related to the size of the visual cortex. However, such a direct interpretation was not demonstrated. Moreover, they did not measured any anatomical areas on the endocasts of *H. neanderthalensis* or of contemporary *H. sapiens*. All those approximations make any interpretation in terms of behaviors impossible.

This example should not prevent us from analyzing morphological variation among hominins species and exploring functional and behavioral implications. However, this needs to be undertaken on a solid anatomical framework, particularly in the context of interspecies comparisons, and with more caution for the evaluation of the potential link between brain anatomy and suspected function.

4. Perspectives for Future Studies of the Evolution of the Human Brain

4.1. The Future of Neuroimaging

The world of neuroimaging is in perpetual development, constantly fed by technological advances and new concepts aimed at deciphering the organization of the human brain. However, large parts of the brain's functioning remain misunderstood. Despite the wealth of knowledge accumulated on its development, the incredible efficiency of its learning processes remains a mystery; it is probably very different from deep learning. Unlocking the secrets of its evolution still seems to be an unattainable goal, given the limited information available to paleoanthropologists. However, the possibility of almost unlimited advances in technology probably holds surprises for us. The last decade has given rise to extraordinary investments in this respect. The American "Brain Initiative" has thus generated science-fiction-like technologies for the "reverse engineering" of rodent brains: for example, the possibility of simultaneously recording the activity of a million neurons, or of mapping the synaptic connectivity between a large number of neurons. The possibilities for the non-invasive exploration of the human brain are much more limited, but the rise of brain imaging raises many hopes. Large shared research infrastructures dedicated to the exploration of the brain are being created, in the spirit of what happened in physics in the middle of the last century. These infrastructures will house outstanding scientific instruments, unique in terms of sensitivity or resolution, built to open up new "discovery spaces". Moreover, the most important discoveries made with these instruments are often those that had not been foreseen in the initial scientific dossier. For example, the French CEA has decided to exploit the expertise of its physicists, who were behind the magnets at CERN in Geneva, to design a new generation of MRI. The 11.7 Tesla magnet located at Neurospin in the southern suburbs of Paris should, for example, make it possible to zoom in vivo to study the functioning of the brain at the true scale of the organization of its cortex into cortical layers and columns. These deep phenotyping initiatives are complemented by major international phenotyping initiatives to understand the genetic basis of the human brain, which will probably provide important insight into the evolutionary events at the origin of our brains. Molecular analysis of humans, archaic hominins, and non-human primates has allowed the identification of chromosomal regions, showing evolutionary changes at different points of our phylogenetic history, which may be related to the evolution of the endocast-based clues about the cortical folding patterns [106]. The coming decades may see the emergence of a better understanding of the evolution of the genetic building plan behind the human brains [107].

4.2. Endocast Side

Variation is an important concept in paleoanthropology. Paleoneurological approaches try to identify as precisely as possible intraspecific variations, as well as diagnostic features, between species. In turn, the initial mainstream paradigm in brain mapping involved canceling out morphological variability to allow comparative analysis of the functional maps across subjects and experiments. Neuroimaging, however, has widened its scope during the last decades to the modeling of intersubject variability, in order to tackle the discovery of biomarkers of pathology or the stratification of populations of patients. Furthermore, neuroimaging is now widely used to understand brain development and to compare primate species. It is time to consider cross-fertilization with paleoneurology, which has evolved in a niche built upon geometric morphometrics, which has prevented synergies. Broadening our knowledge of brain variability in our species by including a long time dimension will be of great help in defining the brain anatomy of *H. sapiens*. It also opens up perspectives for understanding how our brain works.

One original and exciting perspective will be to reconstruct a fossil hominin brain. A recent study [108] was the first to attempt the reconstruction of a *H. neanderthalensis* brain by deforming a population average brain for modern humans into the shape of the endocast of a reconstituted Neandertal. However, this approach does not consider the differences in brain structure between these species, such as those that we documented [103]. The different approaches detailed here, aiming at the collection of better information on the brain/endocast correspondence in living humans, developing new tools of automatic determination of the sulci on the endocasts, and enlarging our knowledge of fossil hominin variation thanks to a better availability of high-quality endocranial surfaces, will make it possible to obtain more satisfactory results.

Modern Artificial Intelligence could even play a role in the cross-fertilization between paleoneurology and neuroimaging. Provided that dedicated MRI sequences can deliver consistent proxies of endocasts on a large scale, deep learning could be trained to transform endocasts into standard representations of the cortical surface used in the mainstream neuroimaging field. Transfer learning could be tested on extant non-human primates and applied to extinct species in case of success.

5. Conclusions

It undeniably emerges from this perspective that paleoanthropology has much to gain from interacting more with the field of neuroimaging. Improving our understanding of the morphology of endocasts necessarily involves studying the external surface of the brain and the link it maintains with the internal surface of the skull. A fundamental perspective is to describe more fossils among more species in order to better understand the evolution of the human brain. Our discipline must also work towards better data accessibility. This will reinforce the quality of the comparisons and the repeatability of the work on the complex material that is the endocast. This will also contribute to a better definition of the traits that are analyzed. This aspect will be greatly improved by the contribution of neuroimaging, which will allow us to better define the relationship between brain and endocast. Finally, the exchanges between these two disciplines will also be beneficial to our knowledge of the *H. sapiens* brain. Documenting the anatomies of other human species and including the variation over time within our own species are approaches that offer us a new perspective through which to appreciate what really characterizes the brain of humanity today.

Author Contributions: A.B. and J.-F.M.: conceptualization, methodology, writing—original draft preparation. All authors have read and agreed to the published version of the manuscript.

Funding: This research received funding from the Agence National de la Recherche (ANR-20-CE27-0009-01) to AB and from the European Union's Horizon 2020 Research and Innovation Programme under Grant Agreement No. 945539 (HBP SGA3), from the FRM DIC20161236445, from the ANR IFOPASUBA and FOLDDICO, and from the Blaise Pascal Chair of Region Ile de France and Université Paris Saclay to W. Hopkins.

Acknowledgments: The authors thank the MNHN for the data for Cro Magnon 1 and La Chapelle-aux-Saints; the 3D mapping of the asymmetries was calculated by S. Prima for a shared study.

Conflicts of Interest: The authors declare no conflict of interest.

References

1. Falk, D. Interpreting sulci on hominin endocasts: Old hypotheses and new findings. *Front. Hum. Neurosci.* **2014**, *8*, 134. [CrossRef]
2. Gall, F.J. *On the Functions of the Brain and of Each of its Parts: With Observations on the Possibility of Determining the Instincts, Propensities, and Talents, or the Moral and Intellectual Dispositions of Men and Animals, by the Configuration of the Brain and Head*; Marsh, Capen & Lyon: Boston, MA, USA, 1835.
3. Parker Jones, O.; Alfaro-Almagro del, F.; Jbabdi, S. An empirical, 21st century evaluation of phrenology. *Cortex* **2018**, *106*, 26–35. [CrossRef]
4. Dumoncel, J.; Subsol, G.; Durrleman, S.; Bertrand, A.; de Jager, E.; Oettlé, A.C.; Lockhat, Z., Suleman, F.E.; Beaudet, A. Are endocasts reliable proxies for brains? A 3D quantitative comparison of the extant human brain and endocast. *J. Anat.* **2021**, *238*, 480–488. [CrossRef]
5. Tononi, G.; Sporns, O.; Edelman, G.M. A measure for brain complexity: Relating functional segregation and integration in the nervous system. *Proc. Natl. Acad. Sci. USA* **1994**, *91*, 5033–5037. [CrossRef]
6. Broca, P. *Mémoires d'Anthropologie*; Reinwald: Paris, France, 1871.
7. Brodmann, K. Physiologie des Gehirns. In *Allgemeine Chirurgie der Gehirnkrankheiten*; Knoblauch, A., Brodmann, K., Hauptmann, A., Eds.; Verlag von Ferdinand Enke: Stuttgart, Germany, 1914; pp. 86–426.
8. Vogt, C.; Vogt, O. Die vergleichend-architektonische und die vergleichend reizphysiologische Felderung der Großhirnrinde unter besonderer Berucksichtigungder menschlichen. *Naturwissenschaften* **1926**, *14*, 1192–1195. [CrossRef]
9. Amunts, K.; Mohlberg, H.; Bludau, S.; Zilles, K. Julich-Brain: A 3D probabilistic atlas of the human brain's cytoarchitecture. *Science* **2020**, *369*, 988–992. [CrossRef]
10. Van Essen, D.C.; Smith, S.M.; Barch, D.M.; Behrens, T.E.; Yacoub, E.; Ugurbil, K.; Consortium W-MH. The WU-Minn Human Connectome Project: An overview. *Neuroimage* **2013**, *80*, 62–79. [CrossRef]
11. Elliott, L.T.; Sharp, K.; Alfaro-Almagro, F.; Shi, S.; Miller, K.L.; Douaud, G.; Marchini, J.; Smith, S.M. Genome-wide association studies of brain imaging phenotypes in UK Biobank. *Nature* **2018**, *562*, 210–216. [CrossRef]
12. Pääbo, S. The human condition-a molecular approach. *Cell* **2014**, *157*, 216–226. [CrossRef]
13. Tilot, A.K.; Khramtsova, E.A.; Liang, D.; Grasby, K.L.; Jahanshad, N.; Painter, J.; Colodro-Conde, L.; Bralten, J.; Hibar, D.P.; Lind, P.A.; et al. The Evolutionary History of Common Genetic Variants Influencing Human Cortical Surface Area. *Cereb. Cortex* **2020**, *5*, 1873–1887.
14. Grasby, K.L.; Jahanshad, N.; Painter, J.N.; Colodro-Conde, L.; Bralten, J.; Hibar, D.P.; Lind, P.A.; Pizzagalli, F.; Ching, C.R.K.; McMahon, M.A.B.; et al. Enhancing NeuroImaging Genetics through Meta-Analysis Consortium (ENIGMA)—Genetics working group. The genetic architecture of the human cerebral cortex. *Science* **2020**, *367*, 6484.
15. Bozek, J.; Makropoulos, A.; Schuh, A.; Fitzgibbon, S.; Wright, R.; Glasser, M.F.; Coalson, T.S.; O'Muircheartaigh, J.; Hutter, J.; Price, A.N.; et al. Construction of a neonatal cortical surface atlas using Multimodal Surface Matching in the Developing Human Connectome Project. *NeuroImage* **2018**, *179*, 11–29. [CrossRef]
16. Harms, M.P.; Somerville, L.H.; Ances, B.M.; Andersson, J.; Barch, D.M.; Bastiani, M.; Bookheimer, S.Y.; Brown, T.B.; Buckner, R.L.; Burgess, G.C.; et al. Extending the Human Connectome Project across ages: Imaging protocols for the Lifespan Development and Aging projects. *Neuroimage* **2018**, *183*, 972–984. [CrossRef]
17. Hopkins, W.D.; Meguerditchian, A.; Coulon, O.; Bogart, S.; Mangin, J.F.; Sherwood, C.C.; Grabowski, M.W.; Bennett, A.J.; Pierre, P.J.; Fears, S.; et al. Evolution of the central sulcus morphology in primates. *Brain Behav. Evol.* **2014**, *84*, 19–30. [CrossRef]
18. Friedrich, P.; Forkel, S.J.; Amiez, C.; Balsters, J.H.; Coulon, O.; Fan, L.; Goulas, A.; Hadj-Bouziane, F.; Hecht, E.E.; Heuer, K.; et al. Imaging evolution of the primate brain: The next frontier? *NeuroImage* **2021**, *228*, 117685. [CrossRef]
19. Milham, M.; Petkov, C.I.; Margulies, D.S.; Schroeder, C.E.; Basso, M.A.; Belin, P.; Fair, D.A.; Fox, A.; Kastner, S.; Mars, R.; et al. Accelerating the Evolution of Nonhuman Primate Neuroimaging. *Neuron* **2020**, *105*, 600–603. [CrossRef]
20. Trinkaus, E.; Churchill, S.E.; Ruff, C.B. Post-cranial robusticity in *Homo* II: Humeral bilateral asymmetry and bone plasticity. *Am. J. Phys. Anthrop.* **1994**, *93*, 1–34. [CrossRef]
21. Balzeau, A.; Gilissen, E.; Grimaud-Hervé, D. Shared pattern of quantified endocranial shape asymmetries among anatomically modern humans, great apes and fossil hominins. *PLoS ONE* **2012**, *7*, e29581.
22. Shaw, C.N.; Stock, J.T. Extreme mobility in the Late Pleistocene? Comparing limb biomechanics among fossil *Homo*, varsity athletes and Holocene foragers. *J. Hum. Evol.* **2013**, *64*, 242–249. [CrossRef]

23. Davies, T.G.; Stock, J.T. Human variation in the periosteal geometry of the lower limb: Signatures of behaviour among human Holocene populations. In *Reconstructing Mobility: Environmental, Behavioral, and Morphological Determinants*; Carlson, K.J., Marchi, D., Eds.; Springer: New York, NY, USA, 2014; pp. 67–90.

24. Goldberg, E.; Roediger, D.; Kucukboyaci, N.E.; Carlson, C.; Devinsky, O.; Kuzniecky, R.; Cash, S.; Thesen, T. Hemispheric asymmetries of cortical volume in the human brain. *Cortex* **2013**, *49*, 200–210. [CrossRef]

25. Tobias, P.V. The brain of *Homo habilis*: A new level of organization in cerebral evolution. *J. Hum. Evol.* **1987**, *16*, 741–761. [CrossRef]

26. Grimaud-Hervé, D. *L'Evolution de l'Encéphale Chez Homo Erectus et Homo Sapiens: Exemples de l'Asie et de l'Europe*; Cahiers de Paléoanthropologie, CNRS: Paris, France, 1997; 406p.

27. Holloway, R.L.; Broadfield, D.C.; Yuan, M.S. *The Human Fossil Record: Brain Endocasts, Paleoneurological Evidence*; Wiley-Liss: New York, NY, USA, 2004; 315p.

28. Balzeau, A.; Grimaud-Hervé, D.; Holloway, R.L.; Détroit, F.; Combès, B.; Prima, S. First description of the Cro-Magnon 1 endocast and study of brain variation an evolution in anatomically modern *Homo Sapiens*. *Bull. Mémoires Société D'anthropologie Paris* **2013**, *25*, 1–18. [CrossRef]

29. Balzeau, A.; Gilissen, E.; Holloway, R.L.; Prima, S.; Grimaud-Hervé, D. Variations in size, shape and asymmetries of the third frontal convolution in hominids: Paleoneurological implications for hominin evolution and the origin of language. *J. Hum. Evol.* **2014**, *76*, 116–128. [CrossRef]

30. Galaburda, A.M.; Corsiglia, J.; Rosen, G.D.; Sherman, G.F. Planum temporale asymmetry, reappraisal since Geschwind and Levitsky. *Neuropsychologia* **1987**, *25*, 853–868. [CrossRef]

31. Mechelli, A.; Price, C.J.; Friston, K.J.; Ashburner, J. Voxel-based morphometry of the human brain: Methods and applications. *Curr. Med. Imaging* **2005**, *1*, 105–113. [CrossRef]

32. Bookstein, F.L. "Voxel-based morphometry" should not be used with imperfectly registered images. *Neuroimage* **2001**, *14*, 1454–1462. [CrossRef]

33. Ashburner, J.; Friston, K.J. Why voxel-based morphometry should be used. *Neuroimage* **2001**, *14*, 1238–1243. [CrossRef]

34. Mangin, J.F.; Lebenberg, J.; Lefranc, S.; Labra, N.; Auzias, G.; Labit, M.; Guevara, M.; Mohlberg, H.; Roca, P.; Guevara, P.; et al. Spatial normalization of brain images and beyond. *Med. Image Anal.* **2016**, *33*, 127–133. [CrossRef]

35. Bookstein, F.L. *Morphometric Tools for Landmark Data: Geometry and Biology*; Cambridge University Press: Cambridge, UK, 1991.

36. Fischl, B. FreeSurfer. *Neuroimage* **2012**, *62*, 774–781. [CrossRef]

37. Mangin, J.F.; Jouvent, E.; Cachia, A. In-vivo measurement of cortical morphology: Means and meanings. *Curr. Opin. Neurol.* **2010**, *23*, 359–367. [CrossRef]

38. Maier, W.O.; Nkini, A.T. The phylogenetic position of Olduvai Hominid 9, especially as determined from basicranial evidence. In *Ancestors: The Hard Evidence*; Delson, E., Alan, R., Eds.; Wiley-Liss: New York, NY, USA, 1985; pp. 249–254.

39. Ruff, C.B.; Leo, F. Use of computed tomography in skeletal structure research. *Yearb. Phys. Anthrop.* **1986**, *29*, 181–196. [CrossRef]

40. Weber, G.W. Virtual Anthropology (VA): A call for glasnost in paleoanthropology. *Anat. Rec.* **2001**, *265*, 193–201. [CrossRef]

41. Immel, A.; Le Cabec, A.; Bonazzi, M.; Herbig, A.; Temming, H.; Schuenemann, V.J.; Bos, K.I.; Langbein, F.; Harvati, K.; Bridault, A.; et al. Effect of X-ray irradiation on ancient DNA in sub-fossil bones–Guidelines for safe X-ray imaging. *Sci. Rep.* **2016**, *6*, 32969. [CrossRef] [PubMed]

42. Duval, M.; Martín-Francés, L. Quantifying the impact of μCT-scanning of human fossil teeth on ESR age results. *Am. J. Phys. Anthropol.* **2017**, *163*, 205–212. [CrossRef] [PubMed]

43. Bastir, M.; Rosas, A.; Gunz, P.; Peña-Melian, A.; Manzi, G.; Harvati, K.; Kruszynski, R.; Stringer, C.; Hublin, J.-J. Evolution of the base of the brain in highly encephalized human species. *Nat. Commun.* **2011**, *2*, 588. [CrossRef]

44. Fournier, M.; Combès, B.; Roberts, N.; Braga, J.; Prima, S. Mapping the distance between the brain and the inner surface of the skull and their global asymmetries. *Med. Imaging 2011 Image Process.* **2011**, *7962*, 79620Y.

45. de Jager, E.J.; van Schoor, A.N.; Hoffman, J.W.; Oettlé, A.C.; Fonta, C.; Mescam, M.; Risser, L.; Beaudet, A. Sulcal pattern variation in extant human endocasts. *J. Anat.* **2019**, *235*, 803–810. [CrossRef]

46. Bruner, E.; Grimaud-Hervé, D.; Wu, X.; Cuétara, J.M.; Holloway, R. A paleoneurological survey of *Homo erectus* endocranial metrics. *Quat. Int.* **2015**, *368*, 80–87. [CrossRef]

47. Neubauer, S.; Hublin, J.-J.; Gunz, P. The evolution of modern human brain shape. *Sci. Adv.* **2018**, *4*, eaao5961. [CrossRef]

48. Neubauer, S.; Gunz, P.; Scott, N.A.; Hublin, J.-J.; Mitteroecker, P. Evolution of brain lateralization: A shared hominid pattern of endocranial asymmetry is much more variable in humans than in great apes. *Sci. Adv.* **2020**, *6*, eaax9935. [CrossRef]

49. Germanaud, D.; Lefèvre, J.; Fischer, C.; Bintner, M.; Curie, A.; Portes, V.D.; Eliez, S.; Elmaleh-Bergès, M.; Lamblin, D.; Passemard, S.; et al. Simplified gyral pattern in severe developmental microcephalies? New insights from allometric modeling for spatial and spectral analysis of gyrification. *NeuroImage* **2014**, *102*, 317–331. [CrossRef] [PubMed]

50. Falk, D.; Zollikofer, C.P.E.; Ponce de León, M.; Semendeferi, K.; Alatorre Warren, J.L.; Hopkins, W.D. Identification of in vivo Sulci on the External Surface of Eight Adult Chimpanzee Brains: Implications for Interpreting Early Hominin Endocasts. *Brain. Behav. Evol.* **2018**, *91*, 45–58. [CrossRef] [PubMed]

51. Alatorre Warren, J.L.; Ponce de León, M.; Hopkins, W.D.; Zollikofer, C.P.E. Evidence for independent brain and neurocranial reorganization during hominin evolution. *Proc. Natl. Acad. Sci. USA* **2019**, *116*, 22115–22121. [CrossRef] [PubMed]

52. Mangin, J.F.; Auzias, G.; Coulon, O.; Sun, Z.Y.; Rivière, D.; Régis, J. Sulci as landmarks. In *Brain Mapping: An Encyclopedic Reference*; Academic Press: Cambridge, MA, USA, 2015; Volume 2, pp. 45–52.

53. Mangin, J.F.; Le Guen, Y.; Labra, N.; Grigis, A.; Frouin, V.; Guevara, M.; Fischer, C.; Rivière, D.; Hopkins, W.D.; Régis, J.; et al. "Plis de passage" deserve a role in models of the cortical folding process. *Brain Topogr.* **2019**, *32*, 1035–1048. [CrossRef]

54. Llinares-Benadero, C.; Borrell, V. Deconstructing cortical folding: Genetic, cellular and mechanical determinants. *Nat. Rev. Neurosci.* **2019**, *20*, 161–176. [CrossRef] [PubMed]

55. Tallinen, T.; Chung, J.Y.; Rousseau, G.; Girard, N.; Lefèvre, J.; Mahadevan, L. On the growth and form of cortical convolutions. *Nat. Phys.* **2016**, *12*, 588–593. [CrossRef]

56. Van Essen, D.C. A tension-based theory of morphogenesis and compact wiring in the central nervous system. *Nature* **1997**, *385*, 313–318. [CrossRef]

57. Cachia, A.; Borst, G.; Jardri, R.; Raznahan, A.; Murray, G.K.; Mangin, J.F.; Plaze, M. Towards deciphering the fetal foundation of normal cognition and cognitive symptoms from sulcation of the cortex. *Front. Neuroanat.* **2021**, in press. [CrossRef]

58. Sun, Z.Y.; Pinel, P.; Rivière, D.; Moreno, A.; Dehaene, S.; Mangin, J.F. Linking morphological and functional variability in hand movement and silent reading. *Brain Struct. Funct.* **2016**, *221*, 3361–3371. [CrossRef]

59. Weaver, A.H. Reciprocal evolution of the cerebellum and neocortex in fossil humans. *Proc. Natl. Acad. Sci. USA* **2005**, *102*, 3576–3580. [CrossRef]

60. Wu, X.; Pan, L. Identification of Zhoukoudian *Homo erectus* brain asymmetry using 3D laser scanning. *Chin. Sci. Bull.* **2011**, *56*, 2215–2220. [CrossRef]

61. Palmer, A.R.; Strobeck, C. Fluctuating asymmetry analyses revisited. In *Developmental Instability: Causes and Consequences*; Polak, M., Ed.; Oxford University Press: Oxford, UK, 2003; pp. 279–319.

62. Gómez-Robles, A.; Hopkins, W.D.; Sherwood, C.C. Increased morphological asymmetry, evolvability and plasticity in human brain evolution. *Proc. R. Soc. B Biol. Sci.* **2013**, *280*, 20130575. [CrossRef]

63. Weaver, T.D.; Gunz, P. Using geometric morphometric visualizations of directional selection gradients to investigate morphological differentiation. *Evolution* **2018**, *72*, 838–850. [CrossRef]

64. Mitteroecker, P.; Gunz, P. Advances in geometric morphometrics. *Evol. Biol.* **2009**, *36*, 235–247. [CrossRef]

65. Richtsmeier, J.T.; Cole, T.M.; Lele, S.R. An invariant approach to the study of fluctuating asymmetry: Developmental instability in a mouse model for Down syndrome. In *Modern Morphometrics in Physical Anthropology*; Springer: Berlin, Germany, 2005; pp. 187–212.

66. Kent, J.T.; Mardia, K.V. Shape, Procrustes tangent projections and bilateral symmetry. *Biometrika* **2001**, *88*, 469–485. [CrossRef]

67. Klingenberg, C.P.; McIntyre, G.S. Geometric Morphometrics of developmental instability: Analyzing patterns of fluctuating asymmetry with Procrustes methods. *Evolution* **1998**, *52*, 1363–1375. [CrossRef] [PubMed]

68. Klingenberg, C.P.; Barluenga, M.; Meyer, A. Shape analysis of symmetric structures: Quantifying variation among individuals and asymmetry. *Evolution* **2002**, *56*, 1909–1920. [CrossRef]

69. Mardia, K.V.; Bookstein, F.L.; Moreton, I.J. Statistical assessment of bilateral symmetry of shapes. *Biometrika* **2000**, *87*, 285–300. [CrossRef]

70. Combès, B.; Hennessy, R.; Waddington, J.L.; Roberts, N.; Prima, S. Automatic symmetry plane estimation of bilateral objects in point clouds. In Proceedings of the 2008 IEEE Conference on Computer Vision and Pattern Recognition-CVPR'2008, Anchorage, AK, USA, 24–26 June 2008.

71. Combès, B.; Fournier, M.; Kennedy, D.N.; Braga, J.; Roberts, N.; Prima, S. EM-ICP strategies for joint mean shape and correspondences estimation: Applications to statistical analysis of shape and of asymmetry. In Proceedings of the 8th IEEE International Symposium on Biomedical Imaging: From Nano to Macro (ISBI'2011), Chicago, IL, USA, 30 March–2 April 2011; pp. 1257–1263.

72. Abdel Fatad, E.E.; Shirley, N.R.; Mahfouz, M.R.; Auerbach, B.M. A three-dimensional analysis of bilateral directional asymmetry in the human clavicles. *Am. J. Phys. Anthrop.* **2012**, *49*, 547–559. [CrossRef] [PubMed]

73. Balzeau, A.; Gilissen, E. Endocranial shape asymmetries in *Pan paniscus*, *Pan troglodytes* and *Gorilla gorilla* assessed via skull based landmark analysis. *J. Hum. Evol.* **2010**, *59*, 54–69. [CrossRef]

74. Sun, Z.Y.; Klöppel, S.; Rivière, D.; Perrot, M.; Frackowiak, R.; Siebner, H.; Mangin, J.F. The effect of handedness on the shape of the central sulcus. *Neuroimage* **2012**, *60*, 332–339. [CrossRef]

75. Sprung-Much, T.; Eichert, N.; Nolan, E.; Petrides, M. Broca's area and the search for anatomical asymmetry: Commentary and perspectives. *Brain Struct. Funct.* **2021**, 1–9. [CrossRef]

76. LeMay, M. Morphological cerebral asymmetries of modern man, fossil man and nonhuman primate. *Ann. N. Y. Acad. Sci.* **1976**, *280*, 349–366. [CrossRef]

77. LeMay, M. Asymmetries of the skull and handedness. *J. Neurol. Sci.* **1977**, *32*, 243–253. [CrossRef]

78. Holloway, R.L.; De La Coste-Lareymondie, M.C. Brain endocast asymmetry in pongids and hominids: Some preliminary findings on the paleontology of cerebral dominance. *Am. J. Phys. Anthrop.* **1982**, *58*, 101–110. [CrossRef]

79. Hopkins, W.D.; Marino, L. Asymmetries in cerebral width in nonhuman primate brains as revealed by magnetic resonance imaging (MRI). *Neuropsychologia* **2000**, *38*, 493–499. [CrossRef]

80. Pilcher, D.L.; Hammock, E.A.D.; Hopkins, W.D. Cerebral volumetric asymmetries in non-human primates: A magnetic resonance imaging study. *Laterality* **2001**, *6*, 165–179. [CrossRef] [PubMed]

81. Good, C.D.; Johnsrude, I.; Ashburner, J.; Henson, R.N.; Friston, K.J.; Frackowiak, R. Cerebral asymmetry and the effects of sex and handedness on brain structure: A voxel-based morphometric analysis of 465 normal adult human brains. *Neuroimage* **2001**, *14*, 685–700. [CrossRef] [PubMed]

82. Watkins, K.E.; Paus, T.; Lerch, J.; Zijdenbos, A.; Collins, D.L.; Neelin, P.; Taylor, J.; Worsley, K.; Evans, A. Structural asymmetries in the human brain: A voxel-based statistical analysis of 142 MRI scans. *Cereb. Cortex* **2001**, *11*, 868–877. [CrossRef]

83. Hopkins, W.D.; Taglialatela, J.P.; Meguerditchian, A.; Nir, T.; Schenker-Ahmed, N.M.; Sherwood, C.C. Gray matter asymmetries in chimpanzees as revealed by voxel-based morphometry. *Neuroimage* **2008**, *42*, 491–497. [CrossRef]

84. Holloway, R.L. Volumetric and asymmetry determinations on recent hominid endocasts: Spy I and II, Djebel Ihroud I, and the Salé *Homo erectus* specimens. *Am. J. Phys. Anthrop.* **1981**, *5*, 385–393. [CrossRef] [PubMed]

85. LeMay, M.; Kido, D.K. Asymmetries of the cerebral hemispheres on computed tomograms. *J. Comput. Assist. Tomogr.* **1978**, *2*, 471–476.

86. LeMay, M.; Billig, M.S.; Geschwind, N. Asymmetries of the brains and skulls of nonhuman primates. In *Primate Brain Evolution, Methods and Concepts*; Falk, D., Armstrong, E., Eds.; Plenum Press: New York, NY, USA, 1976; pp. 263–277.

87. Galaburda, A.M.; LeMay, M.; Kemper, T.L.; Geschwind, N. Right-left asymmetries in the brain. *Science* **1978**, *199*, 852–856. [CrossRef] [PubMed]

88. Kertesz, A.; Black, S.E.; Polk, M.; Howell, J. Cerebral asymmetries on magnetic resonance imaging. *Cortex* **1986**, *22*, 117–127. [CrossRef]

89. Falk, D.; Hildebolt, C.; Cheverud, J.; Vannier, M.W.; Helmkamp, R.C.; Konigsberg, L. Cortical asymmetries in frontal lobes of rhesus monkeys (*Macaca mulatta*). *Brain Res.* **1990**, *512*, 40–45. [CrossRef]

90. LeMay, M. Asymmetries of the brains and skulls of nonhuman primates. In *Cerebral Lateralization in Nonhuman Species*; Glick, S.D., Ed.; Academic Press: New York, NY, USA, 1985; pp. 233–245.

91. Cain, D.P.; Wada, J.A. An anatomical asymmetry in the baboon brain. *Brain Behav. Evol.* **1979**, *16*, 222–226. [CrossRef]

92. Cheverud, J.M.; Falk, D.; Hildebolt, C.; Moore, A.J.; Helmkamp, R.C.; Vannier, M.W. Heritability and association of cortical petalias in rhesus macaques (*Macaca mulatta*). *Brain Behav. Evol.* **1990**, *35*, 368–372. [CrossRef]

93. Hopkins, W.D.; Phillips, K.; Bania, A.; Calcutt, S.E.; Gardner, M.; Russell, J.; Schaeffer, J.; Lonsdorf, E.V.; Ross, S.R.; Schapiro, S.J. Hand preferences for coordinated bimanual actions in 777 great apes: Implications for the evolution of handedness in hominins. *J. Hum. Evol.* **2011**, *60*, 605–611. [CrossRef]

94. Bogart, S.L.; Mangin, J.-F.; Schapiro, S.J.; Reamer, L.; Bennett, A.J.; Pierre, P.J.; Hopkins, W.D. Cortical sulci asymmetries in chimpanzees and macaques: A new look at an old idea. *Neuroimage* **2012**, *61*, 533–541. [CrossRef]

95. Corballis, M.C.; Badzakova-Trajkov, G.; Häberling, I.S. Right hand, left brain: Genetic and evolutionary bases of cerebral asymmetries for language and manual action. *WIREs Cogn. Sci.* **2012**, *3*, 1–17. [CrossRef]

96. Xiang, L.; Crow, T.; Roberts, N. Cerebral torque is human specific and unrelated to brain size. *Brain Struct. Funct.* **2019**, *224*, 1141–1150. [CrossRef]

97. Keller, S.S.; Crow, T.; Foundas, A.; Amunts, K.; Roberts, N. Broca's area: Nomenclature, anatomy, typology and asymmetry. *Brain Lang.* **2009**, *109*, 29–48. [CrossRef]

98. Balzeau, A.; Grimaud-Hervé, D.; Jacob, T. Internal cranial features of the Mojokerto child fossil (East Java, Indonesia). *J. Hum. Evol.* **2005**, *48*, 535–553. [CrossRef]

99. Gunz, P.; Neubauer, S.; Golovanova, L.; Doronichev, V.; Maureille, B.; Hublin, J.J. A uniquely modern human pattern of endocranial development. Insights from a new cranial reconstruction of the Neandertal newborn from Mezmaiskaya. *J. Hum. Evol.* **2012**, *62*, 300–313. [CrossRef]

100. Gunz, P.; Neubauer, S.; Maureille, B.; Hublin, J.J. Brain development after birth differs between Neanderthals and modern humans. *Curr. Biol.* **2010**, *20*, R921–R922. [CrossRef]

101. Neubauer, S.; Gunz, P.; Hublin, J.J. Endocranial shape changes during growth in chimpanzees and humans: A morphometric analysis of unique and shared aspects. *J. Hum. Evol.* **2010**, *59*, 555–566. [CrossRef] [PubMed]

102. Ponce de León, M.S.; Bienvenu, T.; Akazawa, T.; Zollikofer, C.P.E. Brain development is similar in Neanderthals and modern humans. *Curr. Biol.* **2016**, *26*, R665–R666. [CrossRef] [PubMed]

103. Balzeau, A.; Holloway, R.L.; Grimaud-Hervé, D. Variations and asymmetries in regional brain surface in the genus *Homo*. *J. Hum. Evol.* **2012**, *62*, 696–706. [CrossRef] [PubMed]

104. Coulon, O.; Sein, J.; Auzias, G.; Nazarian, B.; Anton, J.L.; Rousseau, F.; Velly, L.; Girard, N. High temporal resolution longitudinal observation of fetal brain development. A baboon pilot study. In Proceedings of the 26th Annual Meeting of the Organization for Human Brain Mapping, Montreal, QC, Canada, 23 June–3 July 2020.

105. Pearce, E.; Stringer, C.; Dunbar, R.I.M. New insights into differences in brain organization between Neanderthals and anatomically modern humans. *Proc. R. Soc. B Biol. Sci.* **2013**, *280*, 20130168. [CrossRef]

106. Lemaitre, H.; Le Guen, Y.; Tilot, A.K.; Stein, J.L.; Philippe, C.; Mangin, J.-F.; Fisher, S.E.; Frouin, V. Genetic variations within human gained enhancer elements affect human brain sulcal morphology. *bioRxiv* **2021**, in press.

107. Ferran, J.L. Architect genes of the brain a look at brain evolution through genoarchitrecture. *Metode Sci. Stud. J.* **2017**, *7*, 17–23.

108. Kochiyama, T.; Ogihara, N.; Tanabe, H.C.; Kondo, O.; Amano, H.; Hasegawa, K.; Suzuki, H.; De León, M.S.P.; Zollikofer, C.P.E.; Bastir, M.; et al. Reconstructing the Neanderthal brain using computational anatomy. *Sci. Rep.* **2018**, *8*, 6296. [CrossRef] [PubMed]

Article

Asymmetry of Endocast Surface Shape in Modern Humans Based on Diffeomorphic Surface Matching

Sungui Lin [1,2], Yuhao Zhao [1,2] and Song Xing [1,*]

[1] Key Laboratory of Vertebrate Evolution and Human Origins, Institute of Vertebrate Paleontology and Paleoanthropology, Chinese Academy of Sciences, Beijing 100044, China; sglin@ivpp.ac.cn (S.L.); zhaoyuhao@ivpp.ac.cn (Y.Z.)

[2] College of Earth and Planetary Sciences, University of Chinese Academy of Sciences, Beijing 100049, China

* Correspondence: xingsong@ivpp.ac.cn

Abstract: Brain asymmetry is associated with handedness and cognitive function, and is also reflected in the shape of endocasts. However, comprehensive quantification of the asymmetry in endocast shapes is limited. Here, we quantify and visualize the variation of endocast asymmetry in modern humans using diffeomorphic surface matching. Our results show that two types of lobar fluctuating asymmetry contribute most to global asymmetry variation. A dominant pattern of local directional asymmetry is shared in the majority of the population: (1) the left occipital pole protrudes more than the right frontal pole in the left-occipital and right-frontal petalial asymmetry; (2) the left Broca's cap appears to be more globular and bulges laterally, anteriorly, and ventrally compared to the right side; and (3) the asymmetrical pattern of the parietal is complex and the posterior part of the right temporal lobes are more bulbous than the contralateral sides. This study confirms the validity of endocasts for obtaining valuable information on encephalic asymmetries and reveals a more complicated pattern of asymmetry of the cerebral lobes than previously reported. The endocast asymmetry pattern revealed here provides more shape information to explore the relationships between brain structure and function, to re-define the uniqueness of human brains related to other primates, and to trace the timing of the human asymmetry pattern within hominin lineages.

Keywords: cerebrum; cerebellum; petalia; shape asymmetry; diffeomorphic surface matching

Citation: Lin, S.; Zhao, Y.; Xing, S. Asymmetry of Endocast Surface Shape in Modern Humans Based on Diffeomorphic Surface Matching. *Symmetry* **2022**, *14*, 1459. https://doi.org/10.3390/sym14071459

Academic Editors: Sebastian Ocklenburg and Jan Awrejcewicz

Received: 29 May 2022
Accepted: 9 July 2022
Published: 17 July 2022

1. Introduction

The structural and functional asymmetries of the human brain have been extensively studied in the fields of medicine and biology, especially as they relate to handedness and cognitive function [1–9]. The morphological characteristics of the brain surface are pressed into the inner cranium and may be visualized on endocasts, which are casts of the interior portion of a cranium [10,11]. Previous studies have demonstrated that the left and right hemispheres of human endocasts are usually asymmetrical in shape [12]. It is worth mentioning that asymmetries are divided into three categories: directional asymmetry, anti-symmetry, and fluctuating asymmetry [13,14]. The most concerning asymmetry pattern is directional asymmetry, characterized by a consistent directional bias within a population [13,15]. "Anti-symmetry" represents the reverse pattern (direction) of the population's directional asymmetry and is restricted to a small portion of the population. The third type of asymmetry is "fluctuating asymmetry", which means the asymmetry pattern of a character is diverse without a particular direction in a population [15].

A local impression on the internal surface of a skull, resulting from a protrusion of one brain hemisphere relative to the other, has been referred to as "petalia" and may be visible on endocasts [7]. Left occipito-petalias have been frequently associated with right fronto-petalias, whereas the directional asymmetries of parieto-petalias and temporo-petalias are inconsistent in different research [16,17]. Petalial asymmetries have been demonstrated to exist in a wide variety of hominids [3,12,15,18,19]. The particular petalia asymmetry pattern with a right-frontal and left-occipital bias, commonly recognized in modern humans and fossil hominins, is considered to be associated with handedness [2,19–22]. Previous studies have also shown that petalial asymmetries differ between males and females, with males having slightly stronger right-frontal and left-occipital lateralization [1,2].

In many cases, protrusions of brain hemispheres are associated with lobar asymmetries. Previous investigations have revealed that the right frontal lobe and the left occipital lobe are frequently wider than the opposite hemisphere in modern humans and great apes [7,12,23]. Indeed, a prominent geometric distortion of the hemispheres, known as Yakovlevian anti-clockwise torque, is frequently observed in human brains and endocasts [7,24]. Specifically, the Yakovlevian anticlockwise torque includes the left-occipital and right-frontal petalias, with the left occipital lobe extending across the midline over the right and wider/larger right frontal and left occipital regions [22]. More recently, endocasts of the genus *Homo* have revealed that the right hemisphere often has a greater surface area than the left, while the right parieto-temporal lobe and the left occipital lobe have larger surface areas than their contralateral regions [4]. In addition, the asymmetries of the cerebellum and temporal lobe have also attracted attention, though relatively little is known about what their structural asymmetries may reflect [25].

The study of the brain lateralization associated with language was one of the most profound discoveries for neurobiology and linguistics. The leftward asymmetry of Broca's area, as a motor speech area, including the pars triangularis and pars opercularis of the inferior frontal gyrus, was first identified by Broca in 1861 [26]. Functional magnetic resonance imaging (fMRI) studies have provided more evidence for the specialization of the left hemisphere for language [5,27]. Broca's area is referred to as Broca's cap on an endocast, representing a "protrusion of the orbital portion of the inferior frontal gyrus" [28]. Since the emergence of genus *Homo*, it is generally accepted that the left Broca's cap is larger and more prominent than the right [29–31]. Recently, a landmark-based quantitative study of the asymmetry of the third frontal convolution in endocasts suggested that the left Broca's cap of modern humans, although smaller in size, is more globular and better defined than the right side [32]. Wernicke's area is responsible for language comprehension and mainly includes the posterior portion of the superior temporal gyrus and middle temporal gyrus, as well as the inferior parietal lobule, which includes the angular gyrus and supramarginal gyrus [24]. Due to the lack of homologous anatomical markers, Wernicke's area is not well defined on endocast surfaces and displays an unclear pattern of asymmetry [5,33–35]. Generally, though, the planum temporale (the main cortical area of Wernicke's area) is larger on the left hemisphere than on the right [36–38].

As mentioned above, most of the work on the asymmetry of endocasts has been focused on the degree of anterior or posterior protrusion, the lateralization of some regions associated with speech (e.g., Broca's cap), or the relative width and area of frontal and occipital lobes [39]. Moreover, previous research usually compared the human endocast with that of great apes or other primates to reveal a shared directional asymmetry pattern or a particular characteristic rather than conduct a global comparison of the entire brain surface [3,12,32,40]. The breadth of endocast asymmetry patterns within modern humans remains unclear.

Due to methodological limitations, it is difficult to comprehensively quantify the asymmetry of an endocast's entire surface. Previous studies typically relied on morphological descriptions, linear measurements, and geometric morphometrics [41–44]. The development of a landmark-free surface deformation method, diffeomorphic surface matching (DSM), provides interesting research opportunities for evaluating morpho-architectural variation on endocasts [45–47]. Compared with geometric morphometrics, DSM does not rely on the definitions of landmarks and semi-landmarks to capture the shape of the whole anatomical structure and can dynamically display the shape variation among different specimens [45–47]. Analytic results based on DSM indicate that endocasts of *Australopithecus africanus* (Sts 5 and Sts 60) display a more elongated frontal bec and a substantially less elevated parietal area, different from those of genus *Homo* [46]. Visualizations of sulcal patterns have contributed more information to taxonomic identification in Old World monkeys [45]. Additionally, this deformation-based approach has been applied to dental materials and the vestibular apparatus [48,49]. However, no studies focusing on the morphological asymmetry of endocasts in modern humans have yet used this landmark-free method.

In the present study, 58 endocasts of archaeological modern Chinese crania were virtually reconstructed with high-resolution computed tomography and three-dimensional virtual technology. Landmark-free diffeomorphic surface matching analysis was performed to quantify and visualize the shape variation of the endocasts. We aim to quantify individual variation in the asymmetry of the endocast surface shape and analyze the variation of asymmetry patterns between the left and right hemispheres within the modern human population, as well as tentatively discuss the correlations between the structural and functional asymmetry of human brains.

2. Materials and Methods

2.1. Materials

In total, 58 adult endocasts, including 28 females and 30 males, were collected from the same archaeological site in Yunnan Province of Southwestern China, dated to about 300 years ago [50]. The skulls were well preserved or only minorly damaged in a way that would have no significant influence on the endocast reconstruction. The sexual assignment of specimens relied on diagnostic characteristics of the pelvis and cranium [50].

2.2. Endocast Reconstruction and Processing

All of the modern human specimens investigated in this study were scanned by a 450 KV industrial CT scanner with a spatial resolution of 160 μm (designed by the Institute of High Energy Physics, Chinese Academy of Sciences, and housed at the Key Laboratory of Vertebrate Evolution and Human Origins, Institute of Vertebrate Paleontology and Paleoanthropology, Chinese Academy of Sciences) [51]. The virtual reconstruction of each endocast was performed through semi-automatic threshold-based segmentation via the Mimics v. 17.0 software (Materialise, Leuven). A three-dimensional (3D) mesh of each endocast was generated and saved as an STL file in Mimics, and then imported into MeshLab software (Bangalore, India) [52] for 'Cleaning and Repairing' and a resulting 3D surface was obtained.

Considering the presence of Yakovlevian anticlockwise torque, the cerebral longitudinal fissure is not entirely on the midsagittal plane. Thus, the central axis or central plane separating a complete left hemisphere from its right side is difficult to determine. To investigate shape differences of the endocast between the right and left hemispheres, mirrored versions of each original specimen were created via Avizo v. 8.0 (FEI Visualization Sciences Group, Houston, TX, USA) [15,25]. The left side of the original endocast corresponds to the right side of the mirrored version and vice versa, as shown in Figure 1. Therefore, a total of 116 cases of endocast surfaces were included in the following deformation analyses. In this way, we could obtain a symmetrical mean shape by averaging the original and mirrored endocast of each individual in the following steps. The shape asymmetry of endocast surfaces was analyzed by calculating the deformation between each original endocast and its mirrored endocast.

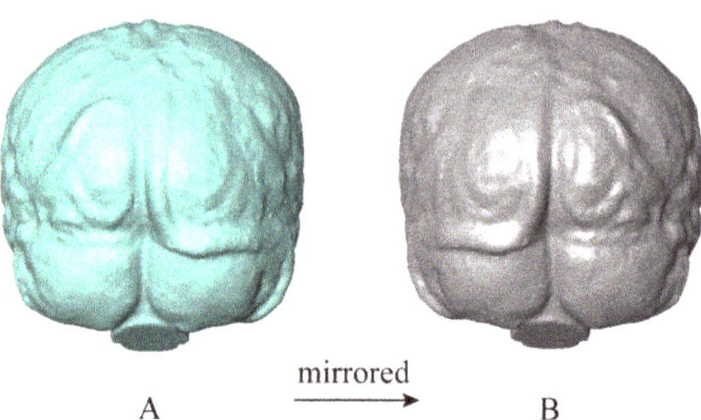

Figure 1. The original endocast (**A**) and the mirrored endocast (**B**) of the same individual in occipital view.

2.3. Diffeomorphic Surface Asymmetry

All surfaces were superimposed and aligned in Avizo through translation, rotation, and dilation (scaling) to eliminate differences, except for shapes. The aligned surfaces were exported as PLY files and then transformed to VTK format by ParaView v. 5.6.0 software (Kitware Inc., Clifton Park, NY, USA). The set of VTK data were imported into Deformetrica v. 4.3 (Paris, France) [53] to carry out the diffeomorphic calculation. The outputs include the spatial coordinates of control points defining the deformable space (9200 in this study), the momenta vectors recording the deformation information of each control point, and a symmetric endocast configuration representing the global mean shape [45,54].

For each endocast specimen, the vector difference at each control point was calculated by subtracting the momenta vector of the mirrored one from its counterpart of the origin one. Then, the surface asymmetry was quantified by an asymmetrical matrix depositing the vector differences at all control points.

A non-center principal component analysis (PCA) using the "RToolsForDeformetrica" [55] and "ade4" v. 1.7-17 [56] packages for R v. 4.0.4 [57] was carried out on an array storing the asymmetrical matrices of all specimens. The "ggplot2" package [58] was used to visualize the result of the PCA. A scatter plot of the second principal component (PC2) against the first principal component (PC1) displayed the distributional relationship of each specimen. The deformations displaying the asymmetric patterns at the four extremes were computed via Deformetrica and visualized in ParaView v. 5.6.0. These deformations were displayed in a form of colormap from dark blue (more constricted compared to the opposite) to red (more expanded compared to the opposite).

The mean matrix averaging the asymmetrical matrices of all specimens was calculated to exhibit the general pattern of surface asymmetry. The deformation for this mean matrix was also computed via Deformetrica and visualized in ParaView v. 5.6.0.

3. Results

3.1. Principal Component Analysis

The first two principal components account for 27.06% and 14.32% of the total shape asymmetry variation. The PC1 indicates that there is no clear directional trend for the two asymmetrical types represented by the shapes at the two extremes (Figure 2). At the positive extreme of PC1 (Figure 3A), the frontal, anterior parietal, and anterior temporal lobes in the right hemisphere are more bulged compared to the left hemisphere, whereas the occipital, posterior temporal, and posterior parietal lobes in the left hemisphere project more posteriorly and laterally than the right side. With increasing values along PC1, the left cerebellar lobe bulges and protrudes more posteriorly, while the right cerebellar lobe exhibits the opposite trend. Comparatively, the asymmetrical pattern in the negative-value end of PC1 shows a contrary trend to that at the positive end (Figure 3B).

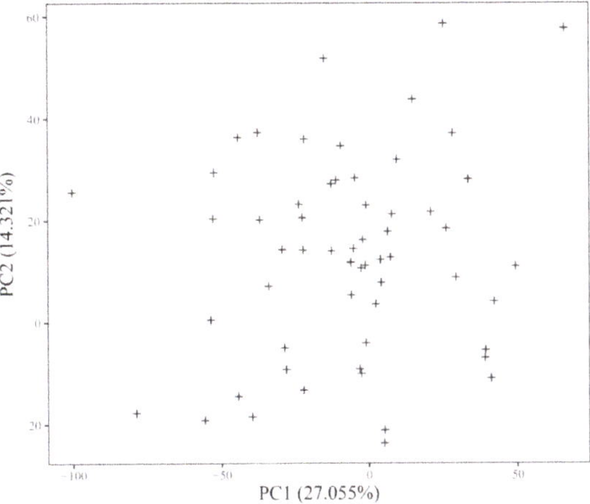

Figure 2. Bivariate plot of PC2 against PC1 based on diffeomorphic surface matching (DSM) analysis for endocast shape asymmetry variation.

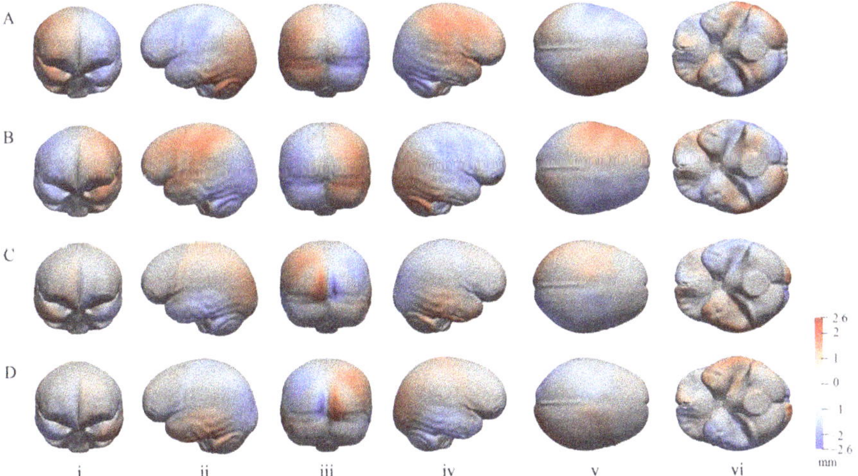

Figure 3. The virtual asymmetric shapes of endocast surfaces at the positive and negative extremes of PC1 (**A,B**) and PC2 (**C,D**) in frontal (**i**), left lateral (**ii**), occipital (**iii**), right lateral (**iv**), dorsal (**v**), and basal (**vi**) views. The colormap shows the degree of asymmetry in endocast surface shape. The dark blue and red represent the most constricted and expanded area of one hemisphere relative to the other half (in millimeter).

As shown in Figure 2, PC2 presents directional asymmetry: about 74% of the individuals are distributed in the positive-value half of PC2 and 26% in the negative-value half. The shape asymmetry pattern of the endocast at the dominant side of the distribution (positive extreme end of PC2, Figure 3C) is mainly shown as left-occipital and right-frontal petalial asymmetries. The superior and middle-frontal convolutions present a slight rightward asymmetry, while the inferior frontal convolution shows a slight leftward asymmetry. The temporal lobe presents a rightward asymmetry. The parietal-occipital lobe presents a leftward asymmetry. The cerebellum shows a double asymmetry with a leftward anterior lobe and rightward posterior lobe, but to a low degree. The negative extreme end of PC2

(Figure 3D) displays the reverse trend of the asymmetrical pattern, namely anti-symmetry of those at the positive extreme end of PC2.

3.2. The General Pattern of Endocast Surface Asymmetry

The mean asymmetric shape of the endocast surface, averaging the asymmetrical deformation of all individuals, is shown in Figure 4. It reveals a directional asymmetry pattern consistent with the positive-value end of PC2 but displays more detailed local asymmetries.

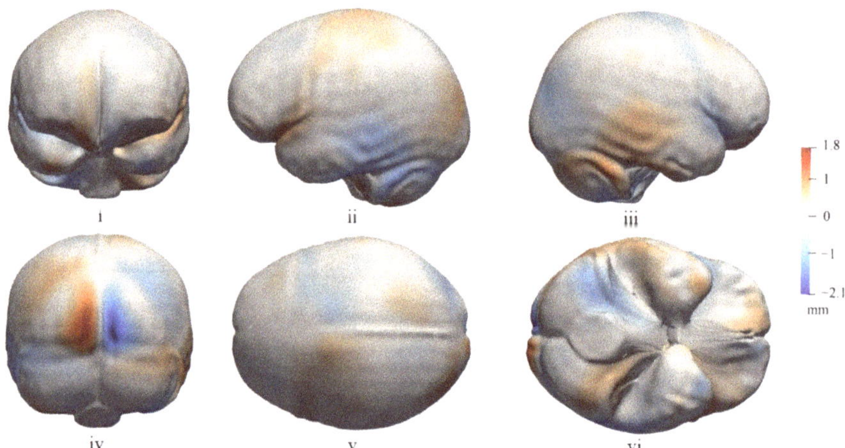

Figure 4. The mean asymmetric shape of the endocast surface in frontal (**i**), left lateral (**ii**), right lateral (**iii**), occipital (**iv**), dorsal (**v**), and basal (**vi**) views. The colormap shows the degree of asymmetry in endocast surface shape. The dark blue and red represent the most constricted and expanded area of one hemisphere relative to the other, and the displacement within 5% from symmetric shape is shown in white (in millimeter).

In the frontal view (Figure 4i), the frontal lobe has a strip adjacent to the longitudinal cerebral fissure, showing expansion on the right hemisphere and contraction on the left side. As a result, the right hemisphere protrudes more anteriorly than the opposite side and slightly bends the anterior interhemispheric fissure towards the left. The left inferior frontal convolution (Figure 4ii), an area that includes Broca's area, is extended more anteriorly, laterally, and ventrally relative to the right side, whereas the right inferior frontal convolution (Figure 4iii) appears more flattened and elongated antero-posteriorly. Additionally, the right frontal bec is more elongated than that of the opposite hemisphere and the ventral surface of the frontal lobes in the left hemisphere is more bulged than the right (Figure 4vi).

In terms of the local shape asymmetry of the temporal lobe (Figure 4ii,iii), the posterior portion, involving all three convolutions, is more inflated in the right hemisphere than the left. The right temporal lobe protrudes more inferiorly than the left one, resulting in a right temporal petalia (Figure 4vi). Additionally, the left temporal lobe appears to be shorter and the left Sylvian fissure is displaced anteriorly to a greater extent than the right. The parietal lobe also shows complex shape asymmetry (Figure 4v): the anterior portion adjacent to the frontal lobe and the posterior portion adjacent to the occipital lobe are more bulged on the left parietal lobe than the right, while an elliptic region, roughly corresponding to the superior parietal lobule, appears to protrude more on the right hemisphere than the left. Asymmetries in the temporal and parietal lobes suggest that the directional asymmetry in the surface of the region corresponding to Wernicke's area is in favor of the right side.

The greatest degree of local asymmetry in the endocast surface was observed in the occipital region (Figure 4iv): the portion of the left occipital lobe near the indentation of the superior sagittal sinus protrudes significantly backward while the relative position of the right hemisphere contracts inward compared to the opposite side. Additionally, the left occipital lobe extends more medially than the right, bending the posterior interhemispheric fissure towards the right.

The surface shape of the cerebellum shows bilateral asymmetry (Figure 4iv,vi): the anterior portion of the cerebellum extends more anteriorly and ventrally in the left hemisphere than the right, whereas the posterior portion of the cerebellum extends more posteriorly and superiorly in the right hemisphere than the left. The cerebellum thus has the appearance of an anticlockwise twisting torque in basal view.

4. Discussion

4.1. Global Endocast Asymmetry

In this study, we performed surface matching using diffeomorphisms to quantify and visualize the asymmetry of human endocasts. We found that two types of lobar asymmetries categorize the majority of the global asymmetry variation. These two types of lobar asymmetries correspond to the clockwise and counterclockwise distortion of the global brain, with the cerebrum and cerebrum being consistent in the deformation trends. Previous studies revealed that the right frontal, right parieto-temporal, and the left occipital lobes have larger surface areas than the contralateral sides [4]. Measurement of lobe volumes using MRI found rightward asymmetries for the frontal and temporal lobes, and leftward asymmetries for the parietal and occipital lobes in right-handed twin pairs [20]. Here, we find that the rightward asymmetries of frontal, anterior parietal, and anterior temporal lobes, and leftward asymmetries of occipital, posterior temporal, and posterior parietal lobes have a roughly equivalent distribution compared to the reverse asymmetry pattern in this population. This indicates that the lobar asymmetry has a more complicated pattern of shape, surface area, and volume asymmetry.

4.2. Local Asymmetries of the Cerebrum

The shape at the positive-value extreme of PC2 and the deformation-based mean asymmetric shape both show a similar pattern of directional asymmetry. This directional asymmetry pattern is about three times as common as its anti-symmetry in the modern population. The general asymmetry pattern illustrated here is similar to the results of the previous study using geometric morphometrics to quantify hominid endocranial asymmetry [15] but reveals more details about the local asymmetries.

The petalia asymmetry pattern of endocasts in modern humans is typically characterized by the right frontal and left occipital lobes protruding outward more than the opposite side. Deformation results in the present study indicate that the leftward occipital petalia is much more prominent than the rightward frontal petalia, which is consistent with analyses of geometric morphometrics and linear measurements in endocasts of humans [15,59]. The temporal lobe presents a right petalia projecting inferiorly, which has not been observed in previous studies.

We found that the asymmetry in the posterior part of the temporal lobe favors the right side and the local asymmetries of the parietal lobe are complex depending on the mean asymmetric shape. In this context, the surface of Wernicke's area is primarily skewed to the right. Previous MRI research showed a complicated asymmetry in the temporal lobe. Kitchell and colleagues [5] have reported that the superior temporal sulcus is rightward-asymmetric while the planum temporale is leftward-asymmetric. A rightward asymmetry in the depth of the superior temporal sulcus ventral to Heschl's gyrus is known to be widely present in modern humans but rare in chimpanzees [60]. Due to the fact that the planum temporale and superior temporal sulcus are internal anatomical structures, it is difficult to define the contour of these area on an endocast. Therefore, the asymmetry of Wernicke's area discussed here is roughly based on the surface shape of the posterior parts of superior and middle-temporal convolutions.

A leftward asymmetry of Broca's area in the inferior frontal convolution has been identified as a feature commonly found in hominins and related to the brain lateralization associated with language [11,29]. Indeed, quantitative measurements revealed that the Broca's cap is larger in the right but more clearly defined in the left in modern humans [25,28,32]. In the present study, the inferior frontal convolution is more flattened and elongated antero-posteriorly in the right hemisphere, while the left inferior frontal convolution extends more laterally, anteriorly, and ventrally relative to the right side. As a result, the left Broca's cap appears to be more globular. In the quantitative study of *H. sapiens* endocast by Balzeau and colleagues [32], the left Broca's cap was also found to be more globular than the right side, but the length and the size of the third frontal convolution displayed a rightward asymmetry. This observation is supported and upheld by the results of the present study. Wada and colleagues [37] measured the visible cortical area on the frontal operculum (including both the pars opercularis and a posterior portion of the pars triangularis) and found that the left side was smaller than the right. However, Falzi and colleagues [61] measured the cortical surface area of Broca's area (including both the extra-sulcal and intra-sulcal cortex) and revealed that the left Broca's area was significantly larger than the right one. The difference in these results is due to the deeper fissure of the cortex in the left Broca's area [62] and perhaps leads to the larger size on the right but more globular shape on the left Broca's area.

4.3. Asymmetry of the Cerebellum

The cerebellum is responsible for controlling movement and coordinating balance, as well as for regulating cognition and emotion through information circuits with the non-motor cortex in the prefrontal and posterior parietal [63–65]. Previous studies have found a leftward asymmetry of the anterior cerebellum and a rightward asymmetry of the posterior cerebellum [25]. Here, the cerebellum shows a double asymmetry in which the right posterior cerebellar lobe extends more posteriorly and superiorly than the left, and across the midline, whereas the left anterior cerebellar lobe extends more anteriorly and ventrally. Therefore, the surface shape of the cerebellum appears as a twisting effect in the opposite direction relative to the Yakovlevian anticlockwise torque of the cerebrum. There is evidence that the motor and non-motor cortex in the left and right hemispheres of the cerebrum show strong preferential correlations with the related functional areas in the contralateral cerebellum [66–68]. In addition, the region and degree of functional lateralization in the cerebellum are correlated with that of lateralization in the cerebrum [6]. With that in mind, the cerebellum may possess roughly similar asymmetrical patterns of function and structure to the cerebrum [6].

5. Significance and Conclusions

Here, we quantified and visualized the asymmetry of endocast surface shapes in a modern human population using landmark-free DSM. Like previous studies, we have found the dominant asymmetry pattern to have left-occipital and right-frontal petalias, a more globular left Broca's area compared to the right, and a double asymmetry in the cerebellum. In addition, our results reveal more information of the asymmetry pattern in parietal and temporal lobes. Brain structural asymmetry is extensively involved in previous studies and often associated with functional lateralization. For example, the left hemisphere is generally dominant for language, with a more prominent Broca's area and a larger planum temporale [26,69]; besides, right-handed individuals often exhibit a more pronounced left-occipital and right-frontal petalias asymmetry than non-right-handed individuals [2]. Additionally, the cortical thickness asymmetry in a specific region of the postcentral gyrus correlated with hand preference, where right-handers dominated by the left hemisphere had a less rightward/more leftward shift of neural resources [9].

Our findings support previous MRI studies and confirm the validity of endocasts for obtaining valuable information on encephalic asymmetries [1,10,15,20,37,59]. Specifically, we find that the surface shape of the temporal language comprehension area (i.e., Wernicke's area) presents a rightward asymmetry, which is different from the asymmetrical pattern of the motor speech area (i.e., Broca's area). Thus, the hemisphere dominance for language in terms of shape asymmetries, reflected by the endocast surface, might not be completely leftward. Furthermore, a rightward temporal petalia and a complex asymmetry pattern of the parietal lobe were also revealed. Whether these asymmetry features in the endocast surface can be correlated with a function of the brain needs to be investigated in combination with MRI studies as well as other morphological and functional studies in the future.

The evolution of the unique structure of the human brain has long been explored by analyzing the evolutionary sequence of endocasts and comparing humans and other primates [3,40,70,71]. A detailed understanding of the asymmetric patterns of the brain structure in modern humans is central to this topic. The PCA results presented in this study demonstrate that modern humans present a fluctuating asymmetry on the endocast surface, as represented by two balanced components of left parietal/right occipital lobes or right parietal/left occipital lobes. On the other hand, the non-center PCA results and the average asymmetric shapes show that most individuals exhibit a prevalent directional asymmetry, as discussed earlier in this paper. According to previous studies, the brain of great apes shows a similar right-frontal and left-occipital directional asymmetry in the width of lobes, and a similar but less variable and low-degree fluctuating asymmetry in components of the petalias [3,15,23]. Moreover, Balzeau and colleagues [32] have found that *Pan paniscus* shares a common pattern of asymmetries in the third frontal convolution with *Homo sapiens* through the quantitative study of endocasts. Great apes have also been shown to have leftward asymmetries in the size of the planum temporale and Broca's area [72,73]. Additionally, previous studies have revealed that levels of brain asymmetry varied in the evolution process of *Homo* species, especially when the common ancestor of *Homo heidelbergensis*, *Homo neanderthalensis*, and *Homo sapiens* emerged [40]. This study reveals that modern humans have a more complex pattern of endocast asymmetry than previously understood, which involves both fluctuating and directional asymmetry in different lobes. Considering this new understanding of endocast asymmetry, it is necessary to assess whether the asymmetry pattern in modern humans is also present in non-human primates and whether it is present in particular stages of hominin evolution.

Author Contributions: Conceptualization, S.X. and S.L.; methodology, Y.Z. and S.X.; software, S.L.; validation, S.L., Y.Z. and S.X.; formal analysis, S.L.; investigation, S.L.; resources, S.X.; data curation, S.L. and S.X.; writing—original draft preparation, S.L.; writing—review and editing, S.X.; visualization, S.L.; supervision, S.X.; project administration, S.X.; funding acquisition, S.X. All authors have read and agreed to the published version of the manuscript.

Funding: This work was supported by the Strategic Priority Research Program of the Chinese Academy of Sciences (no. XDB26000000) and the National Natural Science Foundation of China (41872030).

Institutional Review Board Statement: Not applicable.

Informed Consent Statement: Not applicable.

Data Availability Statement: The data supporting this study are available from the corresponding author on reasonable request.

Acknowledgments: The authors thank Yemao Hou and Jing Zuo for their help in CT scanning and 3D reconstruction processing. Xiujie Wu provided the Mimics files of 3D reconstruction. Mackie O'Hara-Ali helped us in revising the manuscript.

Conflicts of Interest: The authors declare no conflict of interest.

References

1. Zilles, K.; Dabringhaus, A.; Geyer, S.; Amunts, K.; Qü, M.; Schleicher, A.; Gilissen, E.; Schlaug, G.; Steinmetz, H. Structural asymmetries in the human forebrain and the forebrain of non-human primates and rats. *Neurosci. Biobehav. Rev.* **1996**, *20*, 593–605. [CrossRef]
2. Bear, D.; Schiff, D.; Saver, J.; Greenberg, M.; Freeman, R. Quantitative analysis of cerebral asymmetries: Fronto-occipital correlation, sexual dimorphism and association with handedness. *Arch. Neurol.* **1986**, *43*, 598–603. [CrossRef] [PubMed]
3. Balzeau, A.; Gilissen, E.; Grimaud-Hervé, D. Shared pattern of endocranial shape asymmetries among great apes, anatomically modern humans, and fossil hominins. *PLoS ONE* **2012**, *7*, e29581. [CrossRef] [PubMed]
4. Balzeau, A.; Holloway, R.L.; Grimaud-Hervé, D. Variations and asymmetries in regional brain surface in the genus *Homo*. *J. Hum. Evol.* **2012**, *62*, 696–706. [CrossRef]
5. Kitchell, L.M.; Schoenemann, P.T.; Loyet, M. Structural asymmetries in the human brain assessed via MRI. *Am. J. Phys. Anthropol.* **2013**, 167.
6. Wang, D.; Buckner, R.L.; Liu, H. Cerebellar asymmetry and its relation to cerebral asymmetry estimated by intrinsic functional connectivity. *J. Neurophysiol.* **2012**, *109*, 46–57. [CrossRef]
7. LeMay, M. Morphological cerebral asymmetries of modern man, fossil man, and nonhuman primate. *Ann. N. Y. Acad. Sci.* **1976**, *280*, 349–366. [CrossRef]
8. Kuo, F.; Massoud, T.F. Structural asymmetries in normal brain anatomy: A brief overview. *Ann. Anat.-Anat. Anz.* **2022**, *241*, 151894. [CrossRef]
9. Sha, Z.; Pepe, A.; Schijven, D.; Carrión-Castillo, A.; Roe, J.M.; Westerhausen, R.; Joliot, M.; Fisher, S.E.; Crivello, F.; Francks, C. Handedness and its genetic influences are associated with structural asymmetries of the cerebral cortex in 31,864 individuals. *Proc. Natl. Acad. Sci. USA* **2021**, *118*, e2113095118. [CrossRef]
10. Dumoncel, J.; Subsol, G.; Durrleman, S.; Bertrand, A.; de Jager, E.; Oettle, A.C.; Lockhat, Z.; Suleman, F.E.; Beaudet, A. Are endocasts reliable proxies for brains? A 3D quantitative comparison of the extant human brain and endocast. *J. Anat.* **2021**, *238*, 480–488. [CrossRef]
11. Holloway, R.L.; Broadfield, D.C.; Yuan, M.S.; Schwartz, J.H.; Tattersall, I. *The Human Fossil Record, Brain Endocasts: The Paleoneurological Evidence*; John Wiley & Sons: Hoboken, NJ, USA, 2004; Volume 3.
12. Holloway, R.L.; De La Costelareymondie, M.C. Brain endocast asymmetry in pongids and hominids: Some preliminary findings on the paleontology of cerebral dominance. *Am. J. Phys. Anthropol.* **1982**, *58*, 101–110. [CrossRef]
13. Palmer, A.R.; Strobeck, C. Fluctuating asymmetry: Measurement, analysis, patterns. *Annu. Rev. Ecol. Syst.* **1986**, *17*, 391–421. [CrossRef]
14. Klingenberg, C.P.; McIntyre, G.S. Geometric morphometrics of developmental instability: Analyzing patterns of fluctuating asymmetry with procrustes methods. *Evolution* **1998**, *52*, 1363–1375. [CrossRef]
15. Neubauer, S.; Gunz, P.; Scott, N.A.; Hublin, J.-J.; Mitteroecker, P. Evolution of brain lateralization: A shared hominid pattern of endocranial asymmetry is much more variable in humans than in great apes. *Sci. Adv.* **2020**, *6*, eaax9935. [CrossRef]
16. Gundara, N.; Zivanovic, S. Asymmetry in East African skulls. *Am. J. Phys. Anthropol.* **1968**, *28*, 331–337. [CrossRef]
17. Hadžiselimović, H.; Čuš, M. The appearance of internal structures of the brain in relation to configuration of the human skull. *Acta Anat.* **1966**, *63*, 289–299. [CrossRef]
18. Holloway, R.L. Volumetric and asymmetry determinations on recent hominid endocasts-Spy-I and Spy-II, Djebel lhroud-I, and the Sale *Homo-erectus* specimens, with some notes on Neandertal brain size. *Am. J. Phys. Anthropol.* **1981**, *55*, 385–393. [CrossRef]
19. Balzeau, A.; Gilissen, E. Endocranial shape asymmetries in *Pan paniscus*, *Pan troglodytes* and *Gorilla gorilla* assessed via skull based landmark analysis. *J. Hum. Evol.* **2010**, *59*, 54–69. [CrossRef]
20. Geschwind, D.H.; Miller, B.L.; DeCarli, C.; Carmelli, D. Heritability of lobar brain volumes in twins supports genetic models of cerebral laterality and handedness. *Proc. Natl. Acad. Sci. USA* **2002**, *99*, 3176–3181. [CrossRef]
21. Good, C.D.; Johnsrude, I.; Ashburner, J.; Henson, R.N.A.; Friston, K.J.; Frackowiak, R.S.J. Cerebral asymmetry and the effects of sex and handedness on brain structure: A voxel-based morphometric analysis of 465 normal adult human brains. *NeuroImage* **2001**, *14*, 685–700. [CrossRef]
22. Zhao, L.; Matloff, W.; Shi, Y.; Cabeen, R.P.; Toga, A.W. Mapping complex brain torque components and their genetic architecture and phenomic associations in 24,112 individuals. *Biol. Psychiatry* **2022**, *91*, 753–768. [CrossRef] [PubMed]
23. Hopkins, W.D.; Marino, L. Asymmetries in cerebral width in nonhuman primate brains as revealed by magnetic resonance imaging (MRI). *Neuropsychologia* **2000**, *38*, 493–499. [CrossRef]
24. Toga, A.W.; Thompson, P.M. Mapping brain asymmetry. *Nat. Rev. Neurosci.* **2003**, *4*, 37–48. [CrossRef] [PubMed]
25. Kitchell, L.M. Asymmetry of the Modern Human Endocranium. Ph.D. Thesis, Institute of Archaeology, University College London, London, UK, 2015.
26. Broca, P. Remarques sur le siège de la faculté du langage articulé, suivies d'une observation d'aphémie (perte de la parole). *Bull. Mem. Soc. Anat. Paris* **1861**, *6*, 330–357.
27. Shaywitz, B.A.; Shaywltz, S.E.; Pugh, K.R.; Constable, R.T.; Skudlarski, P.; Fulbright, R.K.; Bronen, R.A.; Fletcher, J.M.; Shankweiler, D.P.; Katz, L.; et al. Sex differences in the functional organization of the brain for language. *Nature* **1995**, *373*, 607–609. [CrossRef] [PubMed]

28. Kitchell, L.M. Evaluating Language-Related Asymmetry in Endocasts Using Non-Rigid Diffeomorphic Image Registration. 2017. Available online: https://www.researchgate.net/publication/314299269_Evaluating_Language-Related_Asymmetry_in_Endocasts_using_Non-rigid_Diffeomorphic_Image_Registration (accessed on 5 October 2021).

29. Tobias, P.V. The brain of *Homo habilis*: A new level of organization in cerebral evolution. *J. Hum. Evol.* **1987**, *16*, 741–761. [CrossRef]

30. Holloway, R.L. Human paleontological evidence relevant to language behavior. *Hum. Neurobiol.* **1983**, *2*, 105–114.

31. De Sousa, A.; Cunha, E. Hominins and the emergence of the modern human brain. *Prog. Brain Res.* **2012**, *195*, 293–322. [CrossRef]

32. Balzeau, A.; Gilissen, E.; Holloway, R.L.; Prima, S.; Grimaud-Hervé, D. Variations in size, shape and asymmetries of the third frontal convolution in hominids: Paleoneurological implications for hominin evolution and the origin of language. *J. Hum. Evol.* **2014**, *76*, 116–128. [CrossRef]

33. Barrick, T.R.; Mackay, C.E.; Prima, S.; Maes, F.; Vandermeulen, D.; Crow, T.J.; Roberts, N. Automatic analysis of cerebral asymmetry: An exploratory study of the relationship between brain torque and planum temporale asymmetry. *NeuroImage* **2005**, *24*, 678–691. [CrossRef]

34. Foundas, A.L.; Leonard, C.M.; Hanna-Pladdy, B. Variability in the anatomy of the planum temporale and posterior ascending ramus: Do right- and left handers differ? *Brain Lang.* **2002**, *83*, 403–424. [CrossRef]

35. Watkins, K.E.; Paus, T.; Lerch, J.P.; Zijdenbos, A.; Collins, D.L.; Neelin, P.; Taylor, J.; Worsley, K.J.; Evans, A.C. Structural asymmetries in the human brain: A voxel-based statistical analysis of 142 MRI scans. *Cereb. Cortex* **2001**, *11*, 868–877. [CrossRef]

36. Geschwind, N.; Levitsky, W. Human brain: Left-right asymmetries in temporal speech region. *Science* **1968**, *161*, 186–187. [CrossRef]

37. Wada, J.A.; Clarke, R.; Hamm, A. Cerebral hemispheric asymmetry in humans: Cortical speech zones in 100 adult and 100 infant brains. *Arch. Neurol.* **1975**, *32*, 239–246. [CrossRef]

38. Witelson, S.F.; Pallie, W. Left hemisphere specialization for language in the newborn: Neuroanatomical evidence of asymmetry. *Brain* **1973**, *96*, 641–646. [CrossRef]

39. Wu, X.J.; Zhang, X. Progress in endocast and human brain evolution on Chinese human fossils. *Acta Anthropol. Sin.* **2018**, *37*, 371–383. (In Chinese)

40. Melchionna, M.; Profico, A.; Castiglione, S.; Sansalone, G.; Serio, C.; Mondanaro, A.; Di Febbraro, M.; Rook, L.; Pandolfi, L.; Di Vincenzo, F.; et al. From smart apes to human brain boxes. A uniquely derived brain shape in late hominins clade. *Front. Earth Sci.* **2020**, *8*, 273. [CrossRef]

41. Wu, X.J.; Schepartz, L.A.; Falk, D.; Liu, W. Endocranial cast of Hexian *Homo erectus* from South China. *Am. J. Phys. Anthropol.* **2006**, *130*, 445–454. [CrossRef]

42. Wu, X.J.; Liu, W.; Dong, W.; Que, J.M.; Wang, Y.F. The brain morphology of *Homo* Liujiang cranium fossil by three-dimensional computed tomography. *Sci. Bull.* **2008**, *53*, 2513–2519. [CrossRef]

43. Bruner, E.; Ripani, M. A quantitative and descriptive approach to morphological variation of the endocranial base in modern humans. *Am. J. Phys. Anthropol.* **2008**, *137*, 30–40. [CrossRef]

44. Rohlf, F.J.; Marcus, L.F. A revolution morphometrics. *Trends Ecol. Evol.* **1993**, *8*, 129–132. [CrossRef]

45. Beaudet, A.; Dumoncel, J.; de Beer, F.; Duployer, B.; Durrleman, S.; Gilissen, E.; Hoffman, J.; Tenailleau, C.; Thackeray, J.F.; Braga, J. Morphoarchitectural variation in South African fossil cercopithecoid endocasts. *J. Hum. Evol.* **2016**, *101*, 65–78. [CrossRef]

46. Beaudet, A.; Dumoncel, J.; de Beer, F.; Durrleman, S.; Gilissen, E.; Oettle, A.; Subsol, G.; Thackeray, J.F.; Braga, J. The endocranial shape of *Australopithecus africanus*: Surface analysis of the endocasts of Sts 5 and Sts 60. *J. Anat.* **2018**, *232*, 296–303. [CrossRef]

47. Durrleman, S.; Pennec, X.; Trouvé, A.; Ayache, N.; Braga, J. Comparison of the endocranial ontogenies between chimpanzees and bonobos via temporal regression and spatiotemporal registration. *J. Hum. Evol.* **2012**, *62*, 74–88. [CrossRef]

48. Urciuoli, A.; Zanolli, C.; Beaudet, A.; Dumoncel, J.; Santos, F.; Moyà-Solà, S.; Alba, D.M. The evolution of the vestibular apparatus in apes and humans. *eLife* **2020**, *9*, e51261. [CrossRef]

49. Braga, J.; Zimmer, V.; Dumoncel, J.; Samir, C.; de Beer, F.; Zanolli, C.; Pinto, D.; Rohlf, F.J.; Grine, F.E. Efficacy of diffeomorphic surface matching and 3D geometric morphometrics for taxonomic discrimination of Early Pleistocene hominin mandibular molars. *J. Hum. Evol.* **2019**, *130*, 21–35. [CrossRef]

50. Zhang, X.; Wu, X.J. 3D virtual reconstruction and morphological variations of the mastoid air cells—Take modern skulls from yunnan province for example. *Quat. Sci.* **2017**, *37*, 747–753. (In Chinese)

51. Zhang, X.; Zhang, Y.M.; Wu, X.J. The influence of quality parameter selection on 3D virtual reconstruction model precision based on dry skull. *Acta Anthropol. Sin.* **2020**, *39*, 270–281. [CrossRef]

52. Cignoni, P.; Callieri, M.; Corsini, M.; Dellepiane, M.; Ganovelli, F.; Ranzuglia, G. Meshlab: An open-source mesh processing tool. In Proceedings of the Eurographics Italian Chapter Conference, Salerno, Italy, 2–4 July 2008; Eurographics Association: Aire-la-Ville, Switzerland, 2008; pp. 129–136.

53. Bne, A.; Louis, M.; Martin, B.; Durrleman, S. Deformetrica 4: An open-source software for statistical shape analysis. In Proceedings of the International Workshop on Shape in Medical Imaging, Granada, Spain, 20 September 2018; Lecture Notes in Computer Science. Springer: Cham, Switzerland, 2018; Volume 11167, pp. 3–13. [CrossRef]

54. Durrleman, S. Statistical Models of Currents for Measuring the Variability of Anatomical Curves, Surfaces and Their Evolution. Ph.D. Thesis, Université Nice Sophia Antipolis, Nice, France, 2010.

55. Dumoncel, J. RToolsForDeformetrica. 2020. Available online: https://gitlab.com/jeandumoncel/tools-for-deformetrica/-/tree/master/src/R_script/RToolsForDeformetrica (accessed on 14 December 2021).

56. Dray, S.; Dufour, A.-B. The ade4 package: Implementing the duality diagram for ecologists. *J. Stat. Softw.* **2007**, *22*, 1–20. [CrossRef]
57. Team, R. R: A language and environment for statistical computing. R foundation for statistical computing:Vienna, Austria. *Computing* **2009**, *14*, 12–21.
58. Wickham, H. *ggplot2*; Springer: New York, NY, USA, 2009.
59. Xiang, L.; Crow, T.; Roberts, N. Cerebral torque is human specific and unrelated to brain size. *Brain Struct. Funct.* **2019**, *224*, 1141–1150. [CrossRef] [PubMed]
60. Leroy, F.; Cai, Q.; Bogart, S.L.; Dubois, J.; Coulon, O.; Monzalvo, K.; Fischer, C.; Glasel, H.; Van der Haegen, L.; Bénézit, A.; et al. New human-specific brain landmark: The depth asymmetry of superior temporal sulcus. *Proc. Natl. Acad. Sci. USA* **2015**, *112*, 1208–1213. [CrossRef] [PubMed]
61. Falzi, G.; Perrone, P.; Vignolo, L.A. Right-left asymmetry in anterior speech region. *Arch. Neurol.* **1982**, *39*, 239–240. [CrossRef] [PubMed]
62. Best, C.T. The emergence of cerebral asymmetries in early human development: A literature review and a neuroembryological model. In *Brain Lateralization in Children: Developmental Implications*; Molfese, D.L., Segalowitz, S.J., Eds.; Guilford Press: New York, NY, USA, 1988; pp. 5–34.
63. Strick, P.L.; Dum, R.P.; Fiez, J.A. Cerebellum and nonmotor function. *Annu. Rev. Neurosci.* **2009**, *32*, 413–434. [CrossRef]
64. Schmahmann, J.D. Disorders of the cerebellum: Ataxia, dysmetria of thought, and the cerebellar cognitive affective syndrome. *J. Neuropsych. Clin. Neurosci.* **2004**, *16*, 367–378. [CrossRef]
65. Stoodley, C.J. The cerebellum and cognition: Evidence from functional imaging studies. *Cerebellum* **2012**, *11*, 352–365. [CrossRef]
66. O'Reilly, J.X.; Beckmann, C.F.; Tomassini, V.; Ramnani, N.; Johansen-Berg, H. Distinct and overlapping functional zones in the cerebellum defined by resting state functional connectivity. *Cereb. Cortex* **2010**, *20*, 953–965. [CrossRef]
67. Krienen, F.M.; Buckner, R.L. Segregated fronto-cerebellar circuits revealed by intrinsic functional connectivity. *Cereb. Cortex* **2009**, *19*, 2485–2497. [CrossRef]
68. Buckner, R.L. The cerebellum and cognitive function: 25 years of insight from anatomy and neuroimaging. *Neuron* **2013**, *80*, 807–815. [CrossRef]
69. Foundas, A.L.; Leonard, C.M.; Gilmore, R.; Fennell, E.; Heilman, K.M. Planum temporale asymmetry and language dominance. *Neuropsychologia* **1994**, *32*, 1225–1231. [CrossRef]
70. Beaudet, A. The emergence of language in the hominin lineage: Perspectives from fossil endocasts. *Front. Hum. Neurosci.* **2017**, *11*, 427. [CrossRef]
71. Neubauer, S.; Hublin, J.J.; Gunz, P. The evolution of modern human brain shape. *Sci. Adv.* **2018**, *4*, eaao5961. [CrossRef]
72. Gannon, P.J.; Holloway, R.L.; Broadfield, D.C.; Braun, A.R. Asymmetry of chimpanzee planum temporale: Humanlike pattern of Wernicke's brain language area homolog. *Science* **1998**, *279*, 220–222. [CrossRef]
73. Cantalupo, C.; Hopkins, W.D. Asymmetric Broca's area in great apes. *Nature* **2001**, *414*, 505. [CrossRef]

Article

Asymmetries of Cerebellar Lobe in the Genus *Homo*

Yameng Zhang [1,2] and Xiujie Wu [3,4,*]

1   Joint International Research Laboratory of Environmental and Social Archaeology, Shandong University, Qingdao 266237, China; ymzh@sdu.edu.cn
2   Institute of Cultural Heritage, Shandong University, Qingdao 266237, China
3   Key Laboratory of Vertebrate Evolution and Human Origins of Chinese Academy of Sciences, Institute of Vertebrate Paleontology and Paleoanthropology, Chinese Academy of Sciences, Beijing 100044, China
4   CAS Center for Excellence in Life and Paleoenvironment, Chinese Academy of Sciences, Beijing 100044, China
*   Correspondence: wuxiujie@ivpp.ac.cn

Abstract: The endocast was paid great attention in the study of human brain evolution. However, compared to that of the cerebrum, the cerebellar lobe is poorly studied regarding its morphology, function, and evolutionary changes in the process of human evolution. In this study, we define the major axis and four measurements to inspect possible asymmetric patterns within the genus *Homo*. Results show that significant asymmetry is only observed for the cerebellar length in modern humans and is absent in *Homo erectus* and Neanderthals. The influence of occipital petalia is obscure due to the small sample size for *H. erectus* and Neanderthals, while it has a significant influence over the asymmetries of cerebellar height and horizontal orientation in modern humans. Although the length and height of the Neanderthal cerebellum are comparable to that of modern humans, its sagittal orientation is closer to that of *H. erectus*, which is wider than that of modern humans. The cerebellar morphological difference between Neanderthals and modern humans is suggested to be related to high cognitive activities, such as social factors and language ability.

Keywords: cerebellar lobe; *Homo*; asymmetry; evolutionary changes; cognitive increase

Citation: Zhang, Y.; Wu, X. Asymmetries of Cerebellar Lobe in the Genus *Homo*. *Symmetry* **2021**, *13*, 988. https://doi.org/10.3390/sym13060988

Academic Editor: Antoine Balzeau

Received: 29 April 2021
Accepted: 31 May 2021
Published: 2 June 2021

Publisher's Note: MDPI stays neutral with regard to jurisdictional claims in published maps and institutional affiliations.

1. Introduction

Endocast, or brain endocast, is the cast made of the interior of the neurocranium of a skull [1]. Endocast is the only agent to investigate how the human brain evolved physically in the process of evolution [2] regarding its volume [3–7], surface features [8–11], or size and shape [12–16].

Asymmetry of the brain, as one of the most debated questions, exhibits at different levels, such as the Broca's area [17,18], perisylvian region [19], central sulcus, cortical and subcortical regions, lobes, and hemispheres [20,21]. Brain asymmetry is often related to functional and evolutionary significance; for example, petalia and the Yakovlevian torque is a geometric distortion of the brain hemisphere, in which the left occipital lobe and the right frontal lobe are wider and longer than the opposite side [21]. While a combination of left occipital and right frontal petalia is common in modern humans and fossilized hominins and is regarded as evidence of right-handedness [22–24], such observations are less exaggerated or rarely consistent in great apes and other primates [15,25]. Also, capuchins display a leftward frontal petalia [26] while macaques show the rightward frontal petalia [27], and both were absent of a left occipital petalia.

Compared to that of the cerebrum, the cerebellum receives less attention and is still poorly understood. However, with the advances of neuroimaging technology and theoretical innovation, there is increasing study of the cerebellum regarding its morphology, function, and evolutionary changes. Moreover, evidence from neuroimaging uncovers the cerebellum as the "missing link" in many cognitive domains [28]. Sereno, et al. [29] revealed in a recent study that the cerebellar cortex covers almost 80% of the surface area of the cerebral cortex, and the expansion of the human cerebellar surface area even exceeds the cerebral cortex when compared to that of monkeys. Also, neuroimaging evidence suggests that the function of the cerebellum is also highly involved in cognitive and social activities [30–33], which is also supported by clinical studies, such as developmental dyslexia [34,35].

The evolutionary changes of the cerebellum were also hotly debated. Studies suggest that the cerebellum shows a similar asymmetric pattern as the cerebrum (the left-occipital, right-frontal petalia), with larger anterior lobules on the right side and larger posterior lobules on the left side [36,37]. Compared to that of humans, the cerebellar torque is opposite in chimpanzees [26], while absent in capuchins. MacLeod, Zilles, Schleicher, Rilling, and Gibson [31] found that hominins have a great increase in the lateral cerebellum compared to that of monkeys. It is also a popular idea that the cerebrum and cerebellum underwent several expansions and reorganizations in the process of human evolution [2]. Weaver [38] came up with a hypothesis that the evolution of human cerebellar/neocortical occurred in three stages; the first stage as early encephalization with an expansion of the neocortex during Early-to-Middle Pleistocene; the second stage as the dramatic encephalization primarily happened to the neocortex in Middle-to-Late Pleistocene humans, accompanied by a proliferation of cultural objects as well as an increase in complex behaviors, such as pyrotechnology and prepared core techniques, and the third stage happened in the late Late Pleistocene and Holocene, with an increase in cognitive efficiency as a result of expanded cerebellar capacity, which is the reason why modern humans can do more without an increase in net brain volume. Cerebellar specialization is thought to be an important component in the evolution of humanity's advanced technological capacities and languages [39,40].

The cerebellum is classically divided into three lobes, namely, the anterior lobe, the posterior lobe, and the flocculonodular lobe; it also has three major surfaces, with the superior surface toward the tentorium cerebelli, the posterior surface toward the internal occipital bone, and the anterior surface toward the petrous pyramid [41]. The cerebellar lobe reconstructed from the cerebellar fossa is surrounded by transverse sinus, sigmoid sinus, and occasionally occipital and marginal sinus. Accurate and homologous landmarks are important in the study of endocasts, such as in the case of Taung australopithecine endocast, in which a vague position of the lunate sulcus caused great debate [1,42–48]. Unlike the cerebrum, there are no gyrus or sulcus that exist on the surface of the cerebellum, and hence, it is difficult to identify landmarks to measure.

In this study, we tentatively define the major axis and four measurements on the cerebellum to achieve a more accurate between-group comparison. The main aim of this study is (1) to test whether certain asymmetric patterns existed within the genus *Homo*; (2) to find out possible factors that affected the asymmetry, and to (3) provide morphological evidence for cerebellar reorganization and cognitive increases in the genus *Homo*.

2. Materials and Methods

2.1. Materials

A total of 45 specimens were used in this study (as illustrated in Table 1), including *Homo erectus* (n = 11), *Homo neanderthalensis* (n = 4), and extant *Homo sapiens* (modern humans, n = 30). *H. erectus* specimens were mostly sourced from Asia and Indonesia to avoid significant regional differences. The 3D data of the fossil hominins were laser-scanned from endocast models housed in the American Museum of Natural History (AMNH) and the Institute of Vertebrate Paleontology and Paleoanthropology (IVPP) with a resolution of 0.5 mm or higher. Endocasts of extant *H. sapiens* were reconstructed from CT scans with a resolution of 160 μm. Frontal pole to occipital pole length was also measured in this study and compared to that of those from literature to ascertain whether no great discrepancy existed.

Table 1. List of specimens used in the study.

Populations	Number	Specimens and Source
Homo erectus	11	ZKD III, ZKD XI, Hexian (IVPP); OH 9, WT 15000, Sale, Sangiran 2, Sangiran 17, Ngandong 7, Ngandong 12, Sambungmacan 3 (AMNH)
Homo neanderthalensis	4	La Ferrassie 1, Gibraltar, Spy 1, Spy 2 (AMNH)
Homo sapiens	30	Modern Chinese (IVPP)

2.2. Cerebellar Metrics

Due to lacking prominent anatomical features, homologous landmarks on the cerebellar lobe are difficult to recognize and define. Although previous studies used transverse sinus and sigmoid sinus defining measurements of the cerebellum [1], the great variation suggests that the sinus was not an adequate reference.

Considering the ellipsoid shape of the cerebellar lobe, the size and shape can be best depicted by the major axis and two points it passes through (as illustrated in Figure 1). Here, we define the major axis as a straight line that divides the cerebellar lobe into two halves from both the inferior view and the posterior view. The point that the major axis passes through at the anterior part is defined as the most lateral and inferior point (LI point), which in most cases is medial to the sigmoid sinus. The point that the major axis passes through at the posterior part is defined as the most medial and superior point (MS point), which is close to the internal occipital protuberance point and is often asymmetric because of the occipital petalia. With the endocast at the standard position using the front pole to occipital pole as the horizontal plane [1], four measurements were defined upon the major axis (as illustrated in Figure 1, Table 2), namely, the cerebellar length, cerebellar height, sagittal orientation (the orientation of the major axis relative to the sagittal plane), and horizontal orientation (orientation of the major axis relative to the horizontal plane). Asymmetric parameters were also calculated as the difference between the left and right sides of the same measurement (L–R) [15], namely, the ML.lr, H.lr, Sagi lr, and Hori.lr. The determination of the landmarks and the measurements were performed in Rapidform XOR3.

Table 2. Definition and abbreviation of measurements.

Measurements	Abbreviation (Right/Left)	Definition
Cerebellar length	MLR/MLL	Length of the cerebellar major axis
Cerebellar height	HR/HL	Height of the cerebellum, measured from the MS point to the lowest margin of the cerebellum
Sagittal orientation	SagiR/SagiL	Orientation of the major axis relative to the sagittal plane, depicting how the cerebellar lobe orientated medial-laterally
Horizontal orientation	HoriR/HoriL	Orientation of the major axis relative to the horizontal plane, depicting how the cerebellar lobe orientated superior-inferiorly
Asymmetric parameters	ML.lr, H.lr, Sagi.lr, Hori.lr	Difference between the left and right side of the same measurement, calculated as (L–R)

2.3. Descriptive Statistics of the Asymmetries

To demonstrate overall asymmetries of the cerebellum, all the measurements were summarized regarding different populations and measurements, with mean value, standard deviation, and coefficient of variation (CV) being presented. Paired *t*-test was conducted to test the difference between the two sides with bootstrap considering the small sample size.

2.4. Analysis of Covariance

Cerebellar size and shape may be influenced by occipital petalia and brain size (allometry) [49]. Therefore, those two factors were recorded and analyzed in this study. Endocasts with left occipital petalia, right occipital petalia, or equally bilateral situation are recorded as L, R, and B, respectively. To test how cerebellar asymmetries were influenced by those two factors, analysis of covariance (ANCOVA) was conducted within each population group, with asymmetric parameters as the dependent variable, natural log-transformed cranial capacity as the covariate, and occipital petalia as the independent variable.

The statistical analysis and plotting were carried out in R [50], with packages "plyr", "tidyverse" and "ggplot2" [51].

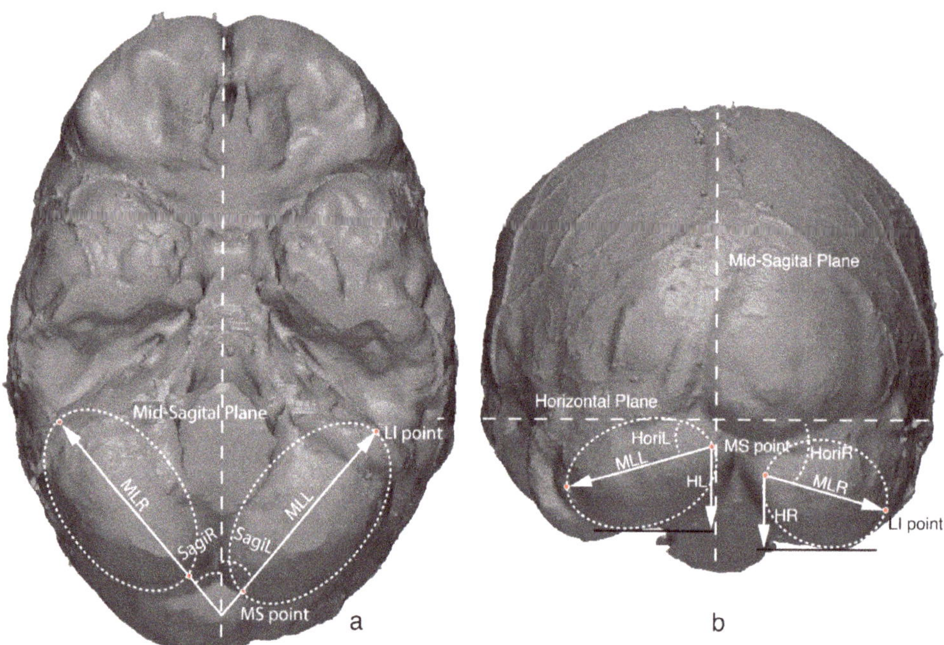

Figure 1. Diagram of major axis and four measurements of cerebellum on endocast from inferior (**a**) and posterior view (**b**). MLL and MLR, cerebellar length on the left and right side; MS point, medial-superior point; LI point, lateral-inferior point; SagiL and SagiR, orientation of major axis relative to sagittal and horizontal plane; HL and HR, cerebellar height on left and right side.

3. Error Evaluation

To avoid interobserver error, all measurements were measured by the same author Y.Z. Also, to assess intraobserver error, three specimens were chosen and measured six times repeatedly on six different days. The coefficient of variation (CV) is calculated with the repeated measurements (as illustrated in Table 3). The intraobserver error is well controlled within most of the measurements except for the horizontal angle. All of the measurements were included in the formal analysis, although the results of the horizontal angle should be taken with great caution.

Table 3. CV calculated from repeated measurements of chosen specimens.

Specimen	Side	ML	H	Sagi	Hori
Ngandong 7	right	2.42	7.95	8.88	16.33
Ngandong 7	left	0.73	2.44	2.01	22.46
Gibraltar	right	1.75	8.34	3.70	40.03
Gibraltar	left	3.52	4.15	4.36	33.63
yno4f	right	0.86	4.12	3.29	24.41
yno4f	left	1.03	1.33	5.49	19.23

4. Results

4.1. Description of Cerebellar Asymmetries

Results of the descriptive statistics and bootstrapped *t*-test were shown in Table 4. No significant asymmetry was observed among the four measurements within *H. erectus* and *H. neanderthalensis*. However, the result of Neanderthals should be taken with caution as only a small sample size is available. Within the *H. sapiens* group, only the cerebellar length is significantly larger on the left side.

Table 4. Descriptive results of measurements and bootstrapped *t*-test.

Population	Side	Metrics	ML	H	Sagi	Hori
	right	mean	48.68	15.18	40.47	7.68
		sd	4.21	1.83	5.49	2.18
		cv	8.65	12.05	13.56	28.42
H. erectus (n = 12)	left	mean	49.95	15.65	38.45	8.81
		sd	3.56	1.30	5.11	3.23
		cv	7.13	8.30	13.29	36.66
		p	0.10	0.45	0.19	0.26
	right	mean	60.33	22.85	41.35	11.20
		sd	4.28	3.39	3.36	4.53
		cv	7.09	14.84	8.13	40.47
H. neanderthalensis (n = 4)	left	mean	61.65	20.38	43.15	10.25
		sd	4.70	4.41	4.66	5.76
		cv	7.63	21.64	10.79	56.19
		p	0.64	0.43	0.25	0.37
	right	mean	58.17	21.78	35.40	9.90
		sd	2.65	3.02	3.27	2.82
		cv	4.55	13.86	9.24	28.52
H. sapiens (n = 30)	left	mean	59.75	22.01	36.50	9.83
		sd	3.33	3.05	3.46	2.77
		cv	5.57	13.85	9.48	28.14
		p	0.01 *	0.64	0.21	0.91

* Significance level is lower than 0.05.

Together with the boxplot (as illustrated in Figure 2), a preliminary evolutionary change of the cerebellar metrics can be summarized. There is an obvious growth of the cerebellar length and cerebellar height for Neanderthals and modern humans over the *H. erectus*. However, the sagittal orientation of the cerebellar is wider for Neanderthals and *H. erectus* while this value is quite small for modern humans. For the horizontal orientation, Neanderthals and modern humans are relatively steep while it is rather flat in *H. erectus*. The CVs of horizontal orientation are high in the process of error evaluation and may contribute to within-group variation here.

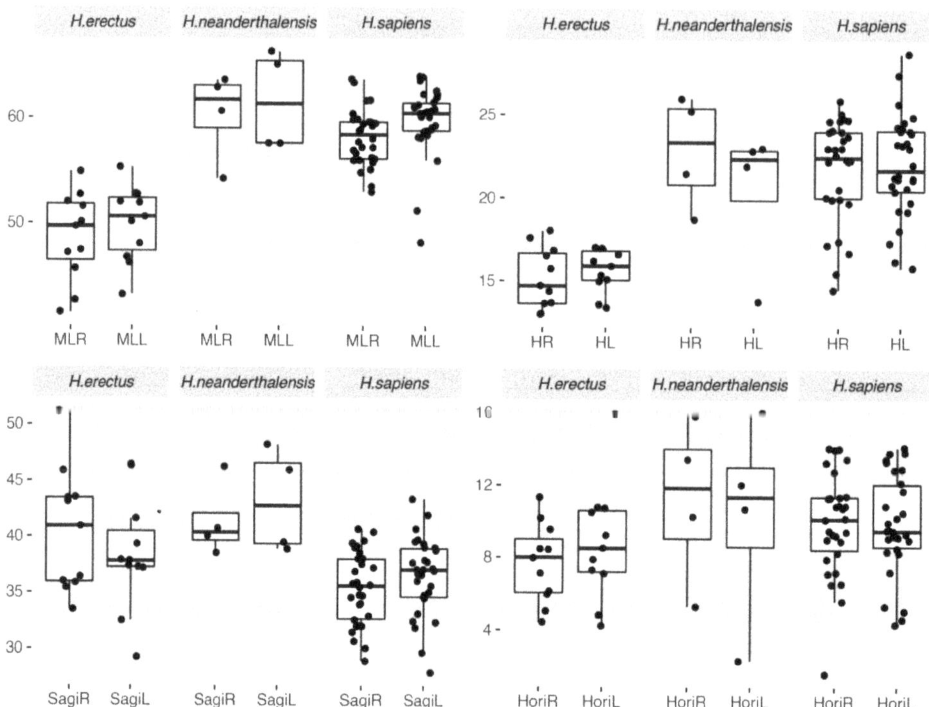

Figure 2. Boxplot of cerebellar metrics.

4.2. ANCOVA

In *H. erectus*, the ANCOVA (as illustrated in Table 5) revealed a significant relationship between cerebellar length asymmetric parameter (ML.lr) and both cranial capacity and occipital petalia while other parameters had no such relationship. From the scatterplot (as illustrated in Figure 3), negative allometry can be observed between ML.lr and cranial capacity in *H. erectus*, indicating that individuals with small brain size will have a larger cerebellar length on the left side. However, only two specimens of *H. erectus* have right occipital petalia (as illustrated in Figure 3), making the relationship between cerebellar length and occipital petalia questionable.

Table 5. *p*-value of asymmetric parameters from ANCOVA.

Asymmetric Parameter	Term	*H. erectus*	*H. sapiens*	*H. neanderthalensis*
ML.lr	ln CC	0.03 *	0.11	0.45
	Occipital Petalia	0.01 *	0.42	
H.lr	ln CC	0.89	0.51	0.21
	Occipital Petalia	0.89	0.01 *	
Sagi.lr	ln CC	0.42	0.26	0.49
	Occipital Petalia	0.06	0.12	
Hori.lr	ln CC	0.14	0.56	0.00 *
	Occipital Petalia	0.51	0.01 *	

* Significance level is lower than 0.05.

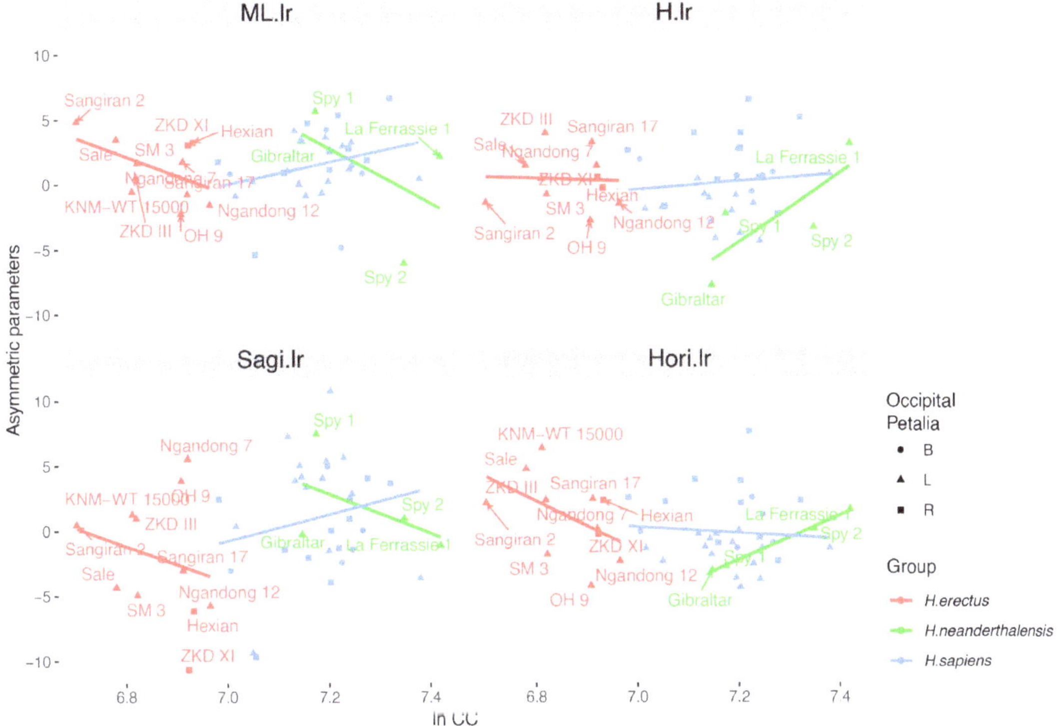

Figure 3. Allometric trend of asymmetric parameters in three populations; B, bilateral; L, left occipital petalia; R, right occipital petalia.

Within *H. sapiens*, cranial capacity does not affect the asymmetric parameter significantly. Occipital petalia had a significant influence on cerebellar height asymmetric parameter (H.lr) and horizontal orientation asymmetric parameter (Hori.lr). Endocasts with right occipital petalia tend to have large positive H.lr and Hori.lr, indicating a tall and steep-orientated cerebellar lobe on the left side.

Because all of the four Neanderthal specimens were left occipital petalia, we performed linear regression instead of the ANCOVA in Neanderthals. Results show that Hori.lr is significantly affected by the cranial capacity. A large brain size would be accompanied by a steep cerebellar lobe, as revealed in the scatterplot (as illustrated in Figure 3).

5. Discussion

5.1. Cerebellar Asymmetric Pattern

The cerebellar asymmetry is only observed in *H. sapiens* while absent in *H. erectus* and possibly *H. neanderthalensis*. In *H. sapiens*, the cerebellar length is significantly longer on the left side, while its height, sagittal orientation, and horizontal orientation do not differ between left and right side (as illustrated in Figure 4).

Further analysis found that the petalia and cranial capacity did not influence the cerebellar length and sagittal orientation in *H. sapiens*. However, the occipital petalia is significantly related to cerebellar height and horizontal orientation in *H. sapiens*. Individuals with right occipital petalia would have high and steep orientated cerebellar on the left side (as illustrated in Figure 4). The prevalence of cerebellum contralateral to the occipital lobe is suggested to be a spatial compensation [52,53], supported by the fact that the occipital petalia can be used as a predictive sign for the transverse sinus [54]. This is not repeated in the cerebellar length and sagittal orientation, indicating that the occipital petalia only affects the superior–inferior dimension instead of the anterior–posterior and medial–lateral dimension of the cerebellum.

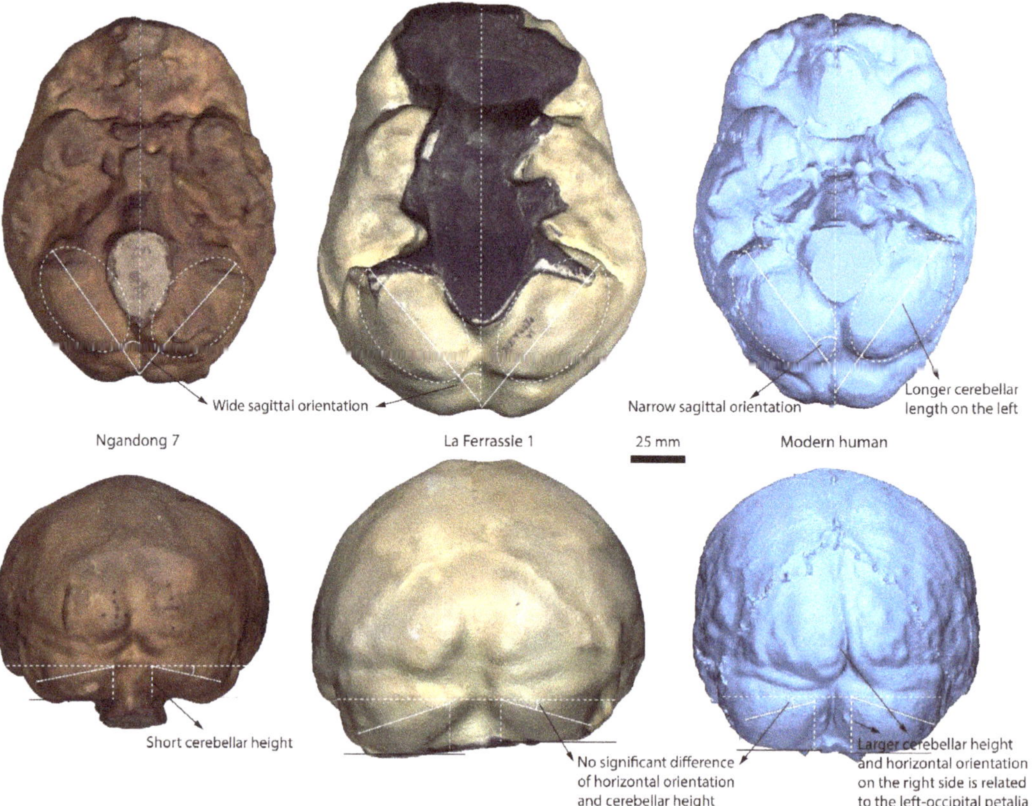

Figure 4. Cerebellar asymmetric pattern represented by Ngandong 7, La Ferrassie 1, and modern human.

5.2. Cerebellar Asymmetry in the Genus Homo

Although cerebellar expansion can be traced back to great apes [38], the asymmetries of the cerebellum appear quite recently in *H. sapiens*, which is absent in *H. erectus* and possibly *H. neanderthalensis*. Also, this differs from the fact that the cerebral laterality is distinguished in early hominins compared to that of the great apes [22,25]. We provide more evidence for the difference of evolutionary trajectories between the cerebrum and cerebellum.

A flat parietal lobe, elongated occipital lobe, and flat cerebellar lobe are thought to be unique features among Neanderthal endocasts [12]. Compared to that of Neanderthals, the endocast of modern humans is much more globular, including the parietal and the cerebellum lobe [55]. The globularity of the endocast of modern humans mainly originates from the parietal expansion rather than the cerebellum [12]. Congruent with previous studies, we find that modern humans do not differ from Neanderthals much at the cerebellar length and height (without size correction), and horizontal orientation. Our results support the idea that the tentorium cerebelli prevented the horizontal dimensions, such as the cerebellar height and horizontal orientation, from diverging greatly between Neanderthals and modern humans [12]. Meanwhile, the sagittal orientation is significantly narrower in modern humans when compared to that of Neanderthals. The small sample size of the Neanderthals aside, this may reflect a species-specific feature. If so, such difference possibly occurs during the "globularization-phase" when critical features of the human brain were established [55]. This might provide new evidence for the different rates and timing of brain development between Neanderthals and modern humans. However, such a statement needs more study and ontogenetic evidence. We suggest this to be related to higher cognitive abilities, such as social factors [56] and language abilities [57], that differ mostly between *H. sapiens* and Neanderthals [58].

5.3. Limitations

Considering the difficulty in obtaining homologous landmarks on the cerebellum, this is still a very preliminary study. The morphological information obtained is limited, and the uncertainty of the landmarks also introduced errors, especially in the measurement of the horizontal orientation. Further studies with landmark-free methods are suggested to reduce measurement error and extract more information from the whole cerebellar surface. Also, the basilar part of the cranium is rarely well preserved in human fossils and led to the small sample size in the analysis, which also prohibited us from obtaining convincing results.

Author Contributions: Conceptualization, Y.Z. and X.W.; methodology, Y.Z.; software, Y.Z. and X.W.; validation, Y.Z. and X.W.; formal analysis, Y.Z.; resources, Y.Z. and X.W., data curation, Y.Z. and X.W.; writing—original draft preparation, Y.Z.; writing—review and editing, Y.Z.; visualization, Y.Z.; funding acquisition, Y.Z. and X.W. All authors have read and agreed to the published version of the manuscript.

Funding: This work was supported by the Key Laboratory of Vertebrate Evolution and Human Origins, IVPP, CAS (NO. LVEHO20005), the Strategic Priority Research Program of Chinese Academy of the Sciences (XDB26000000), and the Shandong University Humanities and Social Science Major Project (No. 19RWZD08).

Institutional Review Board Statement: Not applicable.

Informed Consent Statement: Not applicable.

Data Availability Statement: The data used in the study are available from the authors upon reasonable request.

Conflicts of Interest: The authors declare no conflict of interest.

References

1. Holloway, R.L.; Broadfield, D.C.; Yuan, M.S.; Schwartz, J.H.; Tattersall, I. *The Human Fossil Record, Brain Endocasts–The Paleoneurological Evidence*, 1st ed.; Wiley-Liss: New York, NY, USA, 2004; p. 315.
2. Holloway, R.L. *Evolution of the Human Brain*; Lock, A., Peters, C.R., Eds.; Clarendon Press: Oxford, UK, 1996; pp. 74–125.
3. Holloway, R.L. Cranial capacity, neural reorganization, and hominid evolution: A search for more suitable parameters. *Am. Anthrop.* **1966**, *68*, 103–121.
4. Holloway, R.L. New Endocranial Values for the Australopithecines. *Nature* **1970**, *227*, 199–200. [CrossRef]
5. Holloway, R.L. New Endocranial Values for the East African Early Hominids. *Nature* **1973**, *243*, 97–99. [CrossRef]
6. Conroy, G.C.; Vannier, M.W.; Tobias, P.V. Endocranial features of Australopithecus africanus revealed by 2- and 3-D computed tomography. *Science* **1990**, *247*, 838–841. [CrossRef]
7. Conroy, G.C.; Weber, G.W.; Seidler, H.; Tobias, P.V.; Kane, A.; Brunsden, B. Endocranial capacity in an early hominid cranium from Sterkfontein, South Africa. *Science* **1998**, *280*, 1730–1731. [CrossRef] [PubMed]
8. Falk, D.; Conroy, G.C. The cranial venous sinus system in Australopithecus afarensis. *Nature* **1983**, *306*, 779–781. [CrossRef]
9. Falk, D. Meningeal arterial patterns in great apes: Implications for hominid vascular evolution. *Am. J. Phys. Anthropol.* **1993**, *92*, 81–97. [CrossRef] [PubMed]
10. Weidenreich, F. The ramification of the middle meningeal artery in fossil hominids: And its bearing upon phylogenetic problems. *Palaeontol. Sin.* **1938**, *110*, 1–16.
11. Bruner, E.; Sherkat, S. The middle meningeal artery: From clinics to fossils. *Child's Nerv. Syst.* **2008**, *24*, 1289–1298. [CrossRef]
12. Bruner, E. Geometric morphometrics and paleoneurology: Brain shape evolution in the genus Homo. *J. Hum. Evol.* **2004**, *47*, 279–303. [CrossRef] [PubMed]
13. Pereira-Pedro, A.S.; Bruner, E.; Gunz, P.; Neubauer, S. A morphometric comparison of the parietal lobe in modern humans and Neanderthals. *J. Hum. Evol.* **2020**, *142*, 102770. [CrossRef]
14. Wu, X.-j.; Bruner, E. The endocranial anatomy of maba 1. *Am. J. Phys. Anthropol.* **2016**, *160*, 633–643. [CrossRef] [PubMed]

15. Balzeau, A.; Holloway, R.L.; Grimaud-Hervé, D. Variations and asymmetries in regional brain surface in the genus Homo. *J. Hum. Evol.* **2012**, *62*, 696–706. [CrossRef]
16. Balzeau, A.; Gilissen, E.; Holloway, R.L.; Prima, S.; Grimaud-Hervé, D. Variations in size, shape and asymmetries of the third frontal convolution in hominids: Paleoneurological implications for hominin evolution and the origin of language. *J. Hum. Evol.* **2014**, *76*, 116–128. [CrossRef]
17. Sherwood, C.C.; Broadfield, D.C.; Holloway, R.L.; Gannon, P.J.; Hof, P.R. Variability of Broca's area homologue in African great apes: Implications for language evolution. *Anat. Rec. Part A Discov. Mol. Cell. Evol. Biol.* **2003**, *271*, 276–285. [CrossRef] [PubMed]
18. Bruner, E.; Holloway, R.L. A bivariate approach to the widening of the frontal lobes in the genus Homo. *J. Hum. Evol.* **2010**, *58*, 138–146. [CrossRef] [PubMed]
19. Hou, L.; Xiang, L.; Crow, T.J.; Leroy, F.; Rivière, D.; Mangin, J.-F.; Roberts, N. Measurement of Sylvian Fissure asymmetry and occipital bending in humans and Pan troglodytes. *Neuroimage* **2019**, *184*, 855–870. [CrossRef]
20. LeMay, M. Morphological Cerebral Asymmetries of Modern Man, Fossil Man, and Nonhuman Primate. *Ann. N. Y. Acad. Sci.* **1976**, *280*, 349–366. [CrossRef] [PubMed]
21. Rentería, M.E. Cerebral asymmetry: A quantitative, multifactorial, and plastic brain phenotype. *Twin Res. Hum. Genet.* **2012**, *15*, 401–413. [CrossRef]
22. Holloway, R.L.; De La Costelareymondie, M.C. Brain endocast asymmetry in pongids and hominids: Some preliminary findings on the paleontology of cerebral dominance. *Am. J. Phys. Anthropol.* **1982**, *58*, 101–110. [CrossRef]
23. Galaburda, A.M.; LeMay, M.; Kemper, T.L., Geschwind, N. Right left asymmetries in the brain. *Science* **1978**, *199*, 852–856. [CrossRef]
24. Hervé, P.-Y.; Crivello, F.; Perchey, G.; Mazoyer, B.; Tzourio-Mazoyer, N. Handedness and cerebral anatomical asymmetries in young adult males. *Neuroimage* **2006**, *29*, 1066–1079. [CrossRef]
25. Li, X.; Crow, T.J.; Hopkins, W.D.; Gong, Q.; Roberts, N. Human torque is not present in chimpanzee brain. *Neuroimaging* **2018**, *165*, 285–293. [CrossRef]
26. Phillips, K.A.; Sherwood, C.C. Cerebral petalias and their relationship to handedness in capuchin monkeys (Cebus apella). *Neuropsychologia* **2007**, *45*, 2398–2401. [CrossRef] [PubMed]
27. Falk, D.; Hildebolt, C.; Cheverud, J.; Vannier, M.; Helmkamp, R.C.; Konigsberg, L. Cortical asymmetries in frontal lobes of rhesus monkeys (Macaca mulatta). *J. Brain Res.* **1990**, *512*, 40–45. [CrossRef]
28. MacLeod, C. Chapter 8—The missing link: Evolution of the primate cerebellum. In *Progress in Brain Research*; Hofman, M.A., Falk, D., Eds.; Elsevier: Amsterdam, The Netherlands, 2012; Volume 195, pp. 165–187.
29. Sereno, M.I.; Diedrichsen, J.; Tachrount, M.; Testa-Silva, G.; d'Arceuil, H.; De Zeeuw, C. The human cerebellum has almost 80% of the surface area of the neocortex. *Proc. Natl. Acad. Sci. USA* **2020**, *117*, 19538–19543. [CrossRef]
30. Tanabe, H.C.; Kubo, D.; Hasegawa, K.; Kochiyama, T.; Kondo, O. Cerebellum: Anatomy, Physiology, Function, and Evolution. In *Digital Endocasts: From Skulls to Brains*; Bruner, E., Ogihara, N., Tanabe, H.C., Eds.; Springer: Tokyo, Japan, 2018; pp. 275–289.
31. MacLeod, C.E.; Zilles, K.; Schleicher, A.; Rilling, J.K.; Gibson, K.R. Expansion of the neocerebellum in Hominoidea. *J. Hum. Evol.* **2003**, *44*, 401–429. [CrossRef]
32. Kim, S.; Ugurbil, K.; Strick, P. Activation of a cerebellar output nucleus during cognitive processing. *Science* **1994**, *265*, 949–951. [CrossRef] [PubMed]
33. Raichle, M.E.; Fiez, J.A.; Videen, T.O.; MacLeod, A.-M.K.; Pardo, J.V.; Fox, P.T.; Petersen, S.E. Practice-related changes in human brain functional anatomy during nonmotor learning. *Cereb. Cortex* **1994**, *4*, 8–26. [CrossRef]
34. Rae, C.; Harasty, J.A.; Dzendrowskyj, T.E.; Talcott, J.B.; Simpson, J.M.; Blamire, A.M.; Dixon, R.M.; Lee, M.A.; Thompson, C.H.; Styles, P. Cerebellar morphology in developmental dyslexia. *Neuropsychologia* **2002**, *40*, 1285–1292. [CrossRef]
35. Eckert, M.A.; Leonard, C.M.; Richards, T.L.; Aylward, E.H.; Thomson, J.; Berninger, V.W. Anatomical correlates of dyslexia: Frontal and cerebellar findings. *Brain* **2003**, *126*, 482–494. [CrossRef] [PubMed]
36. Diedrichsen, J.; Balsters, J.H.; Flavell, J.; Cussans, E.; Ramnani, N. A probabilistic MR atlas of the human cerebellum. *Neuroimage* **2009**, *46*, 39–46. [CrossRef]
37. Snyder, P.J.; Bilder, R.M.; Wu, H.; Bogerts, B.; Lieberman, J.A. Cerebellar volume asymmetries are related to handedness: A quantitative MRI study. *Neuropsychologia* **1995**, *33*, 407–419. [CrossRef]
38. Weaver, A.H. Reciprocal evolution of the cerebellum and neocortex in fossil humans. *Proc. Natl. Acad. Sci. USA* **2005**, *102*, 3576–3580. [CrossRef]
39. Barton, R.A. Embodied cognitive evolution and the cerebellum. *Philos. Trans. R. Soc. B Biol. Sci.* **2012**, *367*, 2097–2107. [CrossRef]
40. Barton, R.A.; Venditti, C. Rapid Evolution of the Cerebellum in Humans and Other Great Apes. *Curr. Biol.* **2014**, *24*, 2440–2444. [CrossRef]
41. Naidich, T.P.; Duvernoy, H.M.; Delman, B.N.; Sorensen, A.G.; Kollias, S.S.; Haacke, E.M. *Duvernoy's Atlas of the Human Brain Stem and Cerebellum*; Springer: Vienna, Austria, 2009.
42. Falk, D. Apples, oranges, and the lunate sulcus. *Am. J. Phys. Anthropol.* **1985**, *67*, 313–315. [CrossRef] [PubMed]
43. Falk, D. A reanalysis of the South African australopithecine natural endocasts. *Am. J. Phys. Anthropol.* **1980**, *53*, 525–539. [CrossRef] [PubMed]
44. Falk, D. Ape-like endocast of "ape-man" Taung. *Am. J. Phys. Anthropol.* **1989**, *80*, 335–339. [CrossRef] [PubMed]
45. Falk, D. The Taung endocast: A reply to Holloway. *Am. J. Phys. Anthropol.* **1983**, *60*, 479–489. [CrossRef] [PubMed]

46. Holloway, R.L. Revisiting the South African Taung australopithecine endocast: The position of the lunate sulcus as determined by the stereoplotting technique. *Am. J. Phys. Anthropol.* **1981**, *56*, 43–58. [CrossRef]

47. Holloway, R.L. The taung endocast and the lunate sulcus: A rejection of the hypothesis of its anterior position. *Am. J. Phys. Anthropol.* **1984**, *64*, 285–287. [CrossRef]

48. Holloway, R.L.; Clarke, R.J.; Tobias, P.V. Posterior lunate sulcus in Australopithecus africanus: Was Dart right? *Comptes Rendus Palevol* **2004**, *3*, 287–293. [CrossRef]

49. Rightmire, G.P. Homo erectus and middle pleistocene hominins: Brain size, skull form, and species recognition. *J. Hum. Evol.* **2013**, *65*, 223–252. [CrossRef] [PubMed]

50. R Core Team. *R: A Language and Environment for Statistical Computing*; R Core Team: Vienna, Austria, 2018.

51. Wickham, H. *ggplot2: Elegant Graphics for Data Analysis*; Springer: Dordrecht, The Netherlands, 2016.

52. Enlow, D.H.; Hans, M. *Essentials of Facial Growth*; W.B. Saunders Company: Philadelphia, PA, USA, 1996.

53. Enlow, D.H.; Kuroda, T.; Lewis, A.B. The morphological and morphogenetic basis for craniofacial form and pattern. *Angle Orthod.* **1971**, *41*, 161–188. [PubMed]

54. Arsava, E.Y.; Arsava, E.M.; Oguz, K.K.; Topcuoglu, M.A. Occipital petalia as a predictive imaging sign for transverse sinus dominance. *Neurol. Res.* **2019**, *41*, 306–311. [CrossRef]

55. Gunz, P.; Neubauer, S.; Golovanova, L.; Doronichev, V.; Maureille, B.; Hublin, J.-J. A uniquely modern human pattern of endocranial development. Insights from a new cranial reconstruction of the Neandertal newborn from Mezmaiskaya. *J. Hum. Evol.* **2012**, *62*, 300–313. [CrossRef]

56. Burke, A. Spatial abilities, cognition and the pattern of Neanderthal and modern human dispersals. *Quat. Int.* **2012**, *247*, 230–235. [CrossRef]

57. Lieberman, P. The evolution of human speech-Its anatomical and neural bases. *Curr. Anthropol.* **2007**, *48*, 39–66. [CrossRef]

58. Kochiyama, T.; Ogihara, N.; Tanabe, H.C.; Kondo, O.; Amano, H.; Hasegawa, K.; Suzuki, H.; Ponce de León, M.S.; Zollikofer, C.P.E.; Bastir, M.; et al. Reconstructing the Neanderthal brain using computational anatomy. *Sci. Rep.* **2018**, *8*, 6296. [CrossRef]

 *symmetry*

Article

The Significance of Chimpanzee Occipital Asymmetry to Hominin Evolution

Shawn Hurst [1,*], Ralph Holloway [2], Alannah Pearson [3] and Grace Bocko [1]

[1]   Biology Department, University of Indianapolis, Indianapolis, IN 46227, USA; bockog@uindy.edu
[2]   Anthropology Department, Columbia University, New York, NY 10027, USA; rlh2@columbia.edu
[3]   School of Archaeology and Anthropology, The Australian National University, Canberra 2601, Australia; alannah.pearson@anu.edu.au
*   Correspondence: hursts@uindy.edu; Tel.: +1-317-788-4912

Abstract: Little is known about how occipital lobe asymmetry, width, and height interact to contribute to the operculation of the posterior parietal lobe, despite the utility of knowing this for understanding the relative reduction in the size of the occipital lobe and the increase in the size of the posterior parietal lobe during human brain evolution. Here, we use linear measurements taken on 3D virtual brain surfaces obtained from 83 chimpanzees to study these traits as they apply to operculation of the posterior occipital parietal arcus or bridging gyrus. Asymmetry in this bridging gyrus visibility provides a unique opportunity to study both the human ancestral and human equivalently normal condition in the same individual. Our results show that all three traits (occipital lobe asymmetry, width, and height) are related to this operculation and bridging gyrus visibility but width and not height is the best predictor, against expectations, suggesting that relative reduction of the occipital lobe and exposure of the posterior parietal is a complex phenomenon.

Keywords: chimpanzee; occipital; hominin

Citation: Hurst, S.; Holloway, R.; Pearson, A.; Bocko, G. The Significance of Chimpanzee Occipital Asymmetry to Hominin Evolution. *Symmetry* **2021**, *13*, 1862. https://doi.org/10.3390/sym13101862

Academic Editor: Antoine Balzeau

Received: 18 August 2021
Accepted: 29 September 2021
Published: 3 October 2021

Publisher's Note: MDPI stays neutral with regard to jurisdictional claims in published maps and institutional affiliations.

1. Introduction

In addition to helping us understand the evolution of lateralization [1–3], asymmetries of the brain's surface seen in closely related species such as chimpanzees (*Pan troglodytes*) can also help us to understand the role development plays in brain evolution itself. As an example, a major shape difference in the brains of human (*Homo sapiens*) versus nonhuman primates is that in nonhuman primates the occipital lobe operculates part of the parietal lobe, including a buried annectant gyrus that connects these lobes, known as the 1st parieto-occipital "pli de passage" of Gratiolet or the parieto-occipital arcus [4–6]. The posterior portion or bridge of this gyrus is consistently seen on the brain's surface in humans but is only occasionally seen (often asymmetrically) in chimpanzees [4–8]. Relative reduction of the occipital operculation and expansion of the posterior parietal lobe is a major hallmark in human brain evolution, although debate on when this occurred has been contentious, and currently we have no model of what transitional states between the human ancestral and derived conditions may have looked like. Studying the presence or absence of a visible bridging gyrus in chimpanzees, who are our closest living relatives and who have brains very similar to that of the last common ancestor [7–10] allows us to understand its relationship to the size of the occipital lobe; when this trait is asymmetrical in chimpanzees (who unlike humans still show occasional asymmetry in this region) it allows us to understand this trait developmentally rather than genetically, as it occurs variably in different hemispheres of the same individual, while giving us a greater range of variation in which to build models of transitional states, and to study the evolution of asymmetries and symmetries, since it is asymmetrical in chimpanzees while it is symmetrical in humans. Such an understanding would also be very valuable for the interpretation of hominin endocranial casts, which have morphology that is difficult to interpret in this region due to our lack of transitional models, and so very valuable to the study of brain evolution. If this trait is only associated with occipital lobe height this would suggest that the primary factor in the exposure of the bridging gyrus is posterior movement of the occipital operculation, which retracted inferio-posteriorly during human evolution revealing buried parietal gyri which then expanded; association with asymmetry and/or width in addition to height would suggest a relative change in the size and shape of the entire occipital to the parietal lobe is a more important factor. Using preliminary data, we observed these relationships in a large sample of chimpanzees. The aim of this study is an exploratory assessment of whether the presence or absence of the occipital bridging gyrus is associated with left or right hemispheres, and how hemisphere siding is associated with occipital lobe width and height in the chimpanzee brain. Regression analysis examines the correlation between left and right hemispheres and occipital lobe width and height, where reliable predictions (± 1 s.e.) determined if occipital lobe height or width was a more reliable predictor of hemisphere siding. Ultimately, we found that asymmetry, height, and width are all associated with a visible bridging gyrus, in increasing order.

2. Materials and Methods

This study used three-dimensional surface models of a sample of 83 chimpanzee brains. These brains were reconstructed using MRIs from the National Chimpanzee Brain Resource (https://www.chimpanzeebrain.org (accessed on 1 September 2021)) using BrainVISA software (Pune, India) and measured using MeshLab [11–13]. Although the measurements were able to be collected on the entire sample, the original collectors [12] could not guarantee that the left or right hemisphere siding was correctly labelled. To accommodate this uncertainty, subsample ($n = 15$) was obtained by one of us to allow a comparison and analysis of 'known' and 'unknown' hemisphere siding'. Each brain was rotated such that the lowest points of the left occipital and left temporal lobes both lie on a plane at right angles to the longitudinal fissure. The width of each hemispherical occipital lobe was measured as the distance in millimeters from the longitudinal fissure to the lobe's most lateral extent. Height was measured as the greatest vertical extent between points on each hemispherical lobe, barring its most medial edge if a bridging gyrus was visible; the presence of a visible bridging gyrus between the superior-medial occipital lobe and the parietal-occipital arcus was scored as a Y, while a fully operculated and thus hidden bridging gyrus was scored as an N (see Figure 1).

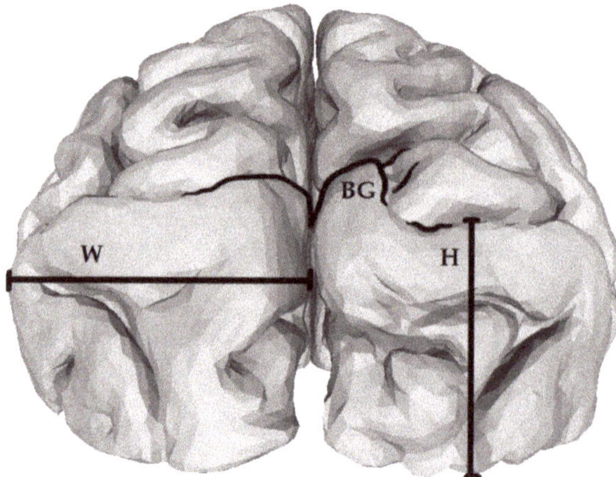

Figure 1. Occipital Measurement Definitions. W = width, H = height. The right hemisphere has a bridging gyrus (BG) not fully operculated by the occipital lobe and was scored as a Y; in the left hemisphere this gyrus is fully operculated, so its condition was scored as an N.

Statistical Analyzes

Preliminary analysis included a measurement error study. All data collection and measurements were conducted by a single operator to prevent the effects on interobserver error. Measurement error was investigated by using an analysis of variance, where measurement error was calculated as the proportion of the mean-squared differences between replicates relative to the total between-group variation [14]. The subsample (n = 15) of known hemisphere siding were measured on two separate occasions and measurement error (ME) calculated as % ME = 100 × MS (within)/MS (within) + MS (among). Measurement error ranged from 0% to 3% (results not shown), and with this low measurement error, we considered intraobserver error had a very minimal effect on further analyzes.

Canonical Correspondence Analysis (CCA) initially examined the potential association between the four metrics: occipital height, both left and right (in mm) and width, both left and right (in mm), and the presence or absence of a left, right, or no occipital bridge (Table 1). CCA is particularly suited to datasets where quantitative variables and presence/absence variables are common, such as ecological datasets [15]. Only recently has this been applied to brain evolution, specifically quantitative variables, and the presence/absence of sulcal patterns [16]. CCA allows a comparison analysis, directly testing a priori hypotheses emphasizing the variance of Y that is related to X, and where CCA combines the properties of both ordination and regression analyses to produce ordinations of Y that are linearly constrained to X [15]. Correlation analysis then tested the strength of the potential correlation between two or more variables using the most common correlation statistic (Pearson's r correlation coefficient), with a two-tailed significance that the variables were uncorrelated and a Monte Carlo permutation (using 9999 iterations) [17].

Table 1. Occipital lobe measurements and bridging pattern type.

Subject	Height [1]		Width		Bridge [2]		
	L	R	L	R	L	R	Both
Abby	36	38	38	37	N	N	N
Agatha	42	44	47	46	N	N	N
Ahni	28	31	35	36	N	N	N
Akimel	42	41	39	41	N	N	N
Alex *	26	27	34	34	Y	Y	Y
Alpha	33	35	36	39	N	N	N
Amanda	41	41	37	37	N	N	N
Angie	27	30	35	35	Y	N	N
Artemus	32	33	35	35	N	Y	N
Arthur	38	37	33	35	N	N	N
Artifee *	39	37	37	36	N	N	N
Augusta	38	35	32	34	N	N	N
Azalea	36	38	33	37	N	N	N
Bahn	35	36	33	33	N	N	N
Barbara	42	43	37	37	N	N	N
Bart	31	29	37	37	N	Y	N
Bashful *	31	32	34	34	N	N	N
Becca	36	38	28	30	N	N	N
Beleka	32	31	28	30	N	N	N
Bernadette	35	39	32	36	N	N	N
Bernie	24	26	27	26	N	N	N
Beta	29	29	29	29	N	N	N
Betty *	44	44	36	38	N	N	N
Billy *	33	39	31	33	N	N	N
Bo *	35	33	33	33	N	N	N
Boka	42	42	38	37	Y	Y	Y
Brandy	35	34	26	29	N	N	N
Bria	34	38	38	40	Y	Y	Y
Brodie	33	33	31	31	N	N	N

Table 1. *Cont.*

Subject	Height [1]		Width		Bridge [2]		
	L	R	L	R	L	R	Both
Callie	40	40	32	32	N	N	N
Carl *	37	32	33	34	Y	Y	Y
Chechkel	43	42	38	41	N	N	N
Cheeta *	45	44	37	39	N	N	N
Cheopi	34	34	31	32	N	N	N
Chester	28	36	37	37	Y	Y	Y
Chinook	35	38	36	36	N	N	N
Chip *	33	34	36	36	Y	Y	Y
Christa	43	43	34	37	N	N	N
Chuhia	37	40	34	34	N	Y	N
Cissie	38	41	35	37	N	N	N
Coco	31	32	37	38	Y	Y	Y
Cybil	27	28	33	34	Y	Y	Y
Dara	36	39	36	34	N	N	N
David *	29	29	37	35	N	Y	N
Drew	37	36	37	40	N	Y	N
Duff	39	39	35	37	N	N	N
Edwina *	31	32	32	32	N	N	N
Eesha	30	32	33	33	N	N	N
Ehsto	42	44	45	45	N	N	N
Elvira	39	39	38	37	Y	Y	Y
Elwood *	39	40	35	35	N	N	N
Emily *	30	32	35	35	N	N	N
Eniga	39	40	35	35	N	N	N
Evelyne	32	29	29	29	N	N	N
Faye	37	38	35	38	N	N	N
Fiona	38	41	38	37	N	N	N
Foxy	37	36	35	35	N	N	N
Frannie	34	35	34	34	N	N	N
Fritz	38	40	34	36	N	N	N
Gaygos	36	35	39	39	N	N	N
Gelb	37	38	31	33	N	N	N
Gigi	34	33	35	35	N	N	N
Gimp	32	33	36	35	Y	N	N
Gisoki	38	40	30	35	N	N	N
Haakid	36	37	38	41	N	N	N
Hannah	35	35	32	33	N	N	N
Helga	30	27	33	35	Y	Y	Y
Heppie	42	42	36	37	N	N	N
Hobbes	30	36	33	32	Y	N	N
Hodari	36	36	37	37	N	N	N
Huey	37	29	37	38	N	Y	N
Hug	31	36	36	36	N	N	N
Huhkalig	38	38	35	36	N	N	N
Iyk	31	35	33	35	N	N	N
Jacqueline	33	31	34	34	N	Y	N
Jadyh	31	33	33	34	N	N	N
Jake	38	40	36	37	N	N	N
Jamie	38	37	37	38	N	N	N
Jane	33	32	38	37	N	N	N
Jarred *	32	33	33	33	N	N	N
Jcarter	35	31	32	34	N	Y	N
Jewelle	28	27	30	29	Y	Y	Y
Jolson *	38	38	39	38	N	N	N

[1] All numbered measurements in left (L) and right (R) height and width in mm. [2] Presence (Y), absence (N), or Both (B) of a visible bridging gyrus. * Indicates the subsample of individuals with known siding.

To estimate the uncertainty due to unknown hemisphere siding, a subsample (n = 15) where the hemisphere siding was known (left and right) was examined with Bivariate ordinary least-squares (OLS) regression to test the strength of association between each of the four variables and occipital lobe side (left and right hemisphere). For regression purposes, and to linearize scaling relationships [18], each variable was converted (from mm) into natural logarithmic units (base e) and a 95% confidence interval fitted to the log–log regressions.

Predicted height and width from both hemispheres was calculated using prediction equations provided by the bivariate OLS regression models, where y = ($a \times$ log[x] + b). The reliability of the predictions was calculated as the percentage of prediction errors (PPE), where PPE = (predicted − observed)/predicted × 100). PPE calculates the uncertainty in an estimate relative to its size [19]. Prediction reliability was determined by applying a bracket of uncertainty produced by the standard error (s.e.) from the bivariate OLS regression models calculating the upper and lower estimates for predicted height or width for each specimen relative to its size, where y = ($a \times$ log[x] + $b \pm$ s.e). This maintained any inherent differences between each variable allowing for changes in the range of uncertainty, where each variable is associated with differences in the standard error [20]. All statistical analyses were conducted in *Past 4.0* [21].

3. Results

Preliminary results from summary statistics (Table 2) detailing the differences between the left and right occipital lobes and the variation between height and width measurements.

Table 2. Summary statistics detailing mean, variance, standard deviations for the subsample (n = 15) with known hemisphere siding.

Summary Statistics (Known Sample)				
	L Height	R Height	L Width	R Width
N	15	15	15	15
Min	26	27	31	32
Max	45	44	39	39
Sum	522	526	522	525
Mean	34.8	35.06667	34.8	35
Std. error	1.40814	1.31	0.57	0.53
Variance	29.74286	25.78095	4.885714	4.285714
Stand. dev	5.453701	5.077495	2.210365	2.070197
Median	33	33	35	35
25 percentile	31	32	33	33
75 percentile	39	39	37	36
Skewness	0.4577742	0.476494	0.108067	0.613097
Kurtosis	−0.4279719	−0.52249	−0.60243	−0.46667
Geom. mean	34.40985	34.73166	34.73453	34.94389
Coeff. var	15.67156	14.47955	6.351624	5.914848

Canonical Correspondence Analysis (CCA) was used to determine the strength of the correlation between different occipital bridge types, and the left (L) and right (R) height or width of the occipital lobe. The presence or absence of bridging patterns requires assessment where the potential correlation between occipital lobe height and width could be assessed against the presence or absence of Left or Right bridging patterns, or whether those with Both patterns were associated more with Occipital lobe width or height. Consistent with CCA, the type of bridging patterns grouped specimens accordingly and the effect of occipital lobe height or width determined. Results indicated that greater occipital width was associated with both Left and Right bridging patterns (Axis 1), while occipital lobe height (Axis 2) was associated more strongly with No Bridging pattern. The correlations between variables indicated by Axis 1 (89% variance) and Axis 2 (11% variance) were statistically significant ($p < 0.002$) with 1000 permutations (Table 3).

Table 3. Canonical Correspondence Analysis values of occipital lobe bridge patterns, with permutation (999 iterations). Statistically significant values are reported in italics.

Axis	Eigenvalue	Percentage	*p*-Value
1	0.2851	89.14	*0.001*
2	0.0347	10.86	*0.002*

Abbreviations: *p*-value is the permutated *p*-value from 1000 iterations.

There were four distinct groups based on the type of bridge patterns observed with a left bridge associated with marginally shorter L lobe height and greater R lobe width, a right bridge was associated with shorter R lobe height and slightly greater R lobe width, where both L and R bridges were present, these were weakly associated with smaller L height, and no bridges was associated with greater R lobe height and width (Figure 2).

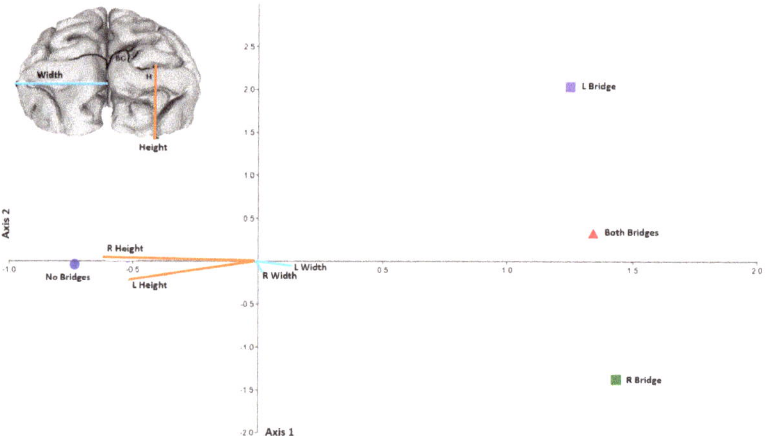

Figure 2. Canonical Correspondence analysis showing the four distinct groups of bridge patterns and a biplot indicating the direction of correlations between variables where longer lines indicate a stronger correlation. Abbreviations: Green square = Right Bridge; Purple square = Left bridge; Blue Sphere = No bridge; Red Triangle = Both bridges; L Height = Left occipital lobe height; R Height = Right occipital lobe height; L Width = Left occipital lobe width; R Width = Right occipital lobe width.

Correlation analysis examined potential correlations between variables using Pearson's *r* correlation coefficient for significance and a Monte Carlo permutation (9999 iterations) with the probability of variables being uncorrelated using a two-tailed significance set to $p < 0.01$. Statistically significant correlations using Monte Carlo permutation are reported (Table 4) for R and L lobe height and width ($p \leq 0.0001$), with slightly less robust correlations for R lobe width and right bridge ($p = 0.0008$), and L lobe height and L bridge ($p = 0.0022$). Correlations between bridging patterns are entirely due to the binary coding and do not reflect a true correlation.

Table 4. Correlation Analysis between occipital lobe metrics and bridging patterns, with Monte Carlo permutation (9999 iterations) and two-tailed significance. Statistically significant values are reported in italics ($p < 0.01$). Correlation values reported in the lower triangle with two-tailed significance that variables are uncorrelated are reported in the upper triangle.

Correlation Table							
	L Height	R Height	L Width	R Width	L Bridge [1]	R Bridge [1]	N Bridge [1]
L Height		0.0001	0.0001	0.0001	0.0022	0.0161	0.0026
R Height	*0.0001*		0.0001	0.0001	0.0101	0.0008	0.0002
L Width	*0.0001*	*0.0001*		0.0001	0.6361	0.3920	0.4240
R Width	*0.0001*	*0.0001*	*0.0001*		0.8226	0.7199	0.9431
L Bridge	*0.0022*	*0.0101*	0.6361	0.8226		0.0001	0.0001
R Bridge	*0.0161*	*0.0008*	0.3920	0.7199	*0.0001*		0.0001
N Bridge	*0.0026*	*0.0002*	0.4240	0.9431	*0.0001*	*0.0001*	

Abbreviations: Correlation in lower triangle of matrix; probability of uncorrelated variables with two-tailed significance ($p < 0.05$) in upper triangle of matrix. L Height = Left occipital lobe height; R Height = Right occipital lobe height; L Width = Left occipital lobe width; R Width = Right occipital lobe width; R Bridge = Right Bridge; L Bridge = Left bridge; No Bridge = Nbridge; [1] = Included as binary values (present/absent scores).

Caution is warranted with these initial findings where uncertainty associated with correct hemisphere siding, and the low number of individuals who possessed a bridging pattern could be obscured by the higher number of those who possessed no bridging pattern and where known siding is uncertain. However, correlation results and those reported from the CCA suggest a likely association between lobe width and bridging patterns.

Ordinary Least Squares (OLS) regression examined a subsample ($n = 15$) of individuals with known right and left hemisphere siding allowing a test of bridging and siding prediction and associated uncertainty. Metrics (in mm) for both right and left width and height were first transformed by natural logarithm (base e) maintaining linearity. Both height and width were predicted using Right from Left and then Left from Right to determine the potential effect of siding on prediction uncertainty. All predictions were made with a 95% confidence interval (CI) with strong correlations ($r \geq 0.86$, $p \leq 0.0001$). However, between the regression models, there was little observable difference whether the left or right hemisphere was used for the predictions (Table 5, Figure 3).

Table 5. Parameters for ordinary least-squares regression detailing the regression statistics for the four metrics both left and right side. Statistically significant results reported in italics.

Right Lobe Regression Statistics					
Metrics	*a*	*b*	s.e	*r*	*p*
R Height	0.82901	0.61434	0.11182	0.90	*0.0001*
R Width	0.79421	0.73609	0.12819	0.86	*0.0001*
Left Lobe Regression Statistics					
Metrics	*a*	*b*	s.e	*r*	*p*
L Height	0.97553	0.07749	0.13158	0.90	*0.0001*
L Width	0.94058	0.20515	0.15181	0.86	*0.0001*

Abbreviations: a = slope; b = intercept; s.e = standard error of the regression estimate; r = Correlation coefficient; p = p-value for significance; L Height = Left occipital lobe height; R Height = Right occipital lobe height; L Width = Left occipital lobe width; R Width = Right occipital lobe width.

All regression models showed a strong prediction overall, calculating the percentage of prediction uncertainty (PPE) allows a better comparison of the uncertainty within each model. Percentage of prediction error (PPE) was calculated for occipital height and width, respectively, and the difference between these left and right predictions compared with robust agreement between the observed and the predicted values (Table 5). Prediction reliability assessed the difference within the regression models and between left and right lobes. Greater prediction uncertainty existed for lobe height, with a disparity of 17%, than for width where the disparity was only 6%. This suggest that occipital lobe width might be a more stable variable with less prediction uncertainty than height, potentially making it more suitable for predicting occipital lobe side and hence, more reliable for assessing bridging pattern associations (Table 6).

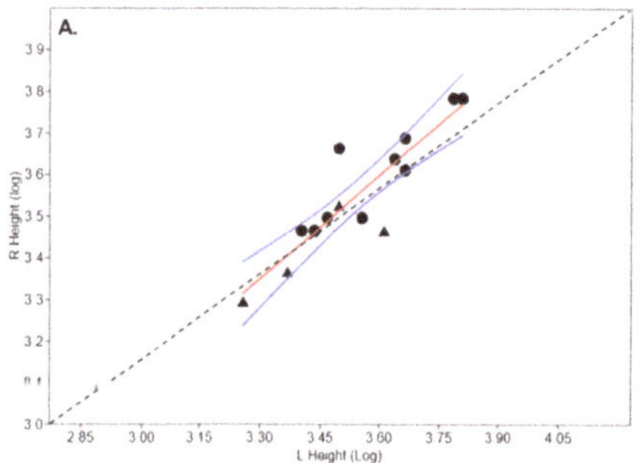
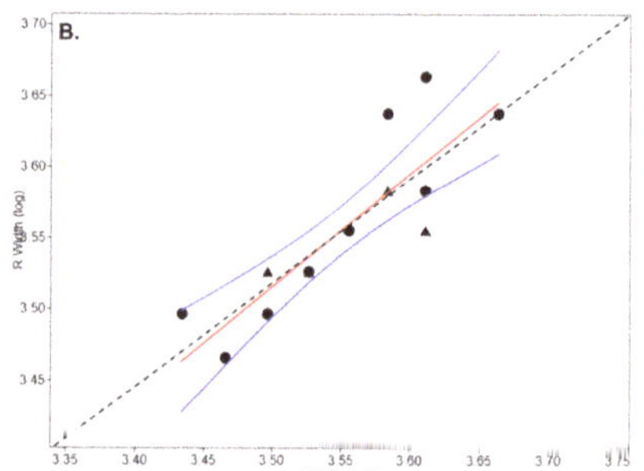

Figure 3. Log-log Ordinary Least Squares (OLS) regression of Occipital lobe fitted with a 95% confidence interval for lobe (**A**) height and (**B**) width where black triangles are specimens with a bridging gyrus and black dashed line to emphasize symmetry and asymmetry (the departure from symmetry).

The predictions for both L and R occipital lobe width and height are provided for both known and unknown sample, with predicted values converted from log-units to metrics (in mm) by taking the inverse-log and the observed values reported in parentheses alongside the predicted values (Table 7, Figure 4). Considering there was no discernible difference in pattern of reliability between the hemispheres, only the prediction of R lobe height and width are provided.

Table 6. Percentage of prediction errors (PPE) for four occipital metrics calculated as the difference between observed and predicted height and width, and percentage of prediction reliability calculated as difference between observed and predicted height and width (in mm) divided by observed height and width. Negative and positive values indicate an increase or decrease, respectively, in the predicted value from the observed.

| | Percentage Prediction Error | | | |
| | Height | | Width | |
Subject	L	R	L	R
Alex	1%	1%	0%	0%
Artifee	−2%	1%	−1%	1%
Bashful	1%	0%	0%	0%
Betty	0%	−1%	1%	−2%
Billy	4%	−4%	2%	−1%
Bo	−2%	2%	0%	0%
Carl	−4%	4%	1%	0%
Cheeta	−1%	0%	1%	−2%
Chip	1%	0%	0%	0%
David	0%	1%	−2%	1%
Edwina	1%	0%	0%	1%
Elwood	0%	−1%	0%	0%
Emily	2%	−1%	0%	0%
Jarred	1%	0%	0%	0%
Jolson	0%	0%	−1%	0%

Table 6. *Cont.*

Relability of Prediction Errors		
Subject	Height	Width
Alex	0%	0%
Artifee	3%	2%
Bashful	−1%	0%
Betty	0%	−3%
Billy	−9%	−3%
Bo	4%	1%
Carl	8%	−1%
Cheeta	1%	−3%
Chip	−1%	0%
David	1%	3%
Edwina	−1%	1%
Elwood	−1%	0%
Emily	−3%	0%
Jarred	−1%	1%
Jolson	0%	1%

Table 7. Prediction of occipital lobe width and height (in mm) listed with the corresponding variable calculated from the bivariate ordinary least-squares equations. Observed values reported beside predicted in parentheses.

Prediction of Height and Width				
	Height [1]		Width [1]	
Subject	R	L	R	L
Alex [2]	28 (27)	26 (26)	34 (34)	34 (34)
Artifee [2]	37 (37)	39 (39)	36 (36)	37 (37)
Bashful [2]	33 (32)	31 (31)	34 (34)	34 (34)
Betty [2]	43 (44)	43 (44)	38 (38)	36 (36)
Billy [2]	39 (39)	33 (34)	34 (33)	31 (31)
Bo [2]	34 (33)	35 (35)	34 (33)	33 (33)
Carl [2]	33 (32)	37 (37)	34 (34)	33 (33)
Cheeta [2]	43 (44)	44 (45)	38 (39)	37 (37)
Chip [2]	34 (34)	33 (33)	36 (36)	36 (36)
David [2]	30 (29)	29 (29)	35 (35)	37 (37)
Edwina [2]	33 (32)	31 (31)	33 (32)	32 (32)
Elwood [2]	39 (40)	39 (39)	35 (35)	35 (35)
Emily [2]	33 (32)	30 (30)	35 (35)	35 (35)
Jarred [2]	34 (33)	32 (32)	34 (33)	33 (33)
Jolson [2]	38 (38)	38 (38)	38 (39)	39 (38)
Abby	38 (38)	36 (36)	37 (37)	38 (39)
Agatha	43 (44)	41 (42)	44 (46)	46 (47)
Ahni	32 (31)	28 (28)	36 (36)	35 (35)
Akimel	40 (41)	41 (42)	40 (41)	39 (39)
Alpha	35 (35)	33 (33)	38 (39)	36 (36)
Amanda	40 (41)	40 (41)	37 (37)	37 (37)
Angie	31 (30)	27 (27)	35 (35)	35 (35)
Artemus	34 (33)	32 (32)	35 (35)	35 (35)
Arthur	37 (38)	38 (37)	35 (35)	33 (33)
Augusta	35 (35)	38 (38)	34 (34)	32 (32)
Azalea	38 (38)	36 (36)	37 (37)	33 (33)
Bahn	36 (36)	35 (35)	34 (33)	33 (33)
Barbara	42 (43)	41 (42)	37 (37)	37 (37)
Bart	30 (29)	31 (31)	37 (37)	37 (37)
Becca	38 (38)	36 (36)	31 (30)	28 (28)

Table 7. *Cont.*

Subject	Prediction of Height and Width			
	Height [1]		Width [1]	
	R	L	R	L
Beleka	32 (31)	32 (31)	31 (30)	28 (28)
Bernadette	39 (39)	35 (35)	36 (36)	32 (32)
Bernie	28 (26)	24 (24)	28 (27)	27 (26)
Beta	30 (29)	29 (29)	30 (29)	29 (29)
Boka	41 (42)	41 (42)	37 (37)	38 (38)
Brandy	34 (34)	35 (35)	30 (29)	26 (26)
Bria	38 (38)	34 (34)	39 (40)	38 (39)
Brodie	34 (33)	33 (33)	32 (31)	31 (31)
Callie	39 (40)	39 (40)	33 (32)	32 (32)
Chechkel	41 (42)	42 (43)	40 (41)	38 (38)
Cheopi	34 (34)	34 (34)	33 (32)	31 (31)
Chester	36 (36)	28 (28)	37 (37)	37 (37)
Chinook	38 (38)	35 (25)	36 (37)	36 (37)
Christa	42 (43)	42 (43)	37 (37)	34 (34)
Chuhia	39 (40)	37 (37)	34 (34)	34 (34)
Cissie	40 (41)	38 (38)	37 (37)	35 (35)
Coco	33 (32)	31 (31)	38 (38)	37 (37)
Cybil	29 (28)	27 (27)	34 (34)	33 (33)
Dara	39 (39)	36 (36)	34 (34)	36 (36)
Drew	36 (36)	37 (37)	39 (40)	37 (37)
Duff	39 (39)	39 (39)	37 (37)	35 (35)
Eesha	33 (32)	30 (30)	34 (33)	33 (33)
Ehsto	43 (44)	41 (42)	43 (45)	44 (45)
Elvira	39 (39)	39 (39)	37 (37)	38 (38)
Eniga	39 (40)	39 (39)	35 (35)	35 (35)
Evelyne	30 (29)	32 (32)	30 (29)	29 (29)
Faye	38 (38)	37 (37)	38 (38)	35 (35)
Fiona	40 (41)	38 (38)	37 (37)	38 (38)
Foxy	36 (36)	37 (37)	35 (35)	35 (35)
Frannie	35 (35)	34 (34)	34 (34)	34 (34)
Fritz	39 (40)	38 (38)	36 (36)	34 (34)
Gaygos	35 (35)	36 (36)	38 (39)	39 (39)
Gelb	38 (38)	37 (37)	34 (33)	31 (31)
Gigi	34 (33)	34 (34)	35 (35)	35 (35)
Gimp	34 (33)	32 (32)	35 (35)	36 (36)
Gisoki	39 (40)	38 (39)	35 (35)	30 (30)
Haakid	37 (37)	36 (36)	40 (41)	38 (38)
Hannah	35 (35)	35 (35)	34 (33)	32 (32)
Helga	28 (27)	30 (30)	35 (35)	33 (33)
Heppie	41 (42)	41 (42)	37 (37)	36 (36)
Hobbes	36 (36)	30 (30)	33 (32)	33 (33)
Hodari	36 (36)	36 (36)	37 (37)	37 (37)
Huey	30 (29)	37 (37)	38 (38)	37 (37)
Hug	36 (36)	31 (31)	36 (36)	36 (36)
Huhkalig	38 (38)	38 (38)	36 (36)	35 (35)
Iyk	35 (35)	31 (31)	35 (35)	33 (33)
Jacqueline	32 (31)	33 (33)	34 (34)	34 (34)
Jadyh	34 (33)	31 (31)	34 (34)	33 (34)
Jake	39 (40)	38 (38)	37 (36)	36 (37)
Jamie	37 (37)	38 (38)	38 (38)	37 (37)
Jane	33 (32)	33 (33)	37 (38)	38 (37)
Jcarter	32 (31)	35 (35)	34 (34)	32 (32)
Jewelle	28 (27)	28 (28)	30 (29)	30 (30)

Abbreviations: [1] Measurements of left (L) and right (R) height and width (in mm), [2] The subsample with known hemisphere siding.

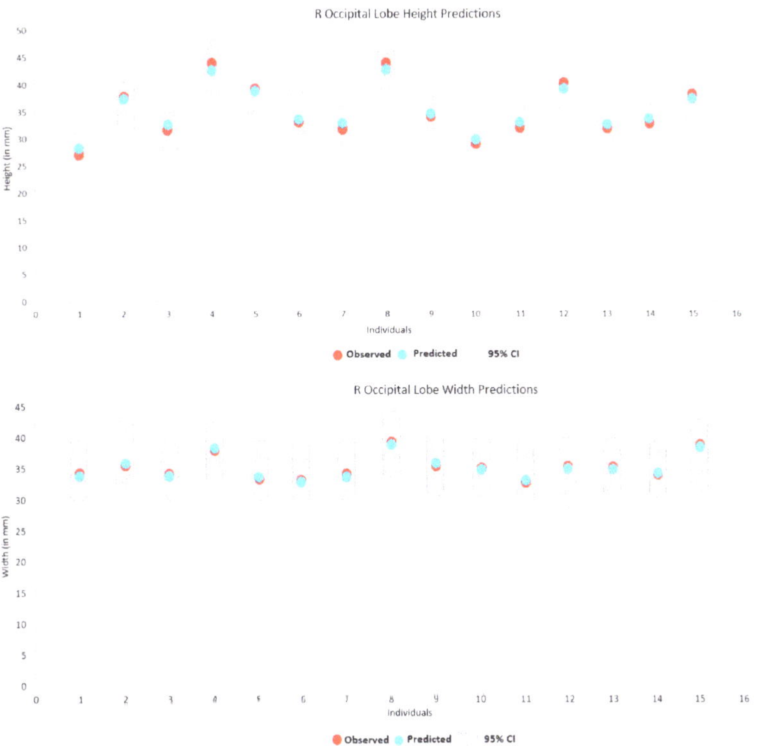

Figure 4. The predicted height and width (in mm) for the R occipital lobe in the known subsample with a confidence interval applied, calculated from the standard error of the regression.

4. Discussion

These findings suggest greater R > L height asymmetry associated with no bridging pattern, moderate R > L height asymmetry for both R and L bridge patterns, smaller L < R height and width asymmetry with a L bridge pattern, and smaller R < L height asymmetry associated with right bridge pattern. Additionally, there was less uncertainty when predicting right and left siding using occipital lobe width rather than occipital lobe height, indicating width is a more reliable predictor than height. This has implications for the suitability of metrics chosen to examine an association with bridging patterns, especially if the sample is unknown where width provides more reliable predictors than height for future research in modelling occipital lobe bridging patterns and possible associations. Although we suggest caution is warranted with the preliminary nature of these results, they also suggest there is a component of asymmetry for chimpanzee occipital lobe bridge patterns, and that increasing width and not simply posterior movement (or reduced height) of the occipital lobe may play an important role in exposure of the occipital-parietal bridge during human evolution, which was unexpected. Future research will compare the size of the parietal to the occipital lobe in these same subjects.

Author Contributions: Conceptualization and methodology, R.H. and S.H.; measurements, G.B.; Statistical analyzes, review and editing, A.P.; original draft preparation, S.H.; review and editing, S.H. All authors have read and agreed to the published version of the manuscript.

Funding: This research received no external funding.

Data Availability Statement: Measurements are contained in the article. MRI data can be obtained from the National Chimpanzee Brain Resource (https://www.chimpanzeebrain.org (accessed on 1 September 2021)).

Acknowledgments: The authors would like to thank Chet Sherwood & Aida Gomez-Robles for providing access to the data for these chimpanzees and Antoine Balzeau for inviting us to participate in this Special Issue.

Conflicts of Interest: The authors declare no conflict of interest.

References

1. LeMay, M. Morphological cerebral asymmetries of modern man, fossil man and nonhuman primate. *Ann. N. Y. Acad. Sci.* **1976**, *280*, 349–366. [CrossRef] [PubMed]
2. Holloway, R.L.; De La Coste-Lareymondie, M.C. Brain endocast asymmetry in pongids and hominids: Some preliminary findings on the paleontology of cerebral dominance. *Am. J. Phys. Anthropol.* **1982**, *58*, 101–110. [CrossRef] [PubMed]
3. Balzeau, A.; Gilissen, E.; Grimaud-Hervé, D. Shared pattern of endocranial shape asymmetries among great apes, anatomically modern humans, and fossil hominins. *PLoS ONE* **2011**, *7*, 29581. [CrossRef] [PubMed]
4. Gratiolet, L.P. *Mémoire sur les Plis Cérébraux de L'homme et des Primates*; Bertrand, A., Ed.; Betrand: Paris, France, 1854.
5. Connolly, J.C. *External Morphology of the Primate Brain*; Thomas, C.C., Ed.; Bannerstone House: Springfield, IL, USA, 1950.
6. Duvernoy, H. *The Human Brain*; Springer: New York, NY, USA, 1991.
7. Holloway, R.L.; Broadfield, D.C.; Yuan, M.S. Morphology and histology of chimpanzee primary visual striate cortex indicate that brain reorganization predated brain expansion in early hominid evolution. *Anat. Rec. A Discov. Mol. Cell Evol. Biol.* **2003**, *273*, 594–602. [CrossRef] [PubMed]
8. Falk, D.; Zollikofer, C.P.E.; Ponce de León, M.; Semendeferi, K.; Alatorre Warren, J.L.; Hopkins, W.D. Identification of in vivo Sulci on the External Surface of Eight Adult Chimpanzee Brains: Implications for Interpreting Early Hominin Endocasts. *Brain Behav. Evol.* **2018**, *91*, 45–58. [CrossRef] [PubMed]
9. Falk, D. A reanalysis of the South African australopithecine natural endocasts. *Am. J. Phys. Anthropol.* **1980**, *53*, 525–539. [CrossRef] [PubMed]
10. Holloway, R. Revisiting the South African Taung australopithecine endocast: The position of the lunate sulcus as determined by the stereoplotting technique. *Am. J. Phys. Anthropol.* **1981**, *56*, 43–58. [CrossRef]
11. Cointepas, Y.; Mangin, J.F.; Garnero, L.; Poline, J.B.; Benali, H. BrainVISA: Software platform for visualization and analysis of multi-modality brain data. *NeuroImage* **2001**, *13*, 98. [CrossRef]
12. Gomez-Robles, A.; Hopkins, W.D.; Sherwood, C.C. Increased morphological asymmetry, evolvability and plasticity in human brain evolution. *Proc. R Soc. B* **2013**, *280*, 20130575. [CrossRef] [PubMed]
13. Cignoni, P.; Callieri, M.; Corsini, M.; Dellepiane, M.; Ganovelli, F.; Ranzuglia, G. MeshLab: An Open-Source Mesh Processing Tool. In Proceedings of the Sixth Eurographics Italian Chapter Conference, Salerno, Italy, 2–4 July 2008; pp. 129–136.
14. Bailey, R.C.; Byrnes, J. A new, old method for assessing measurement error in both univariate and multivariate morphometric studies. *Syst. Zool.* **1990**, *39*, 124–130. [CrossRef]
15. Legendre, P.; Legendre, L. *Numerical Ecology*, 3rd ed.; Elsevier: Oxford, UK, 2012; Volume 24.
16. Pearson, A.; Polly, P.D.; Bruner, E. Making sense of modern human sulcal pattern variation, brain size and temporal lobe boundaries: Implications for fossil Homo. *Am. J. Phys. Anthropol.* **2021**, *174*, 83.
17. Press, W.H.; Teukolsky, S.A.; Vetterling, W.T.; Flannery, B.P. *Numerical Recipes in C*; Cambridge University Press: Cambridge, UK, 1992.
18. Simpson, G.G.; Roe, A.; Lewontin, R.C. *Quantitative Zoology*, Rev. ed.; Dover Publications: New York, NY, USA, 2003.
19. Smith, R.J. Allometric scaling in comparative biology: Problems of concept and method. *Am. J. Phys. Anthropol.* **1984**, *246 Pt 2*, R152–R160. [CrossRef] [PubMed]
20. Pearson, A.; Polly, P.D.; Bruner, E. Is the middle cranial fossa a reliable predictor of temporal lobe volume in extant and fossil anthropoids? *Am. J. Phys. Anthropol.* **2020**, *172*, 698–713. [CrossRef] [PubMed]
21. Hammer, Ø.; Harper, D.A.T.; Ryan, P.D. PAST: Paleontological Statistics Software Package for Education and Data Analysis. *Palaeontol. Electron.* **2001**, *4*, 9.

Article

Retrodeformation of the Steinheim Cranium: Insights into the Evolution of Neanderthals

Costantino Buzi [1,*], Antonio Profico [2], Fabio Di Vincenzo [3,4], Katerina Harvati [1,5], Marina Melchionna [6], Pasquale Raia [6] and Giorgio Manzi [4,7]

[1] DFG Centre for Advanced Studies 'Words, Bones, Genes, Tools', Eberhard Karls University of Tübingen, Rümelinstr. 23, 72070 Tübingen, Germany; katerina.harvati@ifu.uni-tuebingen.de

[2] PalaeoHub, Department of Archaeology, University of York, King's Manor, York YO1 7EP, UK; antonio.profico@york.ac.uk

[3] Natural History Museum, University of Florence, 50121 Firenze, Italy; fabio.divincenzo@unifi.it

[4] Italian Institute of Human Paleontology (Is.I.P.U.), 03012 Anagni, Italy; giorgio.manzi@uniroma1.it

[5] Paleoanthropology, Senckenberg Centre for Human Evolution and Palaeoenvironments, Institute for Archaeological Sciences, Eberhard Karls University of Tübingen, Rümelinstr. 23, 72070 Tübingen, Germany

[6] Dipartimento di Scienze della Terra, dell'Ambiente e delle Risorse, Università di Napoli Federico II, 80138 Napoli, Italy; marina.melchionna@unina.it (M.M.); pasquale.raia@unina.it (P.R.)

[7] Department of Environmental Biology, Sapienza University of Rome, 00185 Roma, Italy

* Correspondence: costantino.buzi@uni-tuebingen.de

Citation: Buzi, C.; Profico, A.; Di Vincenzo, F.; Harvati, K.; Melchionna, M.; Raia, P.; Manzi, G. Retrodeformation of the Steinheim Cranium. Insights into the Evolution of Neanderthals. *Symmetry* **2021**, *13*, 1611. https://doi.org/10.3390/sym13091611

Academic Editor: Chiarella Sforza

Received: 27 July 2021
Accepted: 20 August 2021
Published: 2 September 2021

Publisher's Note: MDPI stays neutral with regard to jurisdictional claims in published maps and institutional affiliations.

Abstract: A number of different approaches are currently available to digitally restore the symmetry of a specimen deformed by taphonomic processes. These tools include mirroring and retrodeformation to approximate the original shape of an object by symmetrisation. Retrodeformation has the potential to return a rather faithful representation of the original shape, but its power is limited by the availability of bilateral landmarks. A recent protocol proposed by Schlager and colleagues (2018) overcomes this issue by using bilateral landmarks and curves as well as semilandmarks. Here we applied this protocol to the Middle Pleistocene human cranium from Steinheim (Germany), the holotype of an abandoned species named *Homo steinheimensis*. The peculiar morphology of this fossil, associated with the taphonomic deformation of the entire cranium and the lack of a large portion of the right side of the face, has given rise to different hypotheses over its phylogenetic position. The reconstruction presented here sheds new light on the taphonomic origin of some features observed on this crucial specimen and results in a morphology consistent with its attribution to the Neanderthal lineage.

Keywords: digital reconstruction; *Homo heidelbergensis*; *Homo neanderthalensis*; *Homo sapiens*; Middle Pleistocene humans; virtual anthropology; Europe

1. Introduction

The study of fossil specimens has been revolutionised by the foundation of modern morphometrics [1]. Symmetry is one prominent feature of biological objects, and possibly the one affected the most by taphonomic processes [2–5]. However, symmetry also offers the possibility to restore the original shapes of fossil remains that are found broken or incomplete [6,7]. This is key to the interpretation of these specimens, since taphonomic alteration affecting diagnostic features may lead to incorrect taxonomic attributions and dubious phylogenetic reconstructions [7–9]. Digital methods for the reconstruction and restoration of broken fossil remains are nowadays available thanks to an ensemble of techniques that commonly fall under the heading of 'virtual anthropology' [10–12]. Specimens can be handled in a safe, virtual environment [7] and undergo restoration protocols that can include the realignment of dislocated fragments [13–16] or the digital removal of the plaster from traditional reconstructions [8,17] without the risk of damaging the original material. These protocols can be associated with symmetrisation, which helps to recreate missing portions or 'undo' the effects of plastic deformation. In the former case, symmetrisation 'fills the gaps' (i.e., missing portions) in one half of the fossil by mirroring the preserved counterparts [7,18–22]. In the latter case, referred to as retrodeformation, the plastic distortion of the original shape is corrected by relying on biological symmetry, as calculated by the acquisition of bilateral landmarks, curves, or patches of semiland-marks [4,7,8,23–28]. Mardia and colleagues [3] defined two types of bilateral symmetry: one referring to structures present as two separate copies on both sides of the specimen as mirror images (matching symmetry), the other defined (in three-dimensional objects) by the midsagittal plane passing through the specimen and thus determining an internal left–right symmetry (object symmetry) [2,3]. One key difference between matching and object symmetry is that genuine asymmetry is ignored by the former, but still apparent under the latter. In the case of the vertebrate skull, which provides an example of object symmetry [3], this implies that retrodeformation preserves genuine asymmetry, whereas mirroring does not. Moreover, mirroring can generate artefacts, or a biased morphology, if the only preserved portion is itself distorted [7]. On the other hand, the application of retrodeformation can be affected by the state of preservation of the object [4].

A perfect example of the combination of missing parts and plastic deformation affecting a single specimen is given by the cranium from Steinheim (hereafter, Steinheim), which is the holotype of the abandoned species *Homo steinheimensis* (Berckhemer, 1936) [29]. This human fossil was found in July 1933 in a gravel pit 70 km north of the town of Steinheim an der Murr, Baden-Württemberg, Germany [30,31] (Figure 1). It was recovered from Pleistocene fluvial deposits along the Murr river, which were well known at the time of the discovery for having yielded well-preserved fossils of Pleistocene mammals [30,32]. Since the discovery came from a well-studied area, the fossil received proper geological contextualisation. It was therefore possible to estimate the specimen's age based on the biochronological dating of the faunal assemblage, roughly corresponding to OIS 9 (i.e., 300–320 ka to 250 ka) [30,32–35].

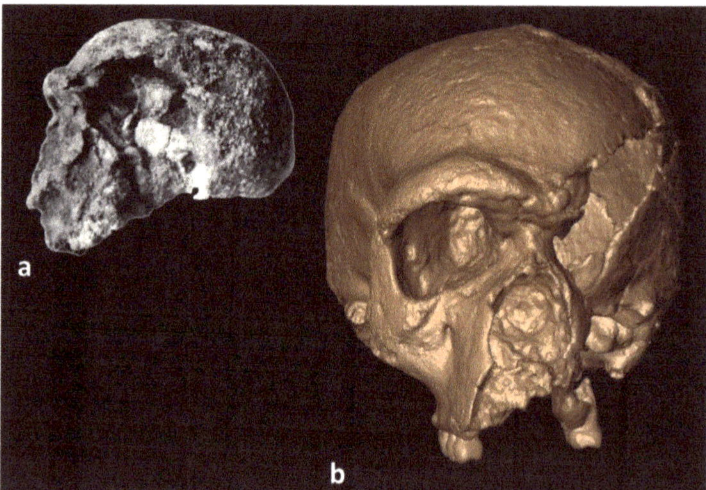

Figure 1. The cranium from Steinheim: (**a**) the cranium (left side) at the moment of recovery (from [30]); (**b**) a digital rendering of the cranium (front side).

The complex pattern of deformation that affected Steinheim, as well as its incomplete status and the presence of extensive incrustations, made it difficult to discern whether its peculiar morphology represents the original shape of the individual, or it is the product of taphonomic deformation [36,37]. This uncertainty contributed to a longstanding debate concerning the Steinheim phylogenetic position [36,38–40]. The cranium is characterised by a peculiar mixture of archaic and derived traits, which originated different proposals about its position within or close to the Neanderthal lineage—as representing a 'pre-Neanderthal stage' along the so-called process of accretion—or even as a specimen somehow related to the origin of *Homo sapiens* [41–45]. However, not only most of the left side of the facial skeleton in Steinheim is missing, but also the cranium presents a peculiar plastic deformation, further complicating the recognition of its features. For example, the highly diagnostic infraorbital plate and orbitomaxillary region are preserved only on the left side. This part of Steinheim's face shows an angled transverse profile, which was interpreted in the past as 'anticipating' the modern human morphology to some degree [30,43], but has been conversely interpreted as the result of the retention of archaic facial morphology, also observed in some Western European earlier taxa (i.e., *Homo antecessor*) [46–48]. The relatively low and long neurocranium of Steinheim, possessing a rather vertical occipital plane, also shows a slightly angled coronal profile, or a 'roofed' appearance [36], with the maximum cranial width occurring in the lower portion [35].

A specific name was initially proposed for this specimen (*Homo steinheimensis* Bereckhemer, 1936) [29], but it is currently considered invalid [40,49], despite that this name has been resurrected at the taxonomic rank of subspecies [50,51]. Steinheim is now generally considered as belonging to the Neanderthal lineage [45,52–54] and possibly related to other Middle Pleistocene populations (e.g., Atapuerca Sima de los Huesos, SH), with which it shares several derived traits in addition to its the geographical and chronological attributions [45,53,54].

2. Materials and Methods

The description of Steinheim's morphology is influenced by the extensive deformation of the skull [32,37]. A representation of the major directions of the deformation has been obtained by observations on the CT scan of the fossil and a review of literature [32,37,55] and is shown in Figure 2.

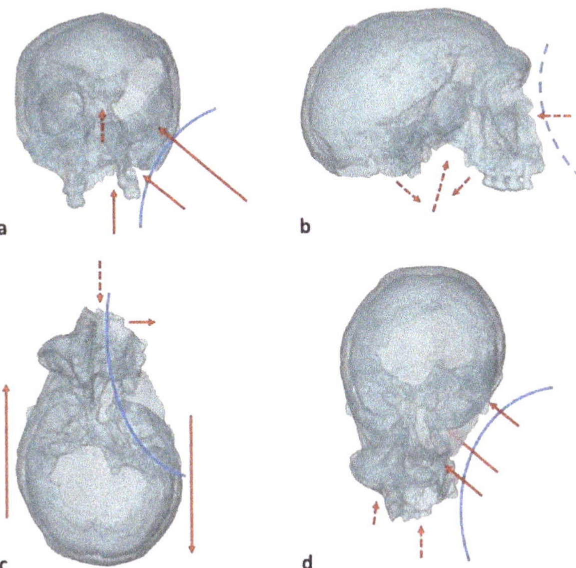

Figure 2. A simplified representation of the deformation of Steinheim. The blue lines resume the extent and area of influence of the deformation; the red arrows resume the directions of the morphological modification, associated with the areas in which the effects are visible. The solid lines point to the more evident effects of the deformation; the dashed lines represent additional possible effects. (**a**): anterior view; (**b**): right-lateral view; (**c**): inferior view; (**d**): superior view.

Prossinger and colleagues [37] performed the first digital segmentation of the cranium, resulting in a model cleared from the encrustations but still heavily affected by taphonomic distortions. Such distortions are observed in the internal structure of the cranium, including a shift to the right of the midsagittal plane of the splanchnocranium, an inward 'inflation' of the left orbital roof, and a rightward rotation of the axis of the *crista galli* in the anterior endocranial surface [37]. Since the left portion of the face is missing, it is difficult to assess how much of this morphology is determined by the deformation [37]. The right orbit is characterized by an angled shape with a sloped inferior margin. The preserved infraorbital plate shows an angled transverse profile with a point of bending roughly corresponding to the infraorbital foramen [33]. Curiously, this is associated with a moderate inflation of the anterior portion of the infraorbital plate, whereas the lateralmost portion appears flattened [35]. Through investigations conducted via CT scan and digital imaging, it was possible to assess the relative size of the frontal sinuses inside the well-developed supraorbital torus [36] and to diagnose a possible meningioma located in the upper part of the neurocranium [56].

To obtain a reconstruction consistent with object symmetry (sensu Mardia and colleagues [3]), we started by applying retrodeformation [4]. The choice of landmarks (Figure 3) was thus constrained by a criterion of symmetry: each landmark chosen on the left side must have a counterpart on the right side [4]. The incomplete state of Steinheim narrowed the choice of possible homologous landmarks and the choice of bilateral curves and surfaces for the definition of semilandmarks (Figure 3).

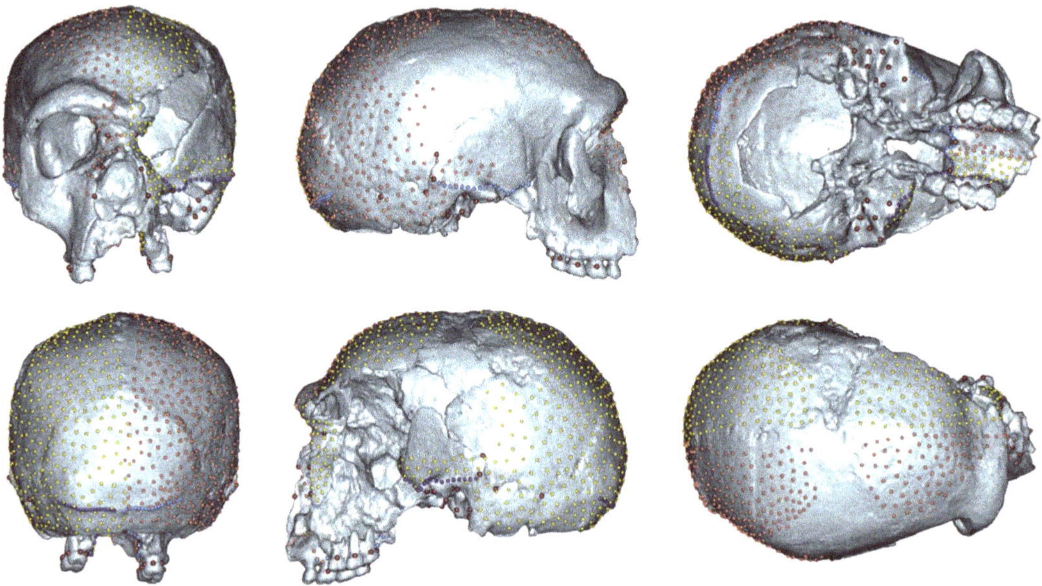

Figure 3. The configurations used: the bilateral landmarks (dark red); the bilateral curves, right (light blue) and left (dark blue); the patches of surface semilandmarks sampled on the left side (yellow) and their projection on the right side (orange).

It was possible to define only a few landmarks on the small preserved portion of the left side of the face, comprising the nasomaxillary region (Figure 3). In defining surface semilandmarks, we excluded the preserved portion of the temporal squama because it is affected by local breakage and subsequent reconstruction (see Figure 1) [36–38]. Similarly, in defining the patches of semilandmarks, the upper part of the left parietal was excluded, as this portion of the neurocranium is more affected by breakage and surface damage (Figure 3). The basioccipital is also damaged, cracked, and partly shifted inside the neurocranium itself, and therefore, no landmarks were placed on this region.

The high-resolution CT scan of Steinheim was kindly provided by Prof. Dr. Christoph P.E. Zollikofer (Department of Anthropology, University of Zurich). The CT data, obtained in the form of a DICOM stack, were processed in Amira [57] to obtain a 3D mesh, subsequently converted into the .ply format. We defined 52 bilateral landmarks on the skull and 8 curves. The curves were later processed in R by the function *equidistantCurve* (*Morpho* R package) [58] to sample evenly spaced semilandmarks along each curve. The coordinates of 500 semilandmarks were obtained by applying a *k*-means clustering algorithm to the vertex coordinates from a portion of the mesh corresponding to the left part of the cranium, from which we excluded the temporal squama and the damaged area of the coronal suture (Figure 3). The set of surface semilandmarks built this way was rotated and projected on the right side. In sum, we defined 8 curves (120 points), 52 bilateral landmarks, and 1000 surface semilandmarks for a total of 1172 paired coordinates (Figure 3). After the retrodeformation, applied according to the protocol in Schlager and colleagues [4], we calculated and visualized the local displacement between the starting and retrodeformed meshes using the function *localmeshdiff* (*Arothron* R package) [59]. The retrodeformed model of Steinheim was eventually subjected to a principal component analysis (PCA) in the shape space, together with the original model and a comparison sample including modern humans (N = 17), Neanderthals (N = 5), and Middle Pleistocene humans (N = 3). The comparison sample for the PCA is reported in Supplementary Table S1. The cranial landmark configuration used for the analysis was built upon the preserved portions of Steinheim and is figured in Supplementary Figure S1.

3. Results

3.1. The Retrodeformation

Most of the retrodeformation procedure intervened on the anteroposterior shift of the two sides of the skull (Figure 2c). In the frontal view (Figure 4a), the shift produces a relative enlargement of the piriform aperture, mainly on the left side, associated with a forward shift of the left rim and a slight retraction of the medial portion of the right rim. A slight 'relaxation' of the nasal profile in the superoinferior direction is apparent, as well as the symmetrisation of the general profile of the neurocranium, which is even more evident in the posterior view (Figure 4b). Symmetrisation of the occlusal plane of the teeth eliminates the unnatural downward displacement of the right maxilla along the midsagittal plane, which is present in the original specimen (Figure 4a,b).

The correction of the anteroposterior shift of the face along the midsagittal plane is also evident from the inferior view (Figure 4c), where the reduction of the slight clockwise rotation of the palate becomes apparent, accompanied by a deflation of the right postorbital portion of the neurocranium. In addition, the basicranium regained a more natural position, appearing straighter and medially placed in comparison with the original specimen, even though a slight deformation remains due to the lack of landmarks to be placed on this badly preserved portion. Preservation similarly affects the retrodeformation process of the flexion of the basicranium and the anteroposterior compression along the coronal suture (Figure 4d). In the lateral view (Figure 4d), the general profile of the neurocranium does not show any major changes. However, it is evident that the retrodeformation produces a retraction of the right portion of the face. Corresponding to the frontal squama, it is possible to see in transparency the previous position of the right side (Figure 4d), which was originally shifted forward according to the deformation directions illustrated in Figure 2.

3.2. Local Displacement

The local displacement between the starting and the retrodeformed meshes (Figure 5) indicates the areas of maximum expansion, which is apparent on almost the entire left side of the skull, with the highest values recorded in the preserved portions of the left maxilla, left parietal, and left orbit. Conversely, the right side of the face is affected almost entirely by contraction, with the highest values recorded at the level of the anterior portion of the maxilla and nasofrontal suture. However, the inferior portion of the right maxilla is expanded in the retrodeformed mesh. A somewhat balanced pattern of deformation occurs on the basicranium. The areas of maximum contraction on the right side are associated with areas of maximum expansion on the opposite side. Similarly, the moderate expansion recorded on the right side of the occipital squama corresponds to an almost symmetrical contraction of the left side. Lastly, along the midsagittal plane an area of contraction appears evident.

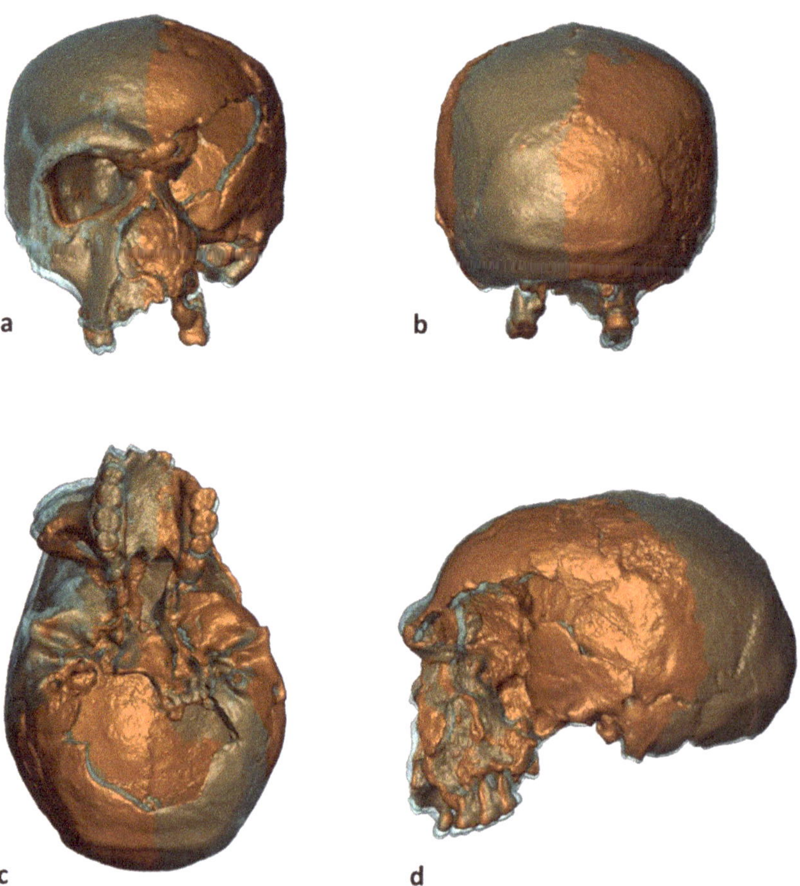

Figure 4. Comparison between the retrodeformed model of Steinheim (brown) and the original specimen (transparent blue). (**a**): anterior view; (**b**): posterior view; (**c**): inferior view; (**d**): left-lateral view.

3.3. Principal Component Analysis

The results of the PCA are reported in Figure 6. The first three PCs explain 62.74% of the total variance in the sample, weighting 41.51%, 12.27% and 8.96%, respectively. In the plot, it is possible to discern a clear separation between modern (*Sap*) and fossil humans both along PC1 and PC3. Along PC2 is visible the separation between the Neanderthals sensu stricto (*Nea*) and a small group of Middle Pleistocene humans (*Mph*). The two models of Steinheim (*Ste*) fall within an intermediate position along PC1, between the *Sap* cluster and the fossil human group. While the original model of Steinheim (*Ori*) clearly diverges from the rest of the sample along both PC1 and PC2, the retrodeformed model of Steinheim (*R.D.*) approaches the fossil human group along the PC1, reaching the limit of the *Nea* cluster along PC2.

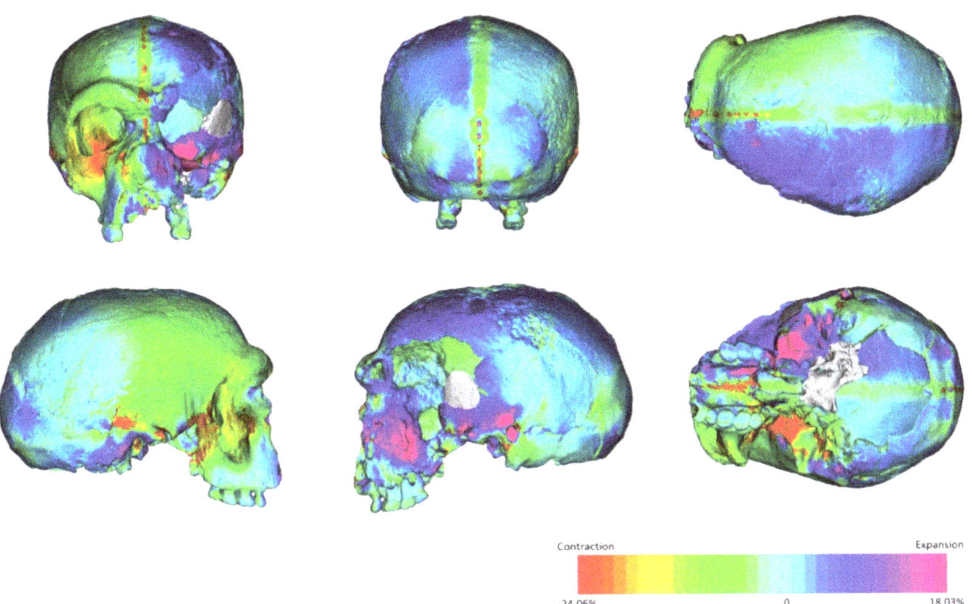

Figure 5. Local displacement (%) in the retrodeformed model of Steinheim, calculated by the function *localmeshdiff*. The white areas represent heavily damaged (basioccipital) or reconstructed (left temporal) portions of the skull that were excluded from this analysis.

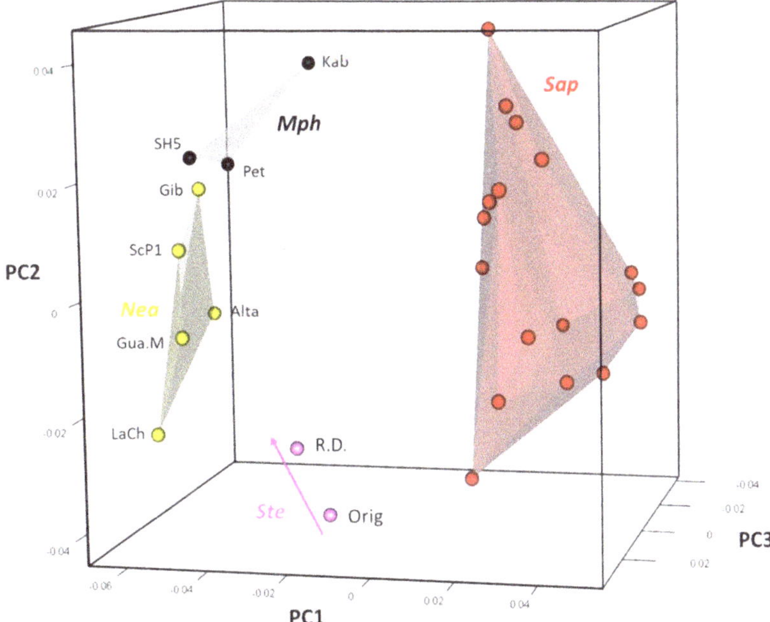

Figure 6. Shape PCA on cranial landmark configuration. In black, Middle Pleistocene humans (*Mph*); in yellow, Neanderthals (*Nea*); in red, modern humans (*Sap*); in violet, Steinheim (*Ste*): original model (*Ori*) and retrodeformed model (*R.D.*). The abbreviations of the fossil samples are reported in Supplementary Table S1.

4. Discussion

We used a retrodeformation protocol to produce a restoration of the Steinheim cranium. The application seemingly restored object symmetry to the specimen [3], despite the poor starting conditions of this incomplete and severely deformed fossil. We relied on the preserved portions, mainly on the left side of the cranium, to drive the reshaping of its counterpart. Even though some directions of the taphonomic deformations were not addressed, our application minimises their effect. The reconstruction (Figure 7), hence, allows us to better contextualise Steinheim among the coeval—or at least chronologically close—Middle Pleistocene *Homo* specimens. The neurocranial shape in the posterior view appears intermediate in morphology between the populations of Sima de los Huesos and early Neanderthals (e.g., Saccopastore 1), in keeping with the slight lateral expansion of the parietal walls after the reconstruction as compared with the more vertical and 'compressed' profile in *Ori*. In this respect, the neurocranial morphology of *R.D.* seems to approach the morphology of the early Neanderthal from Altamura [60]. In the posterior view, the original 'roofed' appearance (as described by Schwartz and Tattersall [36]) weakens in *R.D.* neurocranium, appearing close in morphology to penecontemporaneous individuals such as Skull SH5 from Atapuerca [40,54,61], except for the further laterolateral enlargement of the parietals. This trait, difficult to discern before restoring symmetry, places the maximum width of the skull in a slightly lower position relative to that of the original specimen, and roughly at the level of the temporal squama, similar to the typical Neanderthal condition [62] (Figure 4a,b). It is also possible to see a change in the relative position of the two mastoid processes, which, although partly damaged, after retrodeformation show reduced development compared with those of SH5. Their slight rotation can be interpreted as a trait anticipating the Neanderthal condition of tapering [44,45,62], although high variability in this feature among the Middle Pleistocene humans has been observed [63].

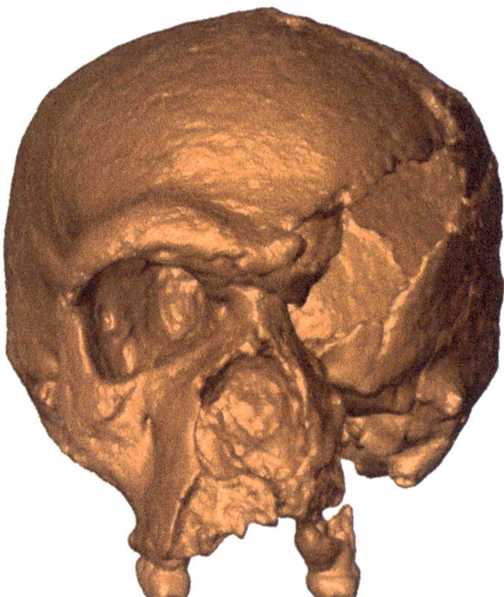

Figure 7. The retrodeformed model of Steinheim.

An almost symmetrical pattern of contraction and expansion is visible at the level of the glenoid fossae (Figure 5), associated with a change in the relative size of the postorbital portion of the neurocranium. This contributes, in turn, to the slight shortening and laterolateral enlargement of the neurocranium. On the other hand, the 'strip' recorded along the midsagittal axis (Figure 5) corresponds to an almost continuous area of contraction, which is a clear indication of the taphonomic deformation that occurred along this axis (see Figure 2c). This 'strip' can probably be traced back to a local expansion along the midsagittal line due to the two 'halves' moving in opposite directions.

As can be seen from Figure 5, the retrodeformation was not able to address the anteroposterior vectors of deformation. This is because such vectors acted in a single straight line, rather than bilaterally. Thus, it is not possible to reach evidence-based assumptions on whether the flexion of the basicranium reflects the original condition or it is the result of taphonomic deformation. Nevertheless, the restored midsagittal profile of the cranium suggests that the anteroposteriorly elongated profile of the neurocranium could possibly be associated with a less flexed basicranium. As we proposed in Figure 2b, the present flexion could be related to a deformation operating along the sagittal axis on the upper midface.

Unfortunately, the almost completely missing left portion of the face made it difficult to correct for some local modifications in this area. Nonetheless, it is still possible to carefully evaluate whether some features are due to taphonomic deformation. As mentioned above, the retrodeformation resulted in a 'proper' midsagittal profile (Figure 4d) by undoing the rotational deformation caused by the anterolateral crushing (Figure 2d). By examining the lateral view of the reconstruction, it is more evident how the 'plica' obliterating the frontonasal suture—which is not found in any other hominin from Middle to Late Pleistocene—is consistent with an anterior crushing of the upper part of the nasal portion (Figure 2a). This, in turn, can be associated with the 'notch' found along the lower-right orbital rim, corresponding to a point of weakness represented by the zygomaxillary suture. We suggest that the peculiar facial morphology of Steinheim is mostly a result of the crushing that occurred in the upper portion of the midface (Figure 2b). In our opinion, the reconstruction showed that the infraorbital plate was in origin possibly less flexed than *Ori* suggests.

As evidenced by the PCA (Figure 6), Steinheim is distinguished from the rest of the sample, and this 'uniqueness' can be traced back to its complex pattern of taphonomic deformation. Nonetheless, when a part of this is corrected by retrodeformation, it is possible to see how the new model approaches the fossil human subsample, towards the Neanderthal cluster. We hypothesize that since some of the deformation vectors—namely, those operating on the anteroposterior axis—cannot be intercepted by the retrodeformation, Steinheim still presents itself with a 'unique' morphology, distinguished from other Middle Pleistocene specimens.

5. Conclusions

The 'mosaic' evolution of the typical Neanderthal cranial morphology (i.e., 'classic' Neanderthal cranial shape [62]) seems to have included an earlier development of some facial traits, combined with the retention of a more primitive condition for the neurocranium [51,61]. The full development of the typical Neanderthal *en-bombe* shape must thus be considered a derived trait. Consequently, the moderate expansion of the parietals of the retrodeformed Steinheim (posterior view, Figure 4b) suggests association with a greater expression of midfacial prognathism than that observed in this individual, consistent with other specimens from the Middle Pleistocene of Europe. In our opinion, the present facial morphology of Steinheim is influenced by the deformation caused on the upper midface by taphonomy. Even though the reconstruction presented here did not correct the whole pattern of deformation, nor did it provide the exact original shape of Steinheim, it contributes to shedding new light on the morphology of this specimen and concurrently to placing Steinheim more firmly in the Neanderthal evolutionary lineage.

Supplementary Materials: The following are available online at https://www.mdpi.com/2073-8994/13/9/1611/s1: Table S1: Comparative samples used in the analysis; Figure S1: Landmark configuration used for the PCA.

Author Contributions: Conceptualization, C.B., A.P. and G.M.; methodology, C.B., A.P. and M.M.; formal analysis, C.B., A.P., F.D.V. and M.M.; investigation, C.B., A.P., F.D.V. and M.M.; data curation,

C.B., K.H. and G.M.; writing—original draft preparation, C.B., A.P., F.D.V. and G.M.; writing—review and editing, C.B., A.P., F.D.V., K.H., M.M., P.R. and G.M.; visualization, C.B. and A.P.; supervision, K.H., P.R. and G.M. All authors have read and agreed to the published version of the manuscript.

Funding: This research received no external funding.

Institutional Review Board Statement: Not applicable.

Informed Consent Statement: Not applicable.

Data Availability Statement: Not applicable.

Acknowledgments: We are extremely grateful to Christoph P.E. Zollikofer (Department of Anthropology, University of Zurich) for kindly providing the CT data for the Steinheim cranium. We thank the Museum of Anthropology 'G. Sergi' (Sapienza University of Rome) and the digital database NESPOS—Pleistocene People and Places for granting access to the comparative sample.

Conflicts of Interest: The authors declare no conflict of interest.

References

1. Bookstein, F.L. The study of shape transformation after D'Arcy Thompson. *Math. Biosci.* **1977**, *34*, 177–219. [CrossRef]
2. Klingenberg, C.P.; Barluenga, M.; Meyer, A. Shape analysis of symmetric structures: Quantifying variation among individuals and asymmetry. *Evolution* **2002**, *56*, 1909–1920. [CrossRef]
3. Mardia, K.V.; Bookstein, F.L.; Moreton, I.J. Statistical assessment of bilateral symmetry of shapes. *Biometrika* **2000**, *87*, 285–300. [CrossRef]
4. Schlager, S.; Profico, A.; Di Vincenzo, F.; Manzi, G. Retrodeformation of fossil specimens based on 3D bilateral semi-landmarks: Implementation in the R package "Morpho". *PLoS ONE* **2018**, *13*, e0194073. [CrossRef]
5. Fernández-Jalvo, Y.; Andrews, P. *Atlas of Taphonomic Identifications*; Springer: Dordrecht, The Netherlands, 2016; p. 4.
6. Adams, D.C.; Rohlf, F.J.; Slice, D.E. A field comes of age: Geometric morphometrics in the 21st century. *Hystrix* **2013**, *24*, 7–14.
7. Gunz, P.; Mitteroecker, P.; Neubauer, S.; Weber, G.W.; Bookstein, F.L. Principles for the virtual reconstruction of hominin crania. *J. Hum. Evol.* **2009**, *57*, 48–62. [CrossRef]
8. Di Vincenzo, F.; Profico, A.; Bernardini, F.; Cerroni, V.; Dreossi, D.; Schlager, S.; Zaio, P.; Benazzi, S.; Biddittu, I.; Rubini, M.; et al. Digital reconstruction of the Ceprano calvarium (Italy), and implications for its interpretation. *Sci. Rep.* **2017**, *7*, 1–11. [CrossRef]
9. White, T. Early hominids—Diversity or distortion? *Science* **2003**, *299*, 1994–1997. [CrossRef]
10. Weber, G.W.; Schäfer, K.; Prossinger, H.; Gunz, P.; Mitteröcker, P.; Seidler, H. Virtual anthropology: The digital evolution in anthropological sciences. *J. Physiol. Anthropol. Appl. Human Sci.* **2001**, *20*, 69–80. [CrossRef]
11. Weber, G.W. Another link between archaeology and anthropology: Virtual anthropology. *DAACH* **2014**, *1*, 3–11. [CrossRef]
12. Weber, G.W. Virtual anthropology. *Am. J. Phys. Anthropol.* **2015**, *156*, 22–42. [CrossRef]
13. Amano, H.; Kikuchi, T.; Morita, Y.; Kondo, O.; Suzuki, H.; Ponce de León, M.S.; Zollikofer, C.P.E.; Bastir, M.; Stringer, C.B.; Ogihara, N. Virtual reconstruction of the Neanderthal Amud 1 cranium. *Am. J. Phys. Anthropol.* **2015**, *158*, 185–197. [CrossRef] [PubMed]
14. Bosman, A.M.; Buck, L.T.; Reyes-Centeno, H.; Mirazón Lahr, M.; Stringer, C.B.; Harvati, K. The Kabua 1 cranium: Virtual anatomical reconstructions. In *Modern Human Origins and Dispersal*; Sahle, Y., Reyes-Centeno, H., Bentz, C., Eds.; Kerns Verlag: Tübingen, Germany, 2019; pp. 137–170.
15. Harvati, K.; Röding, C.; Bosman, A.M.; Karakostis, F.A.; Grün, R.; Stringer, C.B.; Karkanas, P.; Thompson, N.C.; Koutoulidis, V.; Moulopoulos, L.A.; et al. Apidima Cave fossils provide earliest evidence of *Homo sapiens* in Eurasia. *Nature* **2019**, *571*, 500–504. [CrossRef]
16. Profico, A.; Buzi, C.; Davis, C.; Melchionna, M.; Veneziano, A.; Raia, P.; Manzi, G. A new tool for digital alignment in virtual anthropology. *Anat. Rec.* **2019**, *302*, 1104–1115. [CrossRef] [PubMed]
17. Benazzi, S.; Bookstein, F.L.; Strait, D.S.; Weber, G.W. A new OH5 reconstruction with an assessment of its uncertainty. *J. Hum. Evol.* **2011**, *61*, 75–88. [CrossRef] [PubMed]
18. Benazzi, S.; Gruppioni, G.; Strait, D.S.; Hublin, J.J. Virtual reconstruction of KNM-ER 1813 *Homo habilis* cranium. *Am. J. Phys. Anthropol.* **2014**, *153*, 154–160. [CrossRef] [PubMed]
19. Berge, C.; Goularas, D. A new reconstruction of Sts 14 pelvis (*Australopithecus africanus*) from computed tomography and three-dimensional modeling techniques. *J. Hum. Evol.* **2010**, *58*, 262–272. [CrossRef]
20. Gunz, P.; Neubauer, S.; Golovanova, L.; Doronichev, V.; Maureille, B.; Hublin, J.J. A uniquely modern human pattern of endocranial development. Insights from a new cranial reconstruction of the Neandertal newborn from Mezmaiskaya. *J. Hum. Evol.* **2012**, *62*, 300–313. [CrossRef]
21. Hublin, J.J.; Ben-Ncer, A.; Bailey, S.E.; Freidline, S.E.; Neubauer, S.; Skinner, M.M.; Bergmann, I.; Le Cabec, A.; Benazzi, S.; Harvati, K.; et al. New fossils from Jebel Irhoud, Morocco and the pan-African origin of *Homo sapiens*. *Nature* **2017**, *546*, 289–292. [CrossRef]

22. Vialet, A.; Guipert, G.; Alçiçek, M.C. *Homo erectus* found still further west: Reconstruction of the Kocabaş cranium (Denizli, Turkey). *Comptes Rendus Palevol.* **2012**, *11*, 89–95. [CrossRef]

23. Arbour, V.M.; Currie, P.J. Analyzing taphonomic deformation of ankylosaur skulls using retrodeformation and finite element analysis. *PLoS ONE* **2012**, *7*, e39323. [CrossRef] [PubMed]

24. Kazhdan, M.M.; Amenta, N.; Gu, S.; Wiley, D.F.; Hamann, B. Symmetry restoration by stretching. In Proceedings of the 21st Canadian Conference on Computational Geometry (CCCG2009), Vancouver, Canada, 17–19 August 2009; CCCG: Vancouver, BC, Canada, 2009; pp. 37–40.

25. Lautenschlager, S. Reconstructing the past: Methods and techniques for the digital restoration of fossils. *Roy. Soc. Open Sci.* **2016**, *3*, 160342. [CrossRef]

26. Ogihara, N.; Nakatsukasa, M.; Nakano, Y.; Ishida, H. Computerized restoration of nonhomogeneous deformation of a fossil cranium based on bilateral symmetry. *Am. J. Phys. Anthropol.* **2006**, *130*, 1–9. [CrossRef]

27. Tallman, M.; Amenta, N.; Delson, E.; Frost, S.R.; Ghosh, D.; Klukkert, Z.S.; Morrow, A.; Sawyer, G.J. Evaluation of a new method of fossil retrodeformation by algorithmic symmetrization: Crania of papionins (Primates, Cercopithecidae) as a test case. *PLoS ONE* **2014**, *9*, e100833. [CrossRef]

28. Tschopp, E.; Russo, J.; Dzemski, G. Retrodeformation as a test for the validity of phylogenetic characters: An example from diplodocid sauropod vertebrae. *Palaeontol. Electron.* **2016**, *16*, 2T. [CrossRef]

29. Berckhemer, F. Der Urmenschenschädel aus den zwischeneiszeitlichen Fluss-Schottern von Steinheim an der Murr. *Forsch u Fortschr* **1936**, *12*, 349–350.

30. Adam, K.D. The chronological and systematic position of the Steinheim skull. In *Ancestors: The Hard Evidence*; Delson, E., Ed.; Liss: New York, NY, USA, 1985; pp. 272–276.

31. Granat, J.; Peyre, É. L'énigme odontologique du crâne de Steinheim 280 ka-Allemagne. *Actes SFHAD* **2008**, *13*, 14–18.

32. Wahl, J.; König, H.; Ziegler, R. Die Defekt-und Verformungsspuren am Schädel des Urmenschen von Steinheim an der Murr. *Fundber. Baden-Württ.* **2009**, *30*, 7–28.

33. Weinert, H. Der urmenschenschädel von Steinheim. *Z. Morphol. Anthropol.* **1936**, *35*, 463–518.

34. Czarnetzki, A. Steinheim skull: A morphological comparison with Tautavel Man. In Proceedings of the L'Homo Erectus et la Place de l'homme de Tautavel Parmi les Hominidés Fossiles, Actes Congrès International de Paléontologie Humaine, Nice, France, 16–21 October 1982; de Lumley, H., de Lumley, M.A., Eds.; pp. 875–893.

35. Street, M.; Terberger, T.; Orschiedt, J. A critical review of the German Paleolithic hominin record. *J. Hum. Evol.* **2006**, *51*, 551–579. [CrossRef] [PubMed]

36. Schwartz, J.H.; Tattersall, I. *The Human Fossil Record. Vol. 1: Terminology and Craniodental Morphology of Genus Homo (Europe)*; Wiley-Liss: New York, NY, USA, 2002; pp. 347–351.

37. Prossinger, H.; Seidler, H.; Wicke, L.; Weaver, D., Recheis, W.; Stringer, C.B.; Müller, G.B. Electronic removal of encrustations inside the Steinheim cranium reveals paranasal sinus features and deformations and provides a revised endocranial volume estimate. *Anat. Rec. B* **2013**, *273*, 132–142. [CrossRef]

38. Wolpoff, M.H. Cranial remains of middle Pleistocene European hominids. *J. Hum. Evol.* **1980**, *9*, 339–358. [CrossRef]

39. Stringer, C.B.; Hublin, J.J.; Vandermeersch, B. The origin of anatomically modern humans in Western Europe. In *The Origins of Modern Humans: A World Survey of the Fossil Evidence*; Smith, F.H., Spencer, F., Eds.; Liss: New York, NY, USA, 1984; pp. 51–135.

40. Arsuaga, J.L.; Martinón-Torres, M.; Santos, E. *Homo steinheimensis*, a comparison between the Steinheim skull and the Atapuerca Sima de los Huesos fossils. In Proceedings of the European Society for the study of Human Evolution (ESHE) 8, Liège, Belgium, 19–21 September 2019; ESHE: Liège, Belgium, 2019; p. 7.

41. Sergi, S. The palaeanthropi in Italy: The fossil men of Saccopastore and Circeo. Part II: Discussion and interpretation. *Man* **1948**, *48*, 76–79. [CrossRef]

42. Vlček, E. A new discovery of *Homo erectus* in Central Europe. *J. Hum. Evol.* **1978**, *7*, 239–251. [CrossRef]

43. Day, M.H. *Guide to Fossil Man*, 4th ed.; University of Chicago Press: Chicago, IL, USA, 1986.

44. Dean, D.; Hublin, J.J.; Holloway, R.; Ziegler, R. On the phylogenetic position of the pre-Neandertal specimen from Reilingen, Germany. *J. Hum. Evol.* **1998**, *34*, 485–508. [CrossRef] [PubMed]

45. Hublin, J.J. The origin of Neandertals. *Proc. Natl. Acad. Sci. USA* **1998**, *106*, 16022–16027. [CrossRef] [PubMed]

46. Bermúdez de Castro, J.M.; Arsuaga, J.L.; Carbonell, E.; Rosas, A.; Martinez, I.; Mosquera, M. A hominid from the Lower Pleistocene of Atapuerca, Spain: Possible ancestor to Neandertals and modern humans. *Science* **1997**, *276*, 1392–1395. [CrossRef]

47. Freidline, S.E.; Gunz, P.; Harvati, K.; Hublin, J.J. Evaluating developmental shape changes in *Homo antecessor* subadult facial morphology. *J. Hum. Evol.* **2013**, *65*, 404–423. [CrossRef]

48. Trafí, F.R.; Bartual, M.G.; Wang, Q. The affinities of *Homo antecessor*—A review of craniofacial features and their taxonomic validity. *Anthropol. Rev.* **2018**, *81*, 225–251. [CrossRef]

49. Stringer, C. The status of *Homo heidelbergensis* (Schoetensack 1908). *Evol. Anthropol.* **2012**, *21*, 101–107. [CrossRef]

50. Manzi, G. Humans of the Middle Pleistocene: The controversial calvarium from Ceprano (Italy) and its significance for the origin and variability of *Homo heidelbergensis*. *Quat. Int.* **2016**, *411*, 254–261. [CrossRef]

51. Manzi, G. Before the emergence of *Homo sapiens*: Overview on the Early-to-Middle Pleistocene fossil record (with a proposal about *Homo heidelbergensis* at the subspecific level). *Int. J. Evol. Biol.* **2011**, *2011*, 582678. [CrossRef]

52. Stringer, C.B. Secrets of the Pit of the Bones. *Nature* **1993**, *362*, 501–502. [CrossRef]

53. Tattersall, I. Before the Neanderthals: Hominid Evolution in Middle Pleistocene Europe. In *Continuity and Discontinuity in the Peopling of Europe*; Condemi, S., Weniger, G.C., Eds.; Springer: Dordrecht, The Netherlands, 2011; pp. 47–53.
54. Mounier, A.; Caparrós, M. The phylogenetic status of *Homo heidelbergensis*—A cladistic study of Middle Pleistocene hominins. *Bull. Mém. Soc. Anthropol. Paris* **2015**, *27*, 110–134. [CrossRef]
55. Braun, M.; Hublin, J.J.; Bouchet, P. New reconstruction of the Middle Pleistocene skull of Steinheim (Baden-Würtemberg, Germany). *Am. J. Phys. Anthropol.* **1998**, *S26*, 113.
56. Czarnetzki, A.; Schwaderer, E.; Pusch, C.M. Fossil record of meningioma. *Lancet* **2003**, *362*, 408. [CrossRef]
57. Stalling, D.; Westerhoff, M.; Hege, H.C. Amira: A highly interactive system for visual data analysis. In *The Visualization Handbook*; Hansen, C., Johnson, C., Eds.; Elsevier: Amsterdam, The Netherlands, 2005; pp. 749–767.
58. Schlager, S. Morpho and Rvcg—Shape Analysis in R: R-Packages for geometric morphometrics, shape analysis and surface manipulations. In *Statistical Shape and Deformation Analysis*; Zheng, G., Li, S., Székely, G., Eds.; Academic Press: London, UK, 2017; pp. 217–256.
59. Profico, A.; Buzi, C.; Castiglione, S.; Melchionna, M.; Piras, P.; Veneziano, A.; Raia, P. Arothron: An R package for geometric morphometric methods and virtual anthropology applications. *Am. J. Phys. Anthropol.* **2021**, *176*, 144–151.
60. Buzi, C.; Di Vincenzo, F.; Profico, A.; Manzi, G. The pre-modern human fossil record in Italy from the Middle to the Late Pleistocene: An updated reappraisal. *Alp. Mediterr. Quat.* **2021**, *34*, 1–16.
61. Arsuaga, J.L.; Martínez, I.; Arnold, L.J.; Aranburu, A.; Gracia-Téllez, A.; Sharp, W.D.; Quam, R.M.; Falguères, C.; Pantoja-Pérez, A.; Bischoff, J.; et al. Neandertal roots: Cranial and chronological evidence from Sima de los Huesos. *Science* **2014**, *344*, 1358–1363. [CrossRef]
62. Churchill, S.E. *Thin on the Ground: Neandertal Biology, Archeology and Ecology*; Wiley: Hoboken, NJ, USA, 2014; pp. 15–22.
63. Harvati, K.; Hublin, J.J.; Gunz, P. Evolution of middle-late Pleistocene human cranio-facial form: A 3-D approach. *J. Hum. Evol.* **2010**, *59*, 445–464. [CrossRef]

 *symmetry*

Article

A New Integrated Tool to Calculate and Map Bilateral Asymmetry on Three-Dimensional Digital Models

Marina Melchionna [1,*,†], Antonio Profico [2,†], Costantino Buzi [3], Silvia Castiglione [1], Alessandro Mondanaro [4], Antonietta Del Bove [5,6], Gabriele Sansalone [7], Paolo Piras [8] and Pasquale Raia [1]

[1] Dipartimento di Scienze della Terra, dell'Ambiente e delle Risorse, Università di Napoli Federico II, 80126 Napoli, Italy; silvia.castiglione@unina.it (S.C.); pasquale.raia@unina.it (P.R.)
[2] PalaeoHub, Department of Archaeology, Hull York Medical School University of York, Heslington YO10 5DD, UK; antonio.profico@york.ac.uk
[3] DFG Centre for Advanced Studies 'Words, Bones, Genes, Tools', Eberhard Karls University of Tübingen, 72070 Tübingen, Germany; costantino.buzi@uni-tuebingen.de
[4] Dipartimento di Scienze della Terra, Università degli Studi di Firenze, 50121 Firenze, Italy; alessandro.mondanaro@unifi.it
[5] Departament d'Història i Història de l'Art, Universitat Rovira i Virgili, 43002 Tarragona, Spain; adelbove@iphes.cat
[6] Institut Català de Paleoecologia Humana i Evolució Social (IPHES-CERCA), 43007 Tarragona, Spain
[7] Function, Evolution & Anatomy Research Lab, Zoology Division, School of Environmental and Rural Science, University of New England, Armidale, NSW 2351, Australia; gabriele.sansalone@uniroma3.it
[8] Dipartimento di Scienze Cardiovascolari, Respiratorie, Nefrologiche, Anestesiologiche e Geriatriche, Università degli Studi di Roma La Sapienza, 00185 Rome, Italy; paolo.piras@uniroma3.it
* Correspondence: marina.melchionna@unina.it
† Co-first authorship.

Citation: Melchionna, M.; Profico, A.; Buzi, C.; Castiglione, S.; Mondanaro, A., Del Bove, A., Sansalone, G.; Piras, P.; Raia, P. A New Integrated Tool to Calculate and Map Bilateral Asymmetry on Three-Dimensional Digital Models. *Symmetry* **2021**, *13*, 1644. https://doi.org/10.3390/sym13091644

Academic Editor: Antoine Balzeau

Received: 22 July 2021
Accepted: 4 September 2021
Published: 7 September 2021

Publisher's Note: MDPI stays neutral with regard to jurisdictional claims in published maps and institutional affiliations.

Abstract: The observation and the quantification of asymmetry in biological structures are deeply investigated in geometric morphometrics. Patterns of asymmetry were explored in both living and fossil species. In living organisms, levels of directional and fluctuating asymmetry are informative about developmental processes and health status of the individuals. Paleontologists are primarily interested in asymmetric features introduced by the taphonomic process, as they may significantly alter the original shape of the biological remains, hampering the interpretation of morphological features which may have profound evolutionary significance. Here, we provide a new R tool that produces the numerical quantification of fluctuating and directional asymmetry and charts asymmetry directly on the specimens under study, allowing the visual inspection of the asymmetry pattern. We tested this *show.asymmetry* algorithm, written in the R language, on fossil and living cranial remains of the genus *Homo*. *show.asymmetry* proved successful in discriminating levels of asymmetry among sexes in *Homo sapiens*, to tell apart fossil from living *Homo* skulls, to map effectively taphonomic distortion directly on the fossil skulls, and to provide evidence that digital restoration obliterates natural asymmetry to unnaturally low levels.

Keywords: asymmetry; *show.asymmetry*; fossil; virtual anthropology; geometric morphometrics; Arothron

1. Introduction

Most living organisms present bilateral symmetry, meaning that the left and right sides of the body represent an almost perfect reflection of one another about the medial plane. However, perfect symmetry is virtually absent in nature, and minor, localized deviations from perfect symmetry are common. Asymmetry can thus be defined as a deviation of the shape from a perfectly mirrored image of the counter-side of a bilateral object. The observation and quantification of asymmetry patterns in biological structures are keenly studied by evolutionary and developmental biologists, anthropologists, and paleontologists. There are three different types of asymmetries in living organisms: (i) fluctuating asymmetry,

(ii) directional asymmetry, and (iii) antisymmetry. The term fluctuating asymmetry (FA) applies to small left–right differences produced by developmental noises in the form of environmental and/or genetic stress [1]. Several studies identified FA as a good proxy for developmental instability. However, this assertion is still questioned, especially when it comes to the effect of habitat fragmentation, urbanization, and pollution on FA [2–6]. In humans, FA is usually linked to childhood diseases and poor genetic quality [7,8]. Different studies report a possible relationship between a mate's facial attractiveness and symmetry and usually support the notion that FA is higher in males than in females ([9–12], but see [13]). However, FA linkage to developmental disorders in our species remains contentious [14]. As an example, in a study carried out in the early medieval society from the Mikulčice settlement (Czech Republic), the higher degree of FA in females is deemed to be linked to the large variety of the female population due to patrilocality, although environmental effects cannot be ruled out [15].

Directional asymmetry (DA) refers to a skewed distribution of asymmetry when comparing the left to the right side of the body. DA has been largely observed in both vertebrates and invertebrates (i.e., the direction of coiling in gastropod shells, the presence of grossly unequal claws in male fiddler crabs [16]). Major examples of DA in humans pertain to handiness and brain lateralization, which in turn relates to the functioning of Broca's area for speech production [17]. Several investigations of DA in humans focused on differences occurring between males and females and usually support the notion that DA is higher in males [18,19]. DA was also used as an indicator of biomechanical loading in humans [15].

Antisymmetry (AS) is commonly defined as the inversion of the regular pattern of asymmetry, and it is widespread in both animals and plants [20]. The analysis of traits with antisymmetry may present a bimodal distribution in the most extreme manifestation, as in the case of left and right claw size in fiddler crabs. An extreme example of antisymmetry in humans is the condition known as *situs inversus*, which refers to the congenital mirrored position of most of the internal organs [21].

Studying and understanding asymmetry patterns also hold a prominent role in paleontology. Taphonomic and diagenetic processes (i.e., the postburial deformation of the organic material) can heavily affect the physical preservation of biological remains and obliterate their natural symmetry. The majority of fossils thence present damages and missing parts, as well as severe, plastic deformations due to compressive and shear forces. Incorrect identification of the nature of taphonomic distortions may misguide the recognition of diagnostic features, producing taxonomic and evolutionary misinterpretations [22,23]. More than DA and FA, which are virtually impossible to determine in the vast majority of fossil species, paleontologists are interested in quantifying the loss of biological symmetry and in identifying patterns of compression and distortion on the remains to guide the restoration of their original shape and the correct interpretation of diagnostic features. In the last few decades, with the rise of virtual paleontology, several methods of digital restoration were developed. Mirroring procedures [24–27], retrodeformation (i.e., the restoration of specimen's symmetry [28,29]), and target deformation [30] are all examples of digital manipulation procedures aiming to produce the genuine shape the remains had before taphonomy impinged on them. Assessing the reliability of these techniques is therefore crucial for paleontologists and anthropologists interested in virtual restoration.

A number of methodological strategies have been proposed to compute and discriminate between FA, DA, and AS by using geometric morphometrics data [19,31–34]. However, these strategies are generally limited in terms of visual outputs, mostly offering a 2D visualization, and/or require multiple steps to prepare the data before the asymmetry analyses can be conducted. The low visual rendering makes these approaches suboptimal in terms of interpreting the topology and regional variation in the intensity of the patterns of asymmetry and is of little help when the goal is to produce a sensible virtual restoration of the features paleontologists are most interested in.

Herein, we present a new function written in R language, named *show.asymmetry*, that allows users to visualize and measure the left–right differences of bilateral biological objects, while mapping the extent of asymmetry on the object surface and calculating levels of FA and DA where appropriate. To test *show.asymmetry*, we applied the tool to (i) visualize and assess levels of asymmetry in male and female *Homo sapiens* skulls from contemporary populations, (ii) identify patterns of asymmetry in human fossil specimens and compare them to modern humans, and (iii) test the effect of retrodeformation techniques in restoring the original biological symmetry.

2. Materials and Methods

2.1. show.asymmetry

The *show.asymmetry* algorithm is a landmark-based procedure embedded in the Arothron R package [35]. The function works with multiple landmark sets. As the first step, *show.asymmetry* splits each configuration in a left (L) and a right (R) half, following the specified indices for bilateral pairs of landmarks. The two halves are superimposed to each other via generalized Procrustes analysis (GPA) to exclude the non-shape-related differences and compute the rotation matrix to mirror, scale, and align the left side onto the right side or vice versa. By setting the argument *scale.sides*, the user may decide to apply the scaling process of the two halves during the Procrustes superimposition (see Table 1 for the detailed explanation of all arguments). As the default, scaling is not performed. The amount of shape difference that is not removed through the GPA process between the two halves is a measure of the shape differences between both sides. Asymmetry is computed as the square root of the sum of the squared distances between each landmark pair (L and R) as follows:

$$\text{asymmetry} = \sqrt{\sum_{i=1}^{n} (L_i - R_i)^2}$$

where L and R are the superimposed left and right landmark configurations and n is the number of landmarks per side. If the samples differ in terms of dimension (i.e., they belong to different species or genera, or they greatly vary in size), it may be useful to standardize the amount of asymmetry to unit size to compare them directly. Thus, in *show.asymmetry*, the total amount of asymmetry is divided by the maximum interlandmark distance of the sample configuration. This correction is triggered by the function's argument *scale.size*. In the case of specimens under analysis falling under discrete groups, *show.asymmetry* automatically retrieves the mean shapes for the groups indicated by the user.

The asymmetry pattern is automatically visualized on one half of the object surface by using *meshDist* function in 'Morpho' R package, [32], with the asymmetry values used as distance vector. The function also displays the two superimposed surfaces and, eventually, the local area differences between the two halves by using the algorithm embedded in the *localmeshdiff* function ('Arothron' R package, [35]). The area difference range for all the given specimens is rescaled into the 0–1 range to make them comparable. If no reference surface is provided, *show.asymmetry* uses the function *vcgBallPivoting* in 'Morpho' to reconstruct both L and R halves for visualization.

show.asymmetry further gives the possibility to perform a principal component analysis (PCA) on a new set of landmarks obtained by subtracting the mean from the left and mirrored right (or the other way around) side. The output from the PCA is used to decompose the total variance in two components describing the percentage of variation attributed respectively to DA (mean difference between sides) and FA (average differences around mean of asymmetry) (for details, see [31,36]).

The function retrieves the asymmetry vectors, the local area differences vector, the surfaces with levels of asymmetry mapped on a color scale, the PCA results, the asymmetric component of shape variance, and the percentage of DA and FA (see Table 2 for a detailed explanation).

Table 1. Explanation of the arguments of *show.asymmetry*.

Argument Name	Explanation
set	A single matrix k × m or a k × m × n array, where k is the number of points, m is the number of dimensions, and n is the sample size.
x	character: the species/specimens to be analyzed; names specified in the x argument must be included and coincide with the dimnames of the array.
pairs	A two-column data frame containing the indices (row numbers) of the bilateral landmarks.
scale.size	logical: if TRUE, the asymmetry will be corrected with the maximum interlandmark distance.
uniform.range	logical: if TRUE, the color range for the asymmetry visualization will be uniform among all specimens analyzed.
scale.sides	logical: if TRUE, the left and the right side will be scaled during the Procrustes superimposition process.
scale.ranges	logical: if TRUE, the vector of asymmetry values will be scaled from 0 to 1.
PCA	logical: if TRUE, a Principal Component Analysis is performed.
pcx	numeric: first PC axis to be visualized.
pcy	numeric: second PC axis to be visualized.
ref.sur	Reference surfaces to be used for the visualization; if ref.sur is NULL, the surfaces will be automatically reconstructed starting from the landmarks by using the *vcgBallPivoting* algorithm from Morpho R package (Schlager, 2017).
.from	numeric: minimum distance for the asymmetry to be colorized.
.to	numeric: maximum distance for the asymmetry to be colorized.
plot	logical: if TRUE, visualize result for asymmetry as 3D plot.
pal.dist	logical: if TRUE, the mesh area differences are displayed in a second 3D plot.
pal.areas	logical: if TRUE, the names of the species and/or the number of the node are displayed in the 3D plot.

Table 2. *show.asymmetry* value illustration.

Value	Explanation
asym	Vector of asymmetry values.
area.differences	Vector of area differences values.
asymmetry.surfaces	List of objects of mesh3d colorized according to the asymmetry values.
area.diff.surfaces	List of objects of mesh3d colorized according to the area differences values.
PCA	List object containing the mean shape, PC scores, PCs, and the variance table according to the output from Morpho::procSym.
asymmetric.component	The percentage of shape variance explained by asymmetry.
DA	The percentage of directional asymmetry.
FA	The percentage of the fluctuating asymmetry.

2.2. Case Studies

We applied and tested *show.asymmetry* on different case studies. First, we applied the function to a collection of sexed modern human crania. We compared levels of asymmetry in two different groups, male (N = 10) and female (N = 10), to observe if asymmetry patterns differ among sexes. We also computed a PCA on the asymmetric component as described in Section 2.1.

The second case study pertains to two fossil *Homo* skulls we studied to look at the patterns of taphonomic distortion. The two specimens refer to *Homo heidelbergensis* from Petralona and the *Homo neanderthalensis* from Saccopastore 1. The Neanderthal specimen shows extensive asymmetry in the parietotemporal region next to the slight clockwise rotation of the facial complex with respect to the neurocranium [29] (Figure 1).

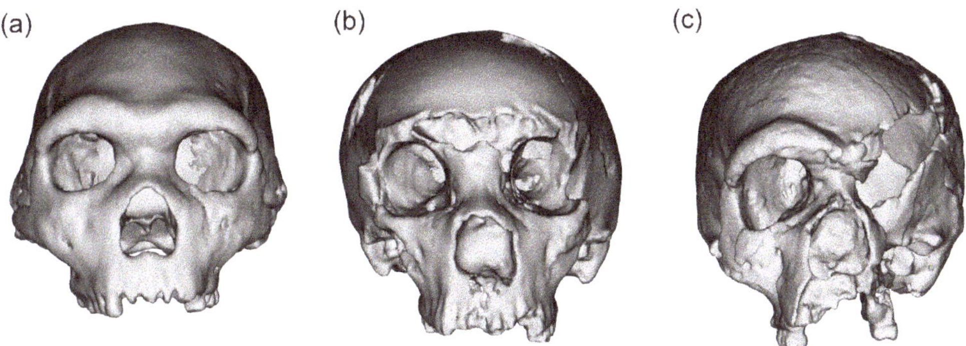

Figure 1. Fossil specimens for *show.asymmetry* case studies. (**a**) Petralona, (**b**) Saccopastore 1, and (**c**) Steinheim skulls.

Petralona represents a well-preserved *Homo heidelbergensis* skull discovered by Malkotsis and colleagues in 1959 in a cave site near the Petralona village (Thessaloniki, Greece). The cranium lacks the right zygomatic arch. The mastoid processes are broken. The upper portion of the sphenoid bone is missing. A wide opening intervenes in between the cranial and nasal cavities and the maxillary sinuses [37]. As highlighted by Rightmire [38], there are slight deformations of the vault. Although the frontal bone is undeformed, the right parietal bulges more than its left counterpart, and the right temporal squama is displaced laterally. The palate is rotated about the sagittal plane of the braincase, indicating some twisting of the facial skeleton towards its right side.

Saccopastore 1 was discovered in 1929 in the aggradational succession of the Aniene River Valley (Rome, Italy) The cranium is almost complete, although it lacks both zygomatic arches and the left orbital region is damaged. Some additional and severe damages were due to its accidental discovery in a gravel pit during construction works. The most extensive damage occurred to the browridge region, which is missing. The neurocranium also presents two pick stroke marks. For the application of *show.asymmetry*, the two holes were closed digitally, while the browridge could not be restored.

The last case study regards the application of *show.asymmetry* to evaluate the effect of retrodeformation on asymmetry. To avoid biases due to the inclusion of deformed specimens in morphometric analysis, or misinterpretation of morphological traits, paleontologists have applied the so-called retodeformation protocol to artificially restore symmetry in digital models [23,26,29,39]. This symmetrization procedure is powerful and effective, yet it cannot discriminate between taphonomic distortion and natural asymmetry. We decided to test *show.asymmetry* on a highly deformed specimen before and after the retrodeformation procedure to see if the retrodeformed specimen shows a lower than expected level of asymmetry (as judged by comparison to living *Homo sapiens* specimens) and whether *show.asymmetry* captures this essential feature of the retrodeformation process. We used the Steinheim skull case-study presented in [40] (this volume). Steinheim cranium was found in 1933 in a gravel pit 70 km north of Steinheim an der Murr (Baden-Württemberg, Germany) and, despite a longstanding debate, is commonly attributed to *Homo heidelbergensis*, or otherwise linked to the Neanderthal lineage [41,42].

2.3. Data Preparation

We acquired 50 anatomical landmarks on each modern human skull specimen on the entire cranial surface (e.g., facial complex, neurocranium, and cranial base). We placed 500 equidistant surface semilandmarks on the left side only of a reference sample. We slid the semilandmark on the entire sample of 20 specimens following the protocol included in 'Morpho' [32]. Then, we mirrored the slid configurations to the other side and projected them on the surfaces after a GPA step rotating semilandmark configuration accordingly to the set of bilateral landmarks.

Concerning fossil specimens, we manually sampled bilateral landmarks on Petralona and Saccopastore by using Amira software (version 5.4.5 [43]) (see Supplementary Materials for the full detailed description of landmarks). We created decimated patches by cutting half of the skull, and then we removed the damaged parts on each surface (i.e., the browridge region from Saccopastore 1). For each patch, we retrieved the coordinates of the vertices and used them as semilandmarks. As we needed bilateral points, we symmetrized the semilandmarks on the opposite side and slid them along the surface by using manually placed landmarks as a reference. The manipulation of landmarks and semilandmarks was performed by using the R Cran software (version 4.0.5).

For the Steinheim case study, we used both the original and the retrodeformed patches from the study presented in [40] (this volume).

3. Results

The comparison between the two modern samples highlighted that male individuals show 40% more asymmetry on average. Such enhanced asymmetry is especially evident in the temporoparietal area, the occipital region, and the maxillary bone (Figure 2a). Student's t-test on asymmetry vectors for male and female mean shapes indicates these differences are significant (t = −9.7703, p values < 0.001). The asymmetric component is 2.7% of total shape variance. This component is primarily made up of FA, which accounts for 93.6% of it, meaning FA represents 2.53% (93.6 times 2.7) of the total shape variance.

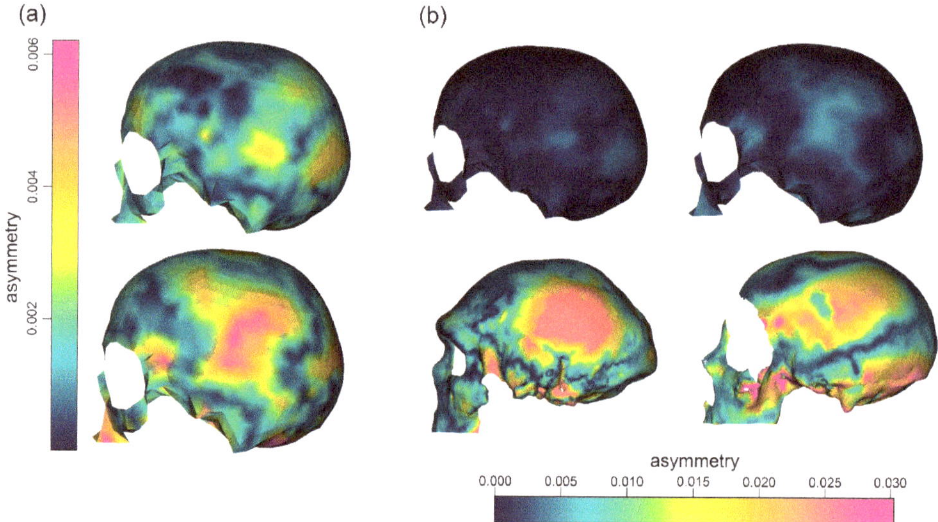

Figure 2. Visualization of the degree of asymmetry obtained with *show.asymmetry*. (**a**) Comparison between the mean shape of female and male modern humans (upper row and bottom row, respectively). The range of asymmetry goes from the minimum to the maximum value of asymmetry between the two samples. (**b**) Comparison between the mean shape of female and male modern humans (upper row left and right, respectively) and Petralona and Saccopastore 1 (lower row, left and right respectively). The range of asymmetry is scaled between samples.

When modern humans are compared to Petralona and Saccopastore 1, their degree of asymmetry appears diminutive (Figure 2b). Petralona shows a marked pattern of asymmetry in the temporoparietal area, while the facial complex appears to be more symmetric than the cranial vault. Saccopastore 1 presents a directional pattern of asymmetry with a peak corresponding to the zygomatic and lateral maxillary areas, due to the bad status of preservation of the left side of the splanchnocranium. Both patterns agree well with what has been reported in the literature regarding these specimens [29,37].

Overall, Steinheim is the most asymmetric specimen (Figures 3 and 4). As expected by the descriptions provided in [44], the skull shows extensive deformations on the splanchnocranium, whereas shape was less affected in its rearmost part. However, in keeping with our hypotheses, after the retrodeformation process, the level of asymmetry is close to zero. Lastly, the modern human crania show a minor level of asymmetry when contrasted with the asymmetry level measured in fossil specimens. (Figures 3 and 4).

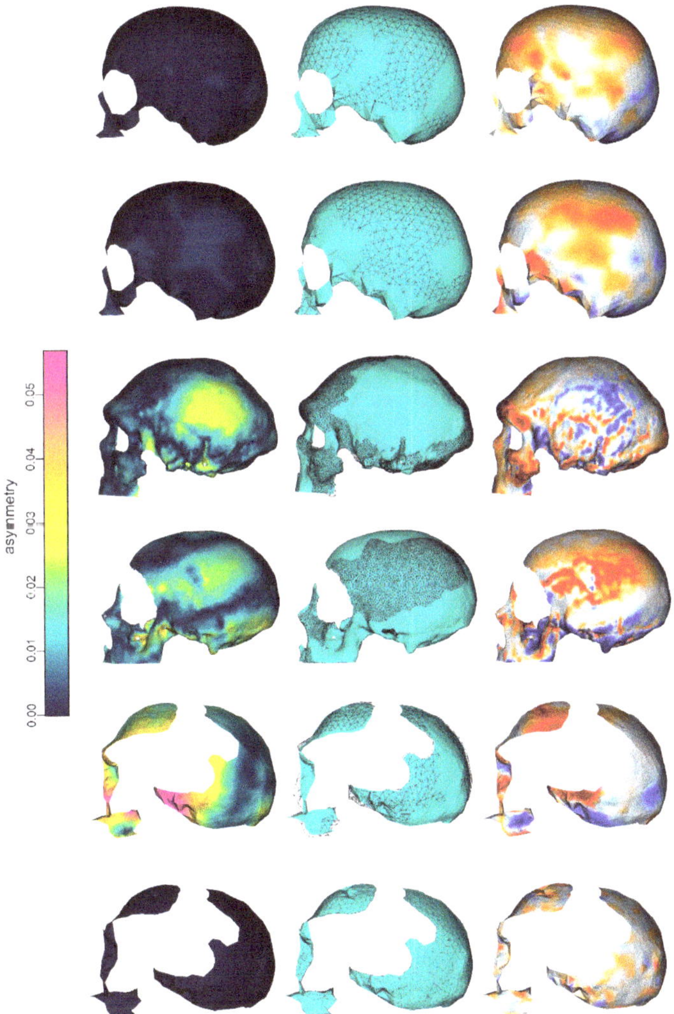

Figure 3. Graphical results produced by *show.asymmetry*. The first column shows the asymmetry pattern in terms of vertex distances between the two superimposed halves (the shared scale is shown on the left). The second column shows the superimposition of the two halves (in this case, the left side is the cyan surfaces, while black wireframe corresponds to the superimposed right side). The third column represents the local area differences between the left and the right side; the scale goes from blue to red (meaning expansion and contraction, respectively), and it is not shared. From top to bottom: mean shape of female individuals, mean shape of male individuals, Petralona, Saccopastore 1, Steinheim before retrodeformation, Steinheim after retrodeformation.

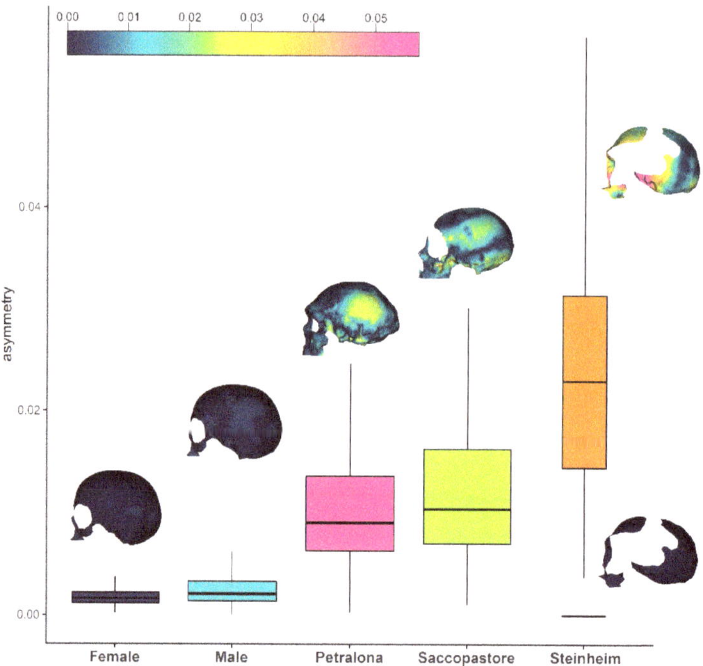

Figure 4. Boxplot of the comparison of asymmetry values for the analyzed samples. From left to right: *Homo sapiens* female mean shape, *Homo sapiens* male mean shape, Petralona, Saccopastore 1, and Steinheim (before and after the retrodeformation procedure). The color gradient for the asymmetry displayed in the crania is shown.

The time requested to run all the four case studies was 3.95 s and in particular: Petralona 0.93 s, Saccopastore 0.81 s, Homo sapiens 1.40 s, Steinheim 0.81 s. Speed tests were run with a laptop Intel Core-i7 10875H (2.30 GHz and 32 GB RAM).

4. Discussion

There are several strategies available to evaluate asymmetry from landmark-based datasets [31,34]. Despite the presence of different methods, none of these offer a fully integrated tool to calculate and especially to map asymmetry from and to mesh-based models. Furthermore, the new function *show.asymmetry* is able to evaluate asymmetry in both multiple datasets (array) and single specimens (matrix), returning colored meshes showing the pattern of asymmetry in two different ways: the 3D map of Euclidean distances and the 3D map of local variations of area.

For example, in the R package 'geomorph' [45], the function *bilat.asymmetry* provides a 3D scatterplot of the distortion of landmarks or the 3D colorless meshes warped according to the detected pattern of asymmetry. Furthermore, landmark clouds can be useful or easy to read when dealing with a small number of points or with relatively simple 3D structures. However, when using complex 3D geometries (such as crania) and/or large numbers of landmarks, other graphical outputs, such as heatmaps, are a more welcome option [34]. Nonetheless, even built-in functions such as *bilat.symmetry* require multiple steps to be performed by the users in order to be implemented. Specifically, the two sets of coordinates defining the two sides are mirror images, and hence they must be reflected for landmark alignment (multiplication of the raw data matrix by -1 is required). On the same page, the approach described by Neubauer et al. [34] requires performing the singular value decomposition (SVD) of the raw data matrix of asymmetric vectors rather than performing a standard PCA of the mean-centered data. These steps must be performed before testing for the presence of any asymmetry pattern, increasing the chances of misuse and lengthening the time to perform the entire set of analyses.

When dealing with fossil specimens, it must be considered that taphonomic processes may sensibly alter the original shape of fossil remains. Whereas cracks and missing parts are undisputable accidents of the preservation process, the compressive and shear stresses acting upon the remains over prolonged periods of time may bring about plastic deformations that could be misinterpreted as 'natural'. This, in turn, may have important consequences on the correct recognition of the phylogenetic position and taxonomy of the remains [22,23]. For instance, the 'roofed' appearance of the neurocranium in the Steinheim skull was interpreted as evidence of its plesiomorphic condition [46] but may be better indicative of taphonomic alteration [40] (Buzi et al., this volume). A similar misinterpretation might have complicated the interpretation of Ceprano *Homo heidelbergensis* calvarium [47]. Digital restorations help in driving the restoration of the original shapes yet obliterate true object symmetry (sensu [31]) and are uninformative as to where and to what extent asymmetry applies in the first place. The algorithm of *show.symmetry* provides exactly this piece of information and therefore helps to understand the processes behind the taphonomic distortions and their total amount. As demonstrated in the first case study, male *Homo sapiens* skulls are on average more asymmetric than female skulls. Whereas this result does not generalize and was not thought to provide an answer to a complicated question as to whether females, as compared to males, really tend to have a lower level of cranial asymmetry [15], it stills shows that *show.asymmety* retrieves even small differences between closely knit individuals. Similarly, *show.asymmetry* confirms that retrodeformation procedures actually reduce cranial symmetry below the natural level, even applied to a highly deformed skull such as Steinheim. Importantly, the tool successfully estimates and maps levels and direction of asymmetric deformations directly on the fossil remains, which may provide critical information when the recognition of the processes behind the deformation and the proper fossil restoration are at the stake.

5. Conclusions

show.asymmetry estimates and charts patterns of asymmetry on three-dimensional digital models. It straightaway performs a correction for size variation within the sample, decomposes asymmetry into its directional and fluctuating components, and maps asymmetry on the three-dimensional surface, allowing users to grasp immediate visual information on the intensity, topology, and direction of departures from perfect symmetry. Through the three different case studies presented here, we showed the functioning of *show.asymmetry* algorithm with different datasets, from recent human individuals to human fossil specimens.

Supplementary Materials: The *show.asymmetry* function, code, and raw data to reproduce the case studies presented here are available at https://www.mdpi.com/2073-8994/13/9/1644/s1.

Author Contributions: Conceptualization, M.M., A.P., S.C., P.R.; methodology and software, M.M., A.P., S.C., P.R., P.P.; formal analysis, M.M., A.P., A.M.; resources, A.P., C.B.; writing, M.M., A.P., P.R., C.B., A.D.B., G.S., A.M., S.C.; funding, A.P. All authors have read and agreed to the published version of the manuscript.

Funding: This project has received funding from the European Union's Horizon 2020 Research and Innovation Programme (H2020 Marie Skłodowska-Curie Actions) under grant agreement H2020-MSCA-IF-2018 No. 835571 to Antonio Profico and Paul O'Higgins.

Institutional Review Board Statement: Not applicable.

Informed Consent Statement: Not applicable.

Data Availability Statement: The data and code and the function *show.asymmetry* are available as Supplementary Materials. The original surface meshes in 3D are available from the senior author upon request.

Acknowledgments: The authors are grateful to Giorgio Manzi for sharing experience and ideas with us.

Conflicts of Interest: The authors declare no conflict of interest.

References

1. Klingenberg, C.P. A developmental perspective on developmental instability: Theory, models and mechanisms. In *Developmental Instability: Causes and Consequences*, 1st ed.; Polak, M., Ed.; Oxford University Press: Oxford, UK, 2003; Volume 1, pp. 14–34.
2. Marchand, H.; Paillat, G.; Montuire, S.; Butet, A. Fluctuating asymmetry in bank vole populations (Rodentia, Arvicolinae) reflects stress caused by landscape fragmentation in the Mont-Saint-Michel Bay. *Biol. J. Linn. Soc.* **2003**, *80*, 37–44. [CrossRef]
3. Oleksyk, T.K.; Novak, J.M.; Purdue, J.R.; Gashchak, S.P.; Smith, M.H. High levels of fluctuating asymmetry in populations of Apodemus flavicollis from the most contaminated areas in Chornobyl. *J. Environ. Radioact.* **2004**, *73*, 1–20. [CrossRef]
4. Lazić, M.M.; Carretero, M.A.; Crnobrnja-Isailović, J.; Kaliontzopoulou, A. Effects of environmental disturbance on phenotypic variation: An integrated assessment of canalization, developmental stability, modularity, and allometry in lizard head shape. *Am. Nat.* **2015**, *185*, 44–58. [CrossRef]
5. Lezcano, A.H.; Rojas Quiroga, M.L.; Liberoff, A.L.; Van der Molen, S. Marine pollution effects on the southern surf crab Ovalipes trimaculatus (Crustacea: Brachyura: Polybiidae) in Patagonia Argentina. *Mar. Pollut. Bull.* **2015**, *91*, 524–529. [CrossRef]
6. Chirichella, R.; Rocca, M.; Brugnoli, A.; Mustoni, A.; Apollonio, M. Fluctuating asymmetry in Alpine chamois horns: An indicator of environmental stress. *Evol. Ecol.* **2020**, *34*, 573–587. [CrossRef]
7. Van Dongen, S.; Gangestad, S.W. Human fluctuating asymmetry in relation to health and quality: A meta-analysis. *Evol. Hum. Behav.* **2011**, *32*, 380–398. [CrossRef]
8. Weisensee, K.E. Assessing the relationship between fluctuating asymmetry and cause of death in skeletal remains: A test of the developmental origins of health and disease hypothesis. *Am. J. Hum. Biol.* **2013**, *25*, 411–417. [CrossRef] [PubMed]
9. Schaefer, K.; Fink, B.; Grammer, K.; Mitteroecker, P.; Gunz, P.; Bookstein, F.L. Female appearance: Facial and bodily attractiveness as shape. *Psychol. Sci.* **2006**, *48*, 187–204.
10. Pflüger, L.S.; Oberzaucher, E.; Katina, S.; Holzleitner, I.J.; Grammer, K. Cues to fertility: Perceived attractiveness and facial shape predict reproductive success. *Evol. Hum. Behav.* **2012**, *33*, 708–714. [CrossRef]
11. Muñoz-Reyes, J.A.; Pita, M.; Arjona, M.; Sanchez-Pages, S.; Turiegano, E. Who is the fairest of them all? The independent effect of attractive features and self-perceived attractiveness on cooperation among women. *Evol. Hum. Behav.* **2014**, *35*, 118–125. [CrossRef]
12. Foo, Y.Z.; Simmons, L.W.; Rhodes, G. Predictors of facial attractiveness and health in humans. *Sci. Rep.* **2017**, *7*, 39731. [CrossRef] [PubMed]
13. Komori, M.; Kawamura, S.; Ishihara, S. Averageness or symmetry: Which is more important for facial attractiveness? *Acta Psychol.* **2009**, *131*, 136–142. [CrossRef]
14. Graham, J.H.; Özener, B. Fluctuating asymmetry of human populations: A review. *Symmetry* **2016**, *8*, 154. [CrossRef]
15. Bigoni, L.; Krajíček, V.; Sládek, V.; Velemínský, P.; Velemínská, J. Skull shape asymmetry and the socioeconomic structure of an early medieval central european society. *Am. J. Phys. Anthropol.* **2013**, *150*, 349–364. [CrossRef]
16. Palmer, A.R. Symmetry breaking and the evolution of development. *Science* **2004**, *306*, 828–833. [CrossRef] [PubMed]
17. Cantalupo, C.; Hopkins, W.D. Asymmetric Broca's area in great apes. *Nature* **2001**, *414*, 505. [CrossRef]
18. Claes, P.; Walters, M.; Shriver, M.D.; Puts, D.; Gibson, G.; Clement, J.; Baynam, G.; Verbeke, G.; Vandermeulen, D.; Suetens, P. Sexual dimorphism in multiple aspects of 3D facial symmetry and asymmetry defined by spatially dense geometric morphometrics. *J. Anat.* **2012**, *221*, 97–114. [CrossRef] [PubMed]
19. Ekrami, O.; Claes, P.; White, J.D.; Weinberg, S.M.; Marazita, M.L.; Walsh, S.; Shriver, M.D.; Van Dongen, S. A multivariate approach to determine the dimensionality of human facial asymmetry. *Symmetry* **2020**, *12*, 348. [CrossRef]
20. Palmer, A.R. Antisymmetry. In *Variation*; Elsevier: Amsterdam, The Netherlands, 2005; p. 359-XXIV.
21. Vingerhoets, G.; Gerrits, R.; Verhelst, H. Atypical brain asymmetry in human situs inversus: Gut feeling or real evidence? *Symmetry* **2021**, *13*, 695. [CrossRef]
22. Sansom, R.S.; Wills, M.A. Fossilization causes organisms to appear erroneously primitive by distorting evolutionary trees. *Sci. Rep.* **2013**, *3*, 2545. [CrossRef]
23. Lautenschlager, S. Reconstructing the past: Methods and techniques for the digital restoration of fossils. *R. Soc. Open Sci.* **2016**, *3*, 160342. [CrossRef]
24. Gunz, P.; Mitteroecker, P.; Neubauer, S.; Weber, G.W.; Bookstein, F.L. Principles for the virtual reconstruction of hominin crania. *J. Hum. Evol.* **2009**, *57*, 48–62. [CrossRef]
25. Benazzi, S.; Bookstein, F.L.; Strait, D.S.; Weber, G.W. A new OH5 reconstruction with an assessment of its uncertainty. *J. Hum. Evol.* **2011**, *61*, 75–88. [CrossRef]
26. Benazzi, S.; Gruppioni, G.; Strait, D.S.; Hublin, J.J. Technical Note: Virtual reconstruction of KNM-ER 1813 *Homo habilis* cranium. *Am. J. Phys. Anthropol.* **2014**, *153*, 154–160. [CrossRef]
27. Zollikofer, C.P.E.; Ponce de León, M.S.; Lieberman, D.E.; Guy, F.; Pilbeam, D.; Likius, A.; Mackaye, H.T.; Vignaud, P.; Brunet, M. Virtual cranial reconstruction of Sahelanthropus tchadensis. *Nature* **2005**, *434*, 755–759. [CrossRef] [PubMed]

28. Tallman, M.; Amenta, N.; Delson, E.; Frost, S.R.; Ghosh, D.; Klukkert, Z.S.; Morrow, A.; Sawyer, G.J. Evaluation of a new method of fossil retrodeformation by algorithmic symmetrization: Crania of papionins (primates, cercopithecidae) as a test case. *PLoS ONE* **2014**, *9*, e100833. [CrossRef] [PubMed]
29. Schlager, S.; Profico, A.; Di Vincenzo, F.; Manzi, G. Retrodeformation of fossil specimens based on 3D bilateral semi-landmarks: Implementation in the R package "Morpho". *PLoS ONE* **2018**, *13*, e0194073. [CrossRef]
30. Cirilli, O.; Melchionna, M.; Serio, C.; Bernor, R.L.; Bukhsianidze, M.; Lordkipanidze, D.; Rook, L.; Profico, A.; Raia, P. Target Deformation of the Equus stenonis Holotype Skull: A Virtual Reconstruction. *Front. Earth Sci.* **2020**, *8*, 247. [CrossRef]
31. Mardia, K.V.; Bookstein, F.L.; Moreton, I.J. Statistical assessment of bilateral symmetry of shapes. *Biometrika* **2000**, *87*, 285–300. [CrossRef]
32. Schlager, S. Morpho and Rvcg—Shape Analysis in R: R-Packages for Geometric Morphometrics, Shape Analysis and Surface Manipulations. In *Statistical Shape and Deformation Analysis: Methods, Implementation and Applications*; Zheng, G., Li, S., Székely, G., Eds.; Academic Press: Cambridge, MA, USA, 2017; pp. 217–256, ISBN 9780128104941.
33. Klingenberg, C.P. Analyzing fluctuating asymmetry with geometric morphometrics: Concepts, methods, and applications. *Symmetry* **2015**, *7*, 843–934. [CrossRef]
34. Neubauer, S.; Gunz, P.; Scott, N.A.; Hublin, J.J.; Mitteroecker, P. Evolution of brain lateralization: A shared hominid pattern of endocranial asymmetry is much more variable in humans than in great apes. *Sci. Adv.* **2020**, *6*, eaax9935. [CrossRef]
35. Profico, A.; Buzi, C.; Castiglione, S.; Melchionna, M.; Piras, P.; Veneziano, A.; Raia, P. Arothron: An R package for geometric morphometric methods and virtual anthropology applications. *Am. J. Phys. Anthropol.* **2021**, *176*, 144–151. [CrossRef]
36. Profico, A.; Zeppilli, C.; Micarelli, I.; Mondanaro, A.; Raia, P.; Marchi, D.; Manzi, G.; O'Higgins, P. Morphometric maps of bilateral asymmetry in the human humerus: An implementation in the R package morphomap. *Symmetry*. under review.
37. Seidler, H.; Falk, D.; Stringer, C.; Wilfing, H.; Müller, G.B.; Zur Nedden, D.; Weber, G.W.; Reicheis, W.; Arsuaga, J.L. A comparative study of stereolithographically modelled skulls of Petralona and Broken Hill: Implications for future studies of middle Pleistocene hominid evolution. *J. Hum. Evol.* **1997**, *33*, 691–703. [CrossRef] [PubMed]
38. Rightmire, G.P. Middle Pleistocene *Homo* Crania from Broken Hill and Petralona: Morphology, metric comparisons, and evolutionary relationships. In *Human Paleontology and Prehistory*; Springer: Berlin/Heidelberg, Germany, 2017; pp. 145–159.
39. Cuff, A.R.; Rayfield, E.J. Retrodeformation andmuscular reconstruction of ornithomimosaurian dinosaur crania. *PeerJ* **2015**, *3*, e1093. [CrossRef]
40. Buzi, C.; Profico, A.; Di Vincenzo, F.; Harvati, K.; Melchionna, M.; Raia, P.; Manzi, G. Retrodeformation of the Steinheim cranium: Insights into the evolution of Neanderthals. *Symmetry* **2021**, *31*, 1611. [CrossRef]
41. Hublin, J.J. The origin of Neandertals. *Proc. Natl. Acad. Sci. USA* **2009**, *106*, 16022–16027. [CrossRef] [PubMed]
42. Mounier, A.; Caparros, M. The phylogenetic status of *Homo heidelbergensis*—A cladistic study of Middle Pleistocene hominins. *Bull. Mem. Soc. Anthropol. Paris* **2015**, *27*, 110–134. [CrossRef]
43. Stalling, D.; Westerhoff, M.; Hege, H.C. Amira: A highly interactive system for visual data analysis. *Vis. Handb.* **2005**, *1*, 749–767. [CrossRef]
44. Prossinger, H.; Seidler, H.; Wicke, L.; Weaver, D.; Recheis, W.; Stringer, C.; Müller, G.B. Electronic removal of encrustations inside the steinheim cranium reveals paranasal sinus features and deformations, and provides a revised endocranial volume estimate. *Anat. Rec. Part B New Anat.* **2003**, *273*, 132–142. [CrossRef] [PubMed]
45. Adams, D.; Collyer, M.; Kaliontzopoulou, A.; Baken, E. Geomorph: Software for Geometric Morphometric Analyses. R Package Version 4.0. Available online: https://cran.r-project.org/package=geomorph (accessed on 21 July 2021).
46. Schwartz, J.H.; Tattersall, I. *The Human Fossil Record*; John Wiley & Sons: New York, NY, USA, 2005; Volume 2, ISBN 0471319287.
47. Bruner, E.; Manzi, G. CT-based description and phyletic evaluation of the archaic human calvarium from Ceprano, Italy. *Anat. Rec. Part A Discov. Mol. Cell. Evol. Biol.* **2005**, *285*, 643–657. [CrossRef]

symmetry

Article

Morphometric Maps of Bilateral Asymmetry in the Human Humerus: An Implementation in the R Package Morphomap

Antonio Profico [1,*,†], Carlotta Zeppilli [2,*,†], Ileana Micarelli [2], Alessandro Mondanaro [3], Pasquale Raia [4], Damiano Marchi [5,6], Giorgio Manzi [2] and Paul O'Higgins [1,7]

[1] PalaeoHub, Department of Archaeology, University of York, York YO10 5DD, UK; Paul.Ohiggins@hyms.ac.uk
[2] Department of Environmental Biology, Sapienza Università di Roma, 00185 Rome, Italy; ileana.micarelli@uniroma1.it (I.M.); giorgio.manzi@uniroma1.it (G.M.)
[3] Dipartimento di Scienze della Terra, Università degli Studi di Firenze, 50121 Florence, Italy; alessandro.mondanaro@unifi.it
[4] Dipartimento di Scienze della Terra, dell'Ambiente e delle Risorse, Università di Napoli Federico II, 80138 Naples, Italy; pasquale.raia@unina.it
[5] Department of Biology, University of Pisa, via Luca Ghini, 56126 Pisa, Italy; damiano.marchi@unipi.it
[6] Centre for the Exploration of the Deep Human Journey, University of the Witwatersrand, Private Bag 3, Wits, Johannesburg 2050, South Africa
[7] Hull York Medical School, University of York, York YO10 5DD, UK
*   Correspondence: antonio.profico@york.ac.uk (A.P.); carlotta.zeppilli@uniroma1.it (C.Z.)
†   Co-first authorship.

Citation: Profico, A.; Zeppilli, C.; Micarelli, I.; Mondanaro, A.; Raia, P.; Marchi, D.; Manzi, G.; O'Higgins, P. Morphometric Maps of Bilateral Asymmetry in the Human Humerus: An Implementation in the R Package Morphomap. *Symmetry* **2021**, *13*, 1711. https://doi.org/10.3390/sym13091711

Academic Editor: Antoine Balzeau

Received: 19 August 2021
Accepted: 10 September 2021
Published: 16 September 2021

Publisher's Note: MDPI stays neutral with regard to jurisdictional claims in published maps and institutional affiliations.

Abstract: In biological anthropology, parameters relating to cross-sectional geometry are calculated in paired long bones to evaluate the degree of lateralization of anatomy and, by inference, function. Here, we describe a novel approach, newly added to the morphomap R package, to assess the lateralization of the distribution of cortical bone along the entire diaphysis. The sample comprises paired long bones belonging to 51 individuals (10 females and 41 males) from The New Mexico Decedent Image Database with known biological profile, occupational and loading histories. Both males and females show a pattern of right lateralization. In addition, males are more lateralized than females, whereas there is not a significant association between lateralization with occupation and loading history. Body weight, height and long-bone length are the major factors driving the emergence of asymmetry in the humerus, while interestingly, the degree of lateralization decreases in the oldest individuals.

Keywords: biological anthropology; biomechanics; cortical thickness; lateralization; modern humans; NMDID; upper limb

1. Introduction

In bioarchaeology and anthropology, it is of interest to infer the physical activities, occupations and behaviours of past populations from skeletal material [1,2]. During life, the distribution of cortical bone is influenced by loading history [3–5], and bone remodelling seems to be significantly associated with high-frequency daily action [6]. Asymmetry of loading, as is common in many physical activities, occupations and behaviours, can be expected to lead to asymmetry of bone form. Thus, to fulfil the goal of inferring past lifestyles often requires the assessment of differences in bone shape and cortical thickness distributions among antimeres [5,7].

Different models have been proposed to explain how bone is remodelled in relation to loading [8–11], although bone adaptation and remodelling has a sizeable physiological and environmental (i.e., nutritional) component. The comparison of antimeric bones from the same individual offers the opportunity to identify asymmetry of loading history [12] while ignoring the confounding, presumed bilaterally equal effects of genetics and nutrition. Yet, even the comparison of paired bone elements is not entirely without issues, since inflammatory processes [13] may trigger osteogenesis in distant regions [14,15], and differences in patterns of asymmetry in the upper limb have been found with ageing [16] and long-term disuse [17], in addition to loading history per se. Despite these caveats, traditional methods that rely on calculations of the percentage change of cross-sectional geometric parameters (total area, cortical area, area moments of inertia) on the humerus have provided useful insights into activity patterns in modern [12] and archaeological populations [18–24], as well as in paleontological samples [25–30]. Studies of professional athletes who play unimanual or bimanual sports, such as tennis [5,31–34], throwing and swimming [34–36], provide an interesting natural experiment. Studies of their long bones allow assessment of the extent to which asymmetry of cortical thickness and whole-bone morphology exists between the dominant (e.g., the racket arm) and non-dominant arm. Younger starters show a higher index of "strength" than older, suggesting that intense activities during adolescence lead to greater subperiosteal expansion [37].

Current methods to evaluate asymmetry in long bones often involve comparison of shape and biomechanical parameters (cross sectional geometry) between antimeres based on a limited number of sections along the diaphysis [19,38,39]. More recently, Wei et al. [40] have extended such analyses to multiple closely placed sections along the entire diaphysis, calculating asymmetry of cortical thicknesses and polar moments of area (J) using the R software tool, morphomap [41].

Here, we assess how asymmetry in the distribution of cortical thickness in the human humerus is related to physical activity level, sex, body mass and weight based on data from recently deceased individuals, with known occupational, lifestyle and medical history curated in the New Mexico Decedent Image Database (NMDID) [42].

We tested the hypotheses that: (a) the degree of asymmetry does not differ among sexes or among three different occupation groups; (b) the difference in distribution of cortical bone and degree of asymmetry are not influenced by age, weight, height, humerus length and occupation. The hypotheses we tested have significant implications for the evaluation of asymmetry in archaeological populations and in extinct human species.

2. Materials and Methods

2.1. Data Preparation and Processing

From the New Mexico Decedent Image Database [42], we selected 51 individuals (41 males, 10 females) with known occupation, ranging in age at death from 21 to 54 years. We selected individuals who had been in the armed forces or worked in building/mining or at a desk job to test the methodology using groups with distinctly different occupational histories (likely high vs. low loading).

We extracted from NMDID metadata associated with occupational history for each individual. Then, we computed occupation scores relating activity to energy cost (see Appendix 4.1 from [43]) and duration in years for each occupation. Missing data for duration in years are estimated by calculating the expected working based on the formula, (age at death −18 years)/ the number of recorded occupations.

A total body CT scan is available for each individual at a slice resolution of 0.5 mm with 16 × 0.75 mm collimation, 120 KVp and 300 mAs. From these scans, we cropped the left and right humerus. In order to create 3D models defined by only bony material, the image stacks were automatically segmented using the Otsu algorithm available in the *morphomap* R package. The 102 resulting 3D models (51 left humeri and 51 right humeri) were aligned following the protocols proposed by Ruff [44].

From each 3D model, we extracted 61 cross sections from 20% to 80% of the biomechanical length along the bone shaft. At each cross section, we defined 24 paired equiangular semilandmarks on the external and internal outlines centred at the barycentre of the cross section. The production of the cross sections is automatically executed in *morphomap* by using the functions *morphomapCore* and *morphomapShape* (Figure 1).

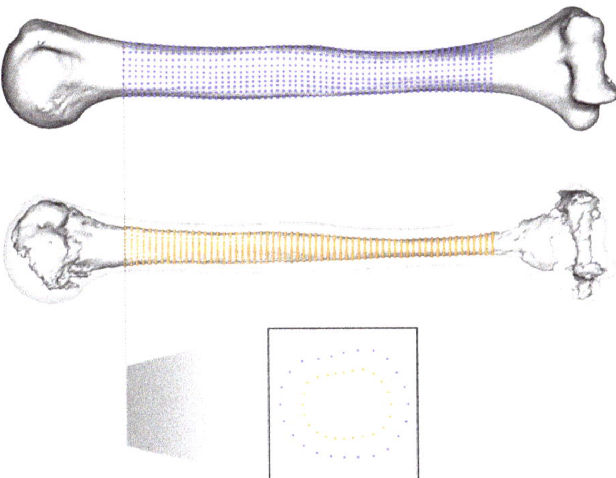

Figure 1. Top: morphomap extracts shape information as equiangular semilandmarks from the periosteal (blue) and endosteal surface (orange). **Bottom**: the cross section at 1% of the biomechanical length.

2.2. Asymmetry and Cross-Sectional Geometry

On each individual, we calculated the polar moment of inertia (J mm^4) at 40% of the biomechanical length on both sides using the function *morphomapCSG*. We avoided standardization of J (on body mass and bone length), because we analyzed the percentage of lateralization (J$_{LAT}$%) using the following equation: J$_{LAT}$% = (| (J$_R$ − J$_L$) | /J$_M$) × 100, where J$_M$ = (J$_R$ + J$_L$)/2.

2.3. Description of the Function MorphomapAsymmetry

The new function, *morphomapAsymmetry*, embedded in *morphomap* facilitates the mapping and analysis of bilateral asymmetry in long bones (Table 1). We provide three strategies to map differences in the distribution of cortical thickness between the two sides: (i) the difference between sides (*type* = "diff"); (ii) the difference from the mean (*type* = "onMean"); (iii) the relative change in thickness (*type* = "relChange") of one side with respect to the other.

Table 1. *morphomapAsymmetry*: description of the main arguments.

Argument	Definition
mshape1	First long bone processed with morphomapShape
mshape2	Second long bone processed with morphomapShape
standandize	If TRUE, the matrices of cortical thickness are standardized using the average biomechanical length between sides.
plot	If TRUE, the map of cortical thickness asymmetry is returned.
type	Defines the method to calculate the differences in cortical thickness between the two long bones: "diff" a map of arithmetic difference between reference and target is computed; "onMean" the morphometric map of asymmetry is defined by computing the differences from the mean for each long bone; "relChange" the morphometric map is computed by calculating the relative change in cortical thickness, expressed as percentage difference between reference and target long bones
reference	If set to 1, mshape1 is defined as reference; if set on 2, mshape2 is defined as reference.
rem.out	If TRUE, outliers are removed from the matrices of cortical thickness.
scale	If TRUE, the matrices of cortical thickness are scaled from 0 to 1.
gamMap	If TRUE, gam smoothing is applied.

The workflow implemented in *morphomap* is as follows:

1. Load the output of the first long bone processed with *morphomapShape*.
2. Load the output of the second long bone processed with *morphomapShape* (Figure 1).
3. Specify if one of the two input objects needs to be mirrored (Figure 2A).
4. Calculate the cortical thickness map of the entire diaphysis in both long bones (Figure 2B).
5. Standardize the cortical thicknesses by dividing the matrices of cortical thickness by the biomechanical length (optional).
6. Choose the method of visualization by setting the argument type to:

    a. *type* = "diff" Calculate the differences between the cortical thickness maps of the two long bones (Figure 2C).

    b. *type* = "onMean" Calculate the differences between the two cortical maps and their mean (Figure 2E).

    c. *type* = "relChange" Compute a cortical map as the percentage change of one side (target) with respect to the other one (reference) (Figure 2D).

7. 2D Plot the map of differences in cortical thickness between the selected specimens. The difference map is displayed after "unrolling" the long-bone shaft to produce a 2D plot, starting and ending at the anterior (A) border passing through the lateral (L), posterior (P), and medial (M) borders (Figure 2C).

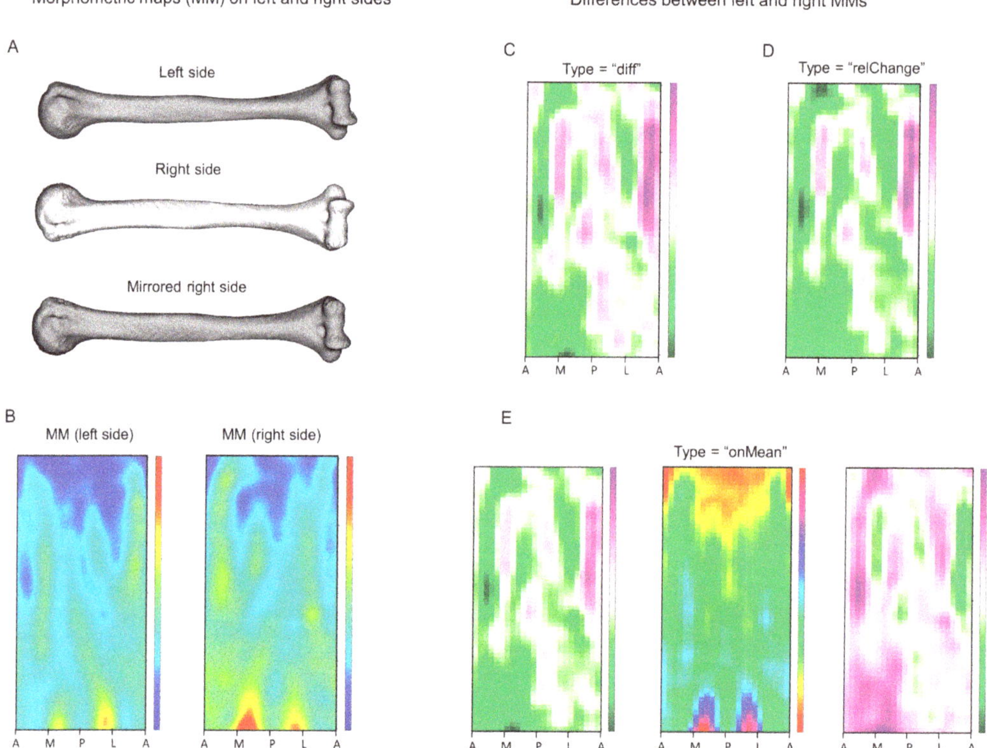

Figure 2. Workflow of the function *morphomapAsymmetry*. In (**A**), one of the two long bones is mirrored, and the two matrices of cortical thicknesses (MM) are computed (**B**). The differences between the two MMs may be computed by calculating (i) the arithmetic difference (**C**), (ii) the percentage change of the target with respect to the reference side (**D**), (iii) the difference between the two MMs and their mean (**E**).

2.4. Description of the Function MorphomapMapPCA

We processed the right and left humerus in 51 individuals selected from the NMDID using the R package *morphomap* (Profico et al. 2021). The individuals belong to three different categories for occupation: "building-mining" (called "building" from now on), "army" and "desk". We extracted 61 cross sections from the humeri and built a multivariate dataset of cortical thicknesses along the entire diaphysis on both sides.

To decompose the total variance of the sample into symmetric and asymmetric components, we performed two different Principal Component Analyses (PCA) on each dataset:

1. PCA of the mean morphometric maps calculated by averaging left and mirrored right side (symmetric component).
2. PCA of the matrices obtained by in each individual subtracting the mean matrix of cortical thicknesses from the matrices of cortical thicknesses of the left and mirrored right sides (asymmetric component).

The function *morphomapPCA* requires two inputs, the left and right sets of long-bone semilandmarks, obtained from *morphomapShape*. The user can select if the calculation of the symmetric and asymmetric component is performed on semilandmark coordinates or on the values of cortical thickness computed from these along the diaphysis.

2.5. Relation between Cortical Thickness Asymmetry Humerus and Biological Variables

Commonly, some limitations apply in evaluating patterns of lateralization (i.e., asymmetry). Analyses are usually limited to a single (e.g., at 40% of the total biomechanical length) or a few cross sections. In addition, the investigation is restricted to the use of univariate and exploratory statistics. Here, we present two different strategies to evaluate the relative contribution to asymmetry of different predictors (i.e., weight, height, age and occupation).

To assess how asymmetry in the distribution of cortical thickness varies in relation to occupation, age, weight, height and biomechanical length, we performed a multiple regression with these variables as independent and maps of differences in cortical thickness between the left and right side as dependent variables. Specifically, the cortical maps of differences between sides are created by subtracting the mean matrix of cortical thicknesses from the matrices of the left and mirrored right sides from each linear regression we computed R^2 and beta coefficients. R^2 quantifies the strength of the relationship between the model and the dependent variable. The beta coefficient describes the rate of change of differences in cortical thickness between sides for every unit of change in the independent variables. In addition, we measured the proportion of variance in total asymmetry related to each independent variable using multivariate regression analysis. Lastly, we applied the variance partitioning method [45] to measure the portions of variance of total asymmetry shared by independent variables. The method calculates the explanatory power of different variables in relation to the same response variable (or matrix). We used redundancy analysis to determine the partial effect of each variable (i.e., weight, height, age and occupation) on the response variable (magnitude of asymmetry of cortical thickness between sides). We used alpha (significance) level = 0.05 for all statistical tests.

3. Results

The asymmetry in polar moment of inertia, J, calculated at 40% of the biomechanical length shows a general trend of lateralization ranging between 0.36% and 10.37%. In all but 4 individuals, J is larger on the right side (Table 2). There are no statistically significant differences between occupation and sex group means (Figure 3), as determined by two-way ANOVA of J among sexes or occupations.

The first two PCs of the symmetric component of cortical thicknesses account for 78% of the total variance (PC1 = 72.33%; PC2 = 5.79%) (Figure 4). On average, the two sexes are separated along PC1 with the females towards the positive limit, and males, the negative ($t_{14.48}$ = 4.66, $p < 0.01$). The three occupation groups largely overlap but with the "building", "army" and "desk" groups approximately distributed in this order from negative to positive limits of PC1.

Table 2. Description of the sample and calculation of lateralization. For all the individuals, we report age, weight, height and occupation. We calculated the biomechanical length of the humerus and the polar moment of inertia, J, at 40% of the biomechanical length on both sides (JL and JR) and the degree of lateralization, $J_{LAT}\%$, between the two sides. Values of J are multiplied by 10^3 (Ruff 2000). D = working at desk job; B = working in building/mining companies; A = working in the armed forces.

ID	Sex	Age	Weight	Height	Occupation	J_L	J_R	$J_{LAT\%}$
100221	male	34	91	188	D	2.84	3.60	5.90
101358	male	21	70	168	B	1.50	1.57	1.07
101510	male	26	86	195	A	2.39	2.96	5.33
102253	male	46	100	183	D	1.62	1.92	4.26
102436	male	37	83	183	D	3.10	3.22	0.91
102602	male	32	86	193	B	2.26	2.40	1.56
103530	male	22	73	168	B	1.51	1.72	3.26
103862	male	31	91	191	A	2.11	2.26	1.76
104373	male	34	86	170	B	2.47	2.56	0.86
108039	male	38	82	188	A	1.65	1.92	3.77
114405	male	27	79	178	B	1.62	1.97	4.81
116546	male	25	86	184	A	2.82	2.43	3.72
116833	male	45	109	183	A	2.40	2.61	2.08
117662	female	24	68	173	A	1.29	1.50	3.85
118646	male	26	84	180	A	1.92	2.84	9.63
121289	male	25	102	184	B	2.74	2.98	2.10
123096	male	24	75	184	B	2.81	3.14	2.78
123240	male	31	82	180	D	1.64	2.05	5.61
125527	female	26	50	157	D	0.69	0.79	3.35
127137	female	54	59	158	D	1.38	1.32	1.20
129131	male	33	79	173	A	1.80	1.87	0.93
129352	male	30	82	180	A	1.56	1.86	4.29
130388	male	37	68	168	A	1.67	1.82	2.16
130964	male	25	68	175	D	2.87	3.42	4.42
132233	male	29	80	166	B	1.77	2.13	4.64
132433	male	27	70	185	D	2.06	2.30	2.72
139871	female	49	61	157	B	1.49	1.56	1.27
140368	male	34	80	185	B	3.17	3.83	4.74
141318	male	32	77	178	B	2.04	2.24	2.32
143365	female	47	72	165	A	1.16	1.10	1.32
143984	male	30	91	183	D	2.51	2.63	1.17
144071	female	35	57	160	A	1.04	1.08	0.94
144977	male	24	75	175	D	2.06	2.24	2.09
146626	male	27	89	163	A	0.98	1.16	4.24
147949	male	29	91	178	A	1.82	2.63	9.15
150608	male	34	63	158	B	1.81	2.01	2.64
152567	male	24	82	185	A	2.46	2.74	2.70
156886	male	29	77	178	B	1.78	1.97	2.51
158402	male	35	109	185	B	2.32	2.37	0.58
162065	male	28	64	170	B	1.56	1.88	4.76
166116	female	36	56	162	B	1.07	1.17	2.22
171170	male	32	106	182	A	2.96	3.75	5.90
171479	male	27	100	185	D	2.50	2.82	3.06
173218	male	22	84	163	D	1.95	2.06	1.36
175725	male	28	86	196	B	3.12	3.46	2.54
176660	female	24	68	173	A	0.79	0.84	1.47
177679	male	38	77	168	A	2.75	2.82	0.68
178078	male	31	73	168	B	2.10	2.31	2.32
180030	male	40	82	180	A	2.06	2.33	3.05
188902	female	26	77	165	D	1.25	1.12	2.82
190756	female	24	48	162	D	0.84	0.98	3.82

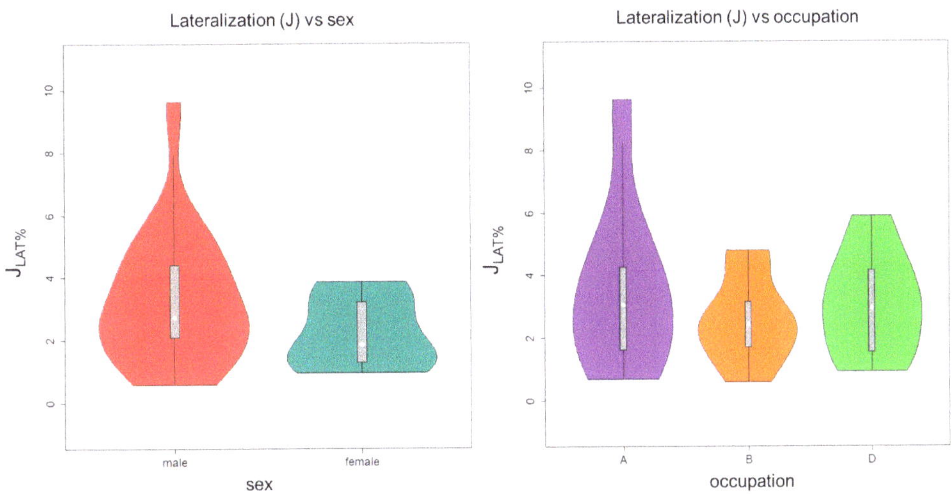

Figure 3. Violin and box plots showing the percentage of right lateralization pooled by sex (**left**) and occupation (**right**, A = "army", B = "building", D = "desk"). Violin plots illustrate the density distribution of the data.

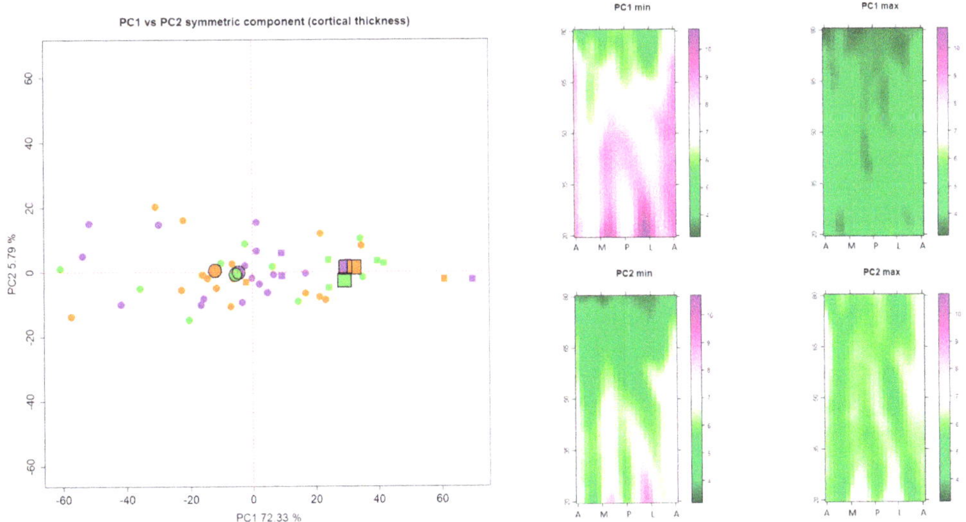

Figure 4. PCA of the symmetric component. Circular and square dots represent male and female individuals, respectively. Violet = "Army", orange = "Building" and green = "Desk". The double-sized circles and squares show the mean values of PC1 and PC2 pooled by sex and occupation. In the morphometric maps, the diaphysis is unrolled from the anterior border (on the left of the x-axis) and follows the medial, posterior, lateral and anterior borders. Violet and green respectively indicate greater or smaller values of cortical thickness. The tops and the bottoms of the morphometric maps correspond to the proximal (80% of the biomechanical length) and distal (20% of the biomechanical length) parts of the diaphysis.

The visualisations of the morphometric maps represented by the extremes of PCs 1 and 2 highlight a general increase in cortical thickness at negative values of PC1. PC2 represents a different pattern of thickening/thinning of the humeral diaphysis. With increasingly negative values of PC2, the cortical bone is thicker in the mid and distal portion of the medial and anterior margins and between the posterior and lateral margins.

Conversely, with more positive values, the proximal portion of the diaphysis is thicker anteriorly (Figure 4).

The PCA of the asymmetric component (Figure 5) indicates how the cortical thickness of the entire diaphysis differs from symmetry. The distance of points from the origin indicates the degree of asymmetry represented by the first two PC scores. Following Mardia et al. [46], we found that directional (mean difference between sides) and fluctuating (average differences from the symmetric mean) components account for 16.00% and 84.00% of the total variance, respectively.

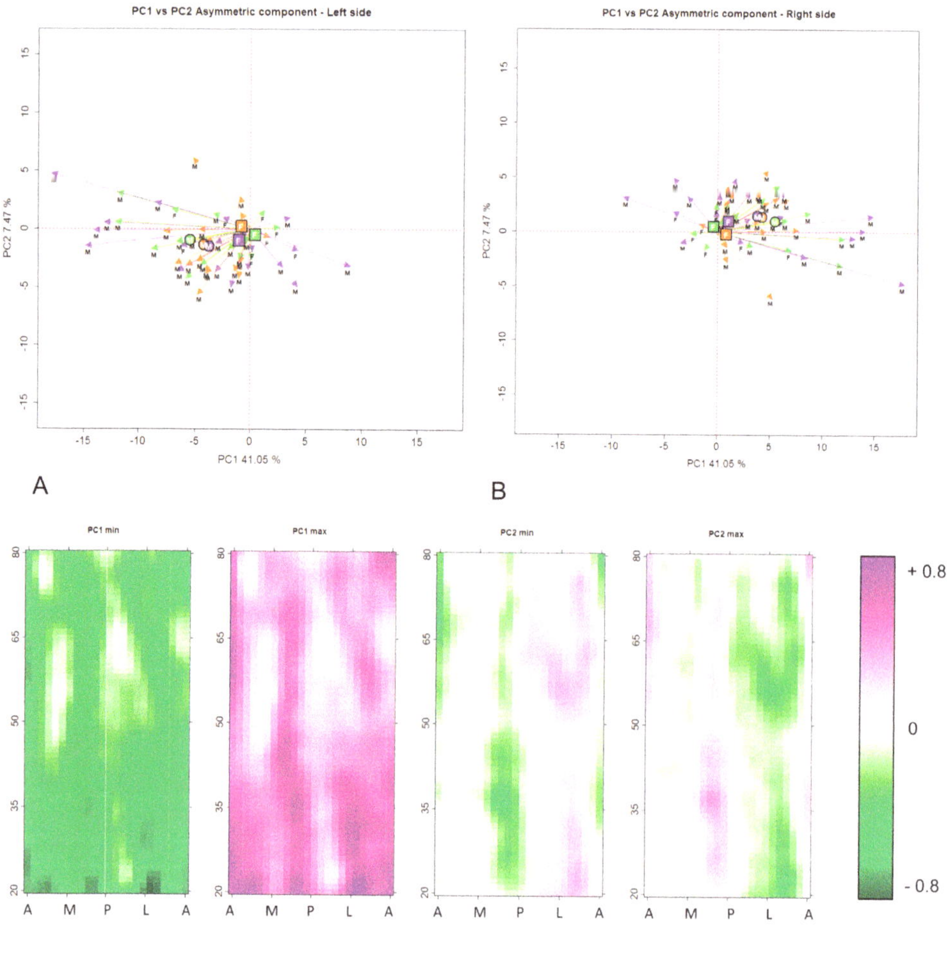

Figure 5. PCA of the asymmetric component shown in two separated plots for the left (**A**) and right (**B**) sides. Arrows connect points representing the left (**A**) and right (**B**) sides of each individual from the origin (i.e., zero asymmetry between left and right side). Violet = "Army", orange = "Building" and green = "Desk". In the morphometric maps (**C**), violet and green palettes respectively indicate larger and smaller values of cortical thickness. In these, the diaphysis is unrolled from the anterior border (on the left of the x-axis) and follow the medial, posterior, lateral and anterior borders. Violet and green respectively indicate greater or smaller values of cortical thickness. The tops and the bottoms of the morphometric maps correspond to the proximal (80% of the biomechanical length) and distal (20% of the biomechanical length) parts of the diaphysis.

PC1 explains 41.05% of the total variance. This axis describes generalised thickening or thinning of the cortex as seen in the morphometric maps representing the extremes of this PC. Scores on PC1 indicate that the right-sided cortex tends to be thicker than the left (with a few exceptions, plausibly explained by handedness). PC2 (7.47% of the total variance) shows a different pattern of asymmetry. The morphometric maps indicate that, from positive to negative limits, this PC represents posterior and anterior thinning of the diaphysis (Figure 5).

While the differences between sexes (F = 18.40, Df = 1, $p < 0.01$) in asymmetry are significant, as indicated by two-way ANOVA (Figure 6), there are no statistically significant differences in asymmetry between the occupation groups.

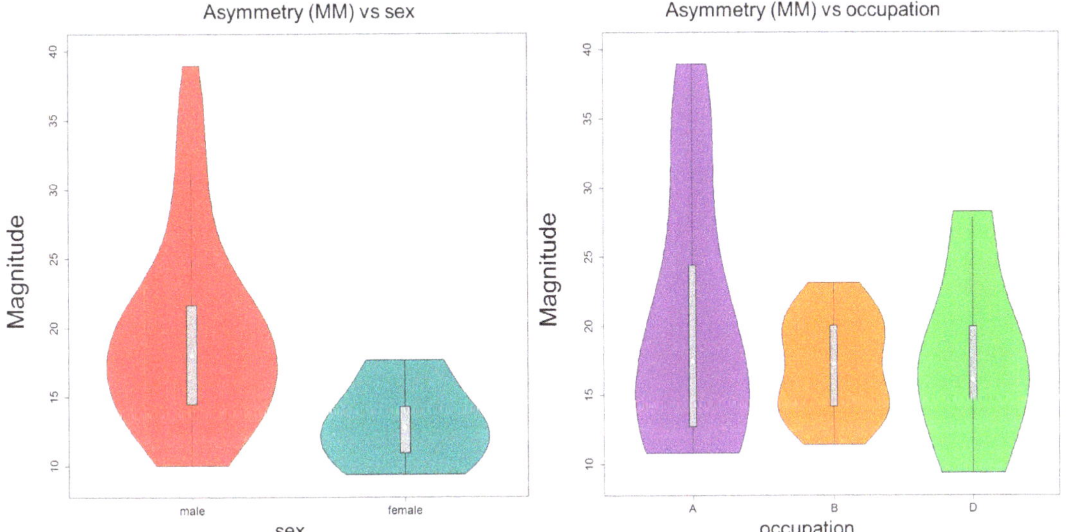

Figure 6. Violin plots of the total length of the displacement vectors of asymmetry in relation to sex (**left**) and occupation (**right**, A = "army", B = "building", D = "desk"). Violin plots illustrate the density distribution of the data.

The lengths of the vectors in Figure 5 indicate the magnitude of asymmetry of cortical thickness, represented by the first two PC scores. The sum of the vectors from the entire matrix of PC scores represents the overall magnitude of asymmetry. In this sample, it is correlated with the index of lateralization (J_{LAT}%), calculated using the polar moment of inertia, J, (correlation = 0.58, $p < 0.01$).

In multivariate regressions, body weight ($R^2 = 0.19$, b = 0.22, $p < 0.01$), height ($R^2 = 0.22$, b = 0.31, $p < 0.01$) and biomechanical length ($R^2 = 0.16$, b = 0.14, $p < 0.01$) significantly predicted asymmetry, while age and occupation score are not statistically significant predictors (Figure 7).

A variance partitioning analysis was performed to evaluate the percentage of variance in asymmetry associated with weight, height, age and occupation score. The combination of all tested variables explained 24.72% of the variance in asymmetry calculated from the PC scores of asymmetric component. Weight (2.09%) and height (2.39%), and their interaction (16.31%) explain the largest portion of asymmetry (Figure 8).

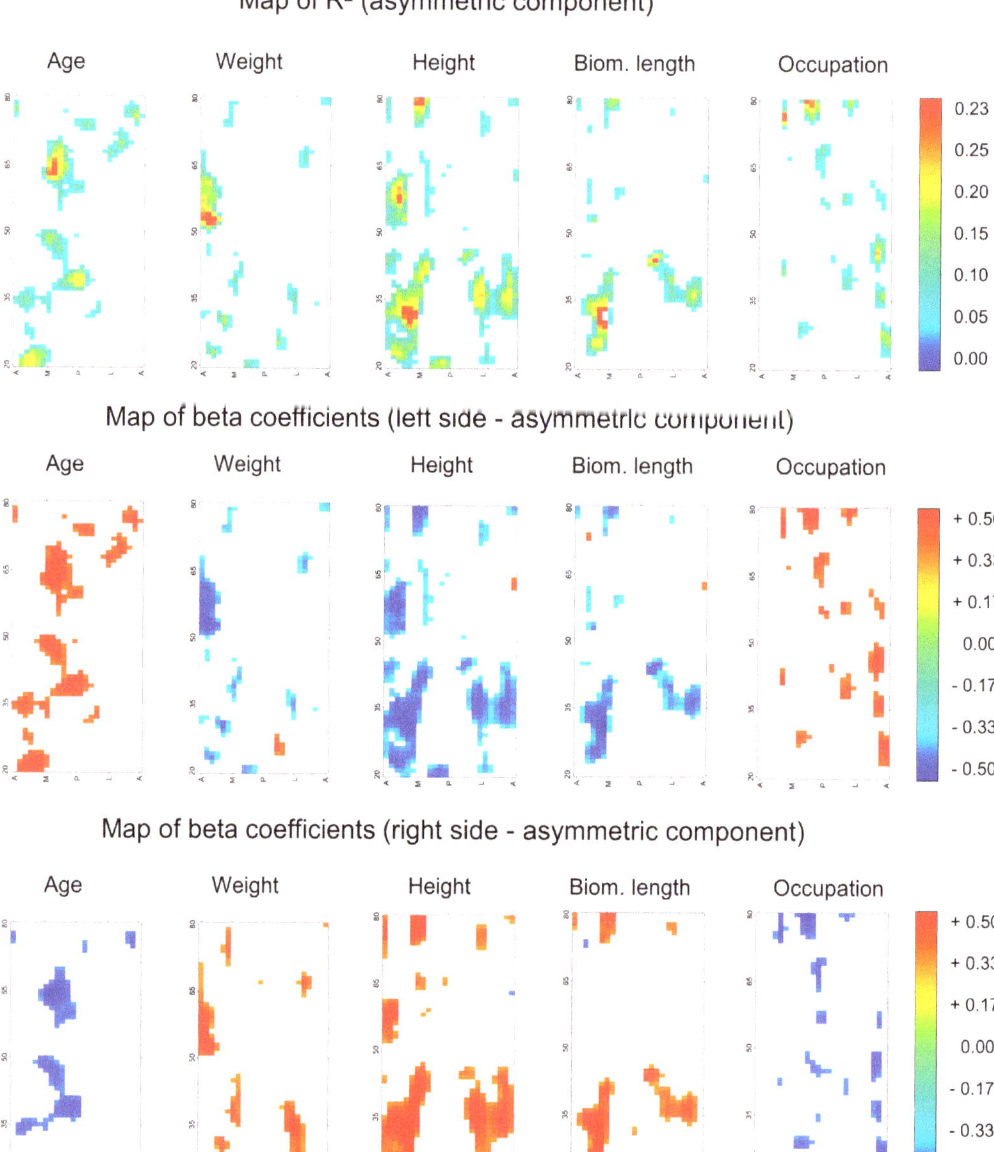

Figure 7. Map of R^2 and beta coefficients calculated from multivariate regression of the asymmetric component on independent variables of interest. Cortical thickness values were rank-transformed. In the first row, maps indicate which regions of the diaphysis show asymmetric variation in thickness with age, weight, height and biomechanical length. The R^2 range is reported using a rainbow palette. White cells indicate statistically insignificant relationships. In the second and third rows, the beta coefficients from the multivariate regressions are mapped on the left and right sides, respectively. Warm colours describe an increase in cortical thickness, and cold colours indicate a reduction.

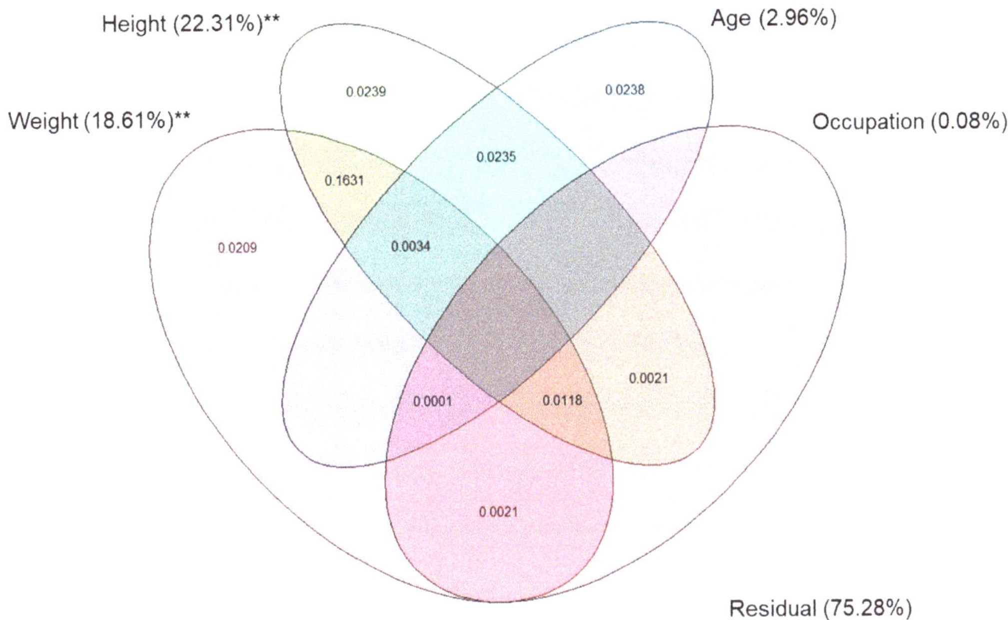

Figure 8. Variation partitioning Venn diagram. Variation in asymmetry expressed as adjusted R^2 explained by weight, height, occupation and age and their intersections. Significance codes: ** = 0.001. Outside the Venn diagram are reported the total explained variances by each variable, taking into account the interactions between them. Note, the sum of the explained variance is not 100% (artefact of the variance partitioning algorithm due to the calculation of negative adjusted R^2).

Multivariate regression was used to assess the relationship between asymmetry in cortical thickness and the independent variable of interest (i.e., weight, age, height, biomechanical length and occupation scores) at each cell of the morphometric maps (i.e., the cortical thickness measured at each semilandmark). At each cell, the explained variance (R^2) and slope (Beta coefficient) can be mapped to visualise the relationship between cortical map asymmetry and the independent variables. Maps can be drawn to represent the (exactly opposite) effects of these relationships on the right or left sides. Such maps are presented in Figure 8. The maps of R^2 indicate that each independent variable is associated with a different pattern of asymmetry, with different localised regions showing an association with each variable. Unsurprisingly, regressions of height and biomechanical length produce the most similar diagrams. On average, the slopes (beta coefficients) of asymmetry of cortical thickness on the independent variables indicate that cortical thicknesses tend to be greater in the right arm (likely this is a mostly right-handed sample). The maps of Figure 8 show values only for those regions where the regression is significant. Weight, height and mechanical length are associated with a larger rate of increase in cortical thickness in the right (usually the dominant) compared to the left humerus. In contrast, with increasing age, cortical thickness decreases more slowly in the dominant arm.

4. Discussion and Conclusions

Studies of patterns of lateralization in archaeological populations often suffer from the lack of a reference sample with known loading history to contextualise the findings. Additionally, the assessment of lateralization is commonly limited to the analysis of one or just a few levels along the diaphysis of paired long bones. The publication of the NMDID [42] offers the prospect of directly relating skeletal form (total body CT-scan) with known biological profile and loading history (metadata with 60 variables). The time and effort in gathering skeletal data is much reduced by the functions available in the R package *morphomap* [41], a recently published toolkit providing functions to extract from CT data, the segmented long bone of interest and based on that. Here, we further extend *morphomap* to visualize and analyze asymmetry in paired long bones. Specifically, the new implementation: (i) performs a PCA on the symmetric and asymmetric component of form variation; (ii) creates morphometric maps of symmetric and asymmetric variation on single individuals or on entire samples from PC scores; (iii) calculates the total magnitude of asymmetry of cortical bone distribution, quantifying deviations from symmetry.

PCAs of both symmetric and asymmetric components indicate that cortical bone distribution differs between sexes, but not between the occupation groups considered in our analyses. On average, male individuals possess thicker and more asymmetric humeri than females. All our measurements of magnitude of asymmetry (cross-sectional geometry and vector lengths from PC scores) present a general pattern of right lateralization. This finding is consistent with previous studies indicating 90% preference for right-handedness in modern humans [47–49]. Further, the analyses of morphometric maps indicate that males are more asymmetric than females. However, since males are larger on average, their more asymmetric cortical thickness might be guided by allometric effects. In contrast, the index of lateralization based on cross-sectional geometry ($J_{LAT\%}$) does not significantly differ between sexes. This contrast may be due to the fact that $J_{LAT\%}$ is calculated at a single level along the diaphysis (40%v of bone biomechanical length), whereas the calculation of absolute lateralization from PC scores takes into account the entire diaphysis. In fact, $J_{LAT\%}$ calculated at 70%, 75%, 76%, 78% and 79% of bone length is statistically different between sexes.

Regression analyses on morphometric maps show that, as body weight, height and longbone biomechanical length increase, so does asymmetry. In contrast, with increasing of age, asymmetry decreases (i.e., the oldest individuals are less asymmetric than the youngest individuals). Interestingly, loading history (occupation scores) does not affect the pattern of asymmetry. The main effect of body proportions (weight and height) as a key factor in determining the degree of asymmetry is also confirmed by the partitioning of variance analysis performed on values of total asymmetry calculated from the PCA.

This study was able to test the hypothesised relationships between loading history and cortical bone distribution in the humeral shaft using the unique and extensive collection curated in the NMDID [42]. Despite the quality of these data, our analyses do not show a significant association between occupation and asymmetry. This analysis was confined to occupations that would be expected to lead to extremely different occupational loading histories (army, building vs. desk) in order to emphasise any effect of occupation. This suggests either that there is no strong difference in the effect of these occupations on loading history, or that occupational history does not reflect the full loading history, and that our categorisation by occupation is inadequate to describe individual loading history. Further studies are required to clarify this finding, which is potentially of great importance in archaeological and forensic contexts.

Author Contributions: Conceptualization, A.P., D.M., G.M. and P.O.; data curation, A.P. and C.Z.; formal analysis, A.P., C.Z., I.M. and A.M.; funding acquisition, A.P. and P.O.; investigation, A.P. and C.Z.; methodology, A.P. and P.O.; project administration, A.P. and P.O.; resources, A.P. and C.Z.; software, A.P., P.R. and P.O.; validation, C.Z.; visualization, A.P. and A.M.; writing—original draft, A.P., C.Z., P.R. and P.O.; writing—review and editing, C.Z., I.M., A.M., P.R., D.M., G.M. and P.O. All authors have read and agreed to the published version of the manuscript.

Funding: This project has received funding from the European Union's Horizon 2020 research and innovation program (H2020 Marie Skłodowska-Curie Actions) under grant agreement H2020-MSCA-IF-2018 No. 835571 to Antonio Profico and Paul O'Higgins.

Institutional Review Board Statement: The study was conducted according to the guidelines of the Declaration of Helsinki, and approved by the Ethics Committee of the University of York, Department of Archaeology.

Informed Consent Statement: Not applicable.

Data Availability Statement: Not applicable.

Acknowledgments: The authors are grateful to two anonymous reviewers for helpful comments.

Conflicts of Interest: The authors declare no conflict of interest.

References

1. Jurmain, R.; Cardoso, F.A.; Henderson, C.; Villotte, S. Bioarchaeology's holy grail: The reconstruction of activity. In *A Companion to Paleopathology*; Grauer, A.L., Ed.; Wiley-Blackwell: Oxford, UK, 2011; pp. 531–552.
2. Larsen, C.S. *Bioarchaeology: Interpreting Behavior from the Human Skeleton*, 2nd ed.; Cambridge Studies in Biological and Evolutionary Anthropology; Cambridge University Press: Cambridge, UK, 2015.
3. Ruff, C.; Holt, B.; Trinkaus, E. Who's afraid of the big bad Wolff? "Wolff's law" and bone functional adaptation. *Am. J. Phys. Anthr.* **2006**, *129*, 484–498. [CrossRef]
4. Ruff, C.B. Biomechanical analyses of archaeological human skeletons. In *Biological Anthropology of the Human Skeleton*, 2nd ed.; Katzenberg, M.A., Saunders, S.R., Eds.; Alan R. Liss: New York, NY, USA, 2008; pp. 183–206.
5. Trinkaus, E.; Churchill, S.E.; Ruff, C.B. Postcranial robusticity inHomo. II: Humeral bilateral asymmetry and bone plasticity. *Am. J. Phys. Anthr.* **1994**, *93*, 1–34. [CrossRef]
6. Miller, B.F.; Hamilton, K.L.; Majeed, Z.R.; Abshire, S.M.; Confides, A.L.; Hayek, A.M.; Hunt, E.R.; Shipman, P.; Peelor, F.F.; Butterfield, T.A.; et al. Enhanced skeletal muscle regrowth and remodelling in massaged and contralateral non-massaged hindlimb. *J. Physiol.* **2017**, *596*, 83–103. [CrossRef] [PubMed]
7. Trinkaus, E.; Churchill, S.E. Diaphyseal cross-sectional geometry of near eastern middle palaeolithic humans: The humerus. *J. Archaeol. Sci.* **1999**, *26*, 173–184. [CrossRef]
8. Frost, H.M. *Orthopaedic Biomechanics*; Charles C. Thomas: Springfield, IL, USA, 1973.
9. Currey, J. The many adaptations of bone. *J. Biomech.* **2003**, *36*, 1487–1495. [CrossRef]
10. Oxnard, C.E. Thoughts on bone biomechanics. *Folia Primatol.* **2004**, *75*, 189–201. [CrossRef] [PubMed]
11. Cardoso, F.A.; Henderson, C. Enthesopathy formation in the humerus: Data from known age-at-death and known occupation skeletal collections. *Am. J. Phys. Anthr.* **2009**, *141*, 550–560. [CrossRef] [PubMed]
12. Weiss, E. Humeral cross-sectional morphology from 18th century Quebec prisoners of war: Limits to activity reconstruction. *Am. J. Phys. Anthr.* **2005**, *126*, 311–317. [CrossRef] [PubMed]
13. Bertram, J.E.A.; Swartz, S.M. The 'law of bone transformation': A case of crying wolff? *Biol. Rev.* **1991**, *66*, 245–273. [CrossRef]
14. Melcher, A.H.; Accursi, G.E. Transmission of an "osteogenic message" through intact bone after wounding. *Anat. Rec. Adv. Integr. Anat. Evol. Biol.* **1972**, *173*, 265–275. [CrossRef]
15. Pearson, O.M.; Lieberman, D.E. The aging of Wolff's "law": Ontogeny and responses to mechanical loading in cortical bone. *Am. J. Phys. Anthr.* **2004**, *125*, 63–99. [CrossRef] [PubMed]
16. Ruff, C.B.; Jones, H.H. Bilateral asymmetry in cortical bone of the humerus and tibia—Sex and age factors. *Hum. Biol.* **1981**, *53*, 69–86. [PubMed]
17. Frost, H.M. An overview: Spinal tissue vital biomechanics for clinicians. In *Spinal Disorders in Growth and Aging*; Takahashi, H.E., Ed.; Springer: Tokyo, Japan, 1995; pp. 95–126.
18. Sparacello, V.; Pearson, O.; Coppa, A.; Marchi, D. Changes in skeletal robusticity in an iron age agropastoral group: The samnites from the Alfedena necropolis (Abruzzo, Central Italy). *Am. J. Phys. Anthr.* **2010**, *144*, 119–130. [CrossRef]
19. Rhodes, J.A.; Knüsel, C.J. Activity-related skeletal change in medieval humeri: Cross-sectional and architectural alterations. *Am. J. Phys. Anthr.* **2005**, *128*, 536–546. [CrossRef]
20. Tur, S. Bilateral asymmetry of long bones in bronze and early iron age pastoralists of the altai. *Archaeol. Ethnol. Anthr. Eurasia* **2014**, *42*, 141–156. [CrossRef]
21. Sládek, V.; Ruff, C.B.; Berner, M.; Holt, B.; Niskanen, M.; Schuplerová, E.; Hora, M. The impact of subsistence changes on humeral bilateral asymmetry in Terminal Pleistocene and Holocene Europe. *J. Hum. Evol.* **2016**, *92*, 37–49. [CrossRef]
22. Kubicka, A.M.; Nowaczewska, W.; Balzeau, A.; Piontek, J. Bilateral asymmetry of the humerus in Neandertals, Australian aborigines and medieval humans. *Am. J. Phys. Anthr.* **2018**, *167*, 46–60. [CrossRef] [PubMed]
23. Churchill, S.E.; Formicola, V. A case of marked bilateral asymmetry in the Upper Limbs of an Upper Palaeolithic Male from Barma Grande (Liguria). *Italy Int. J. Osteoarchaeol.* **1997**, *7*, 18–38. [CrossRef]
24. Perchalski, B.; Placke, A.; Sukhdeo, S.M.; Shaw, C.N.; Gosman, J.H.; Raichlen, D.A.; Ryan, T.M. Asymmetry in the cortical and trabecular bone of the human humerus during development. *Anat. Rec. Adv. Integr. Anat. Evol. Biol.* **2018**, *301*, 1012–1025. [CrossRef]
25. Ben-Itzhak, S.; Smith, P.; Bloom, R.A. Radiographic study of the humerus in Neandertals and Homo sapiens sapiens. *Am. J. Phys. Anthr.* **1988**, *77*, 231–242. [CrossRef]
26. Shaw, C.N.; Hofmann, C.L.; Petraglia, M.D.; Stock, J.T.; Gottschall, J.S. Neandertal humeri may reflect adaptation to scraping tasks, but not spear thrusting. *PLoS ONE* **2012**, *7*, e40349. [CrossRef]
27. Sparacello, V.S.; Villotte, S.; Shackelford, L.L.; Trinkaus, E. Patterns of humeral asymmetry among Late Pleistocene humans. *C. R. Palevol* **2017**, *16*, 680–689. [CrossRef]

28. Vandermeersch, B.; Trinkaus, E. The postcranial remains of the Régourdou 1 Neandertal: The shoulder and arm remains. *J. Hum. Evol.* **1995**, *28*, 439–476. [CrossRef]
29. Pérez-Criado, L.; Rosas, A.; Bastir, M.; Pastor, F. Humeral laterality in modern humans and Neanderthals: A 3D geometric morphometric analysis. *Anthr. Sci.* **2017**, *125*, 117–128. [CrossRef]
30. Uomini, N.T. Handedness in Neanderthals. In *Neanderthal Lifeways, Subsistence and Technology: One Hundred Fifty Years of Neanderthal Study*; Conard, N.J., Richter, J., Eds.; Springer: Dordrecht, The Netherlands, 2011; pp. 139–154.
31. Ireland, A.; Maden-Wilkinson, T.; Mcphee, J.; Cooke, K.; Narici, M.; Degens, H.; Rittweger, J. Upper limb muscle-bone asymmetries and bone adaptation in elite youth tennis players. *Med. Sci. Sports Exerc.* **2013**, *45*, 1749–1758. [CrossRef]
32. Haapasalo, H.; Kontulainen, S.; Sievänen, H.; Kannus, P.; Järvinen, M.; Vuori, I. Exercise-induced bone gain is due to enlargement in bone size without a change in volumetric bone density: A peripheral quantitative computed tomography study of the upper arms of male tennis players. *Bone* **2000**, *27*, 351–357. [CrossRef]
33. Kontulainen, S.; Sievänen, H.; Kannus, P.; Pasanen, M.; Vuori, I. Effect of long-term impact-loading on mass, size, and estimated strength of humerus and radius of female racquet-sports players: A peripheral quantitative computed tomography study between young and old starters and controls. *J. Bone Miner. Res.* **2003**, *18*, 352–359. [CrossRef] [PubMed]
34. Nikander, R.; Sievänen, H.; Uusi-Rasi, K.; Heinonen, A.; Kannus, P. Loading modalities and bone structures at nonweight-bearing upper extremity and weight-bearing lower extremity: A pQCT study of adult female athletes. *Bone* **2006**, *39*, 886–894. [CrossRef]
35. Shaw, C.N.; Stock, J.T. Habitual throwing and swimming correspond with upper limb diaphyseal strength and shape in modern human athletes. *Am. J. Phys. Anthr.* **2009**, *140*, 160–172. [CrossRef]
36. Qu, X. Morphological effects of mechanical forces on the human humerus. *Br. J. Sports Med.* **1992**, *26*, 51–53. [CrossRef]
37. Ruff, C.B.; Walker, A.; Trinkaus, E. Postcranial robusticity inHomo. III: Ontogeny. *Am. J. Phys. Anthr.* **1994**, *93*, 35–54. [CrossRef]
38. Niinimäki, S. The relationship between musculoskeletal stress markers and biomechanical properties of the humeral diaphysis. *Am. J. Phys. Anthr.* **2012**, *147*, 618–628. [CrossRef]
39. Ibáñez-Gimeno, P.; De Esteban-Trivigno, S.; Jordana, X.; Manyosa, J.; Malgosa, A.; Galtés, I. Functional plasticity of the human humerus: Shape, rigidity, and muscular entheses. *Am. J. Phys. Anthr.* **2013**, *150*, 609–617. [CrossRef]
40. Wang, X.; Liu, H.; Zhao, L.; Fei, C.; Feng, X.; Chen, S.; Wang, Y. Structural properties characterized by the film thickness and annealing temperature for La2O3 films grown by atomic layer deposition. *Nanoscale Res. Lett.* **2017**, *12*, 233. [CrossRef]
41. Profico, A.; Bondioli, L.; Raia, P.; O'Higgins, P.; Marchi, D. Morphomap: An R package for long bone landmarking, cortical thickness, and cross-sectional geometry mapping. *Am. J. Phys. Anthr.* **2020**, *174*, 129–139. [CrossRef]
42. Edgar, H.J.H.; Daneshvari-Berry, S.; Moes, E.; Adolphi, N.L.; Bridges, P.; Nolte, K.B. *New Mexico Decedent Image Database*; Office of the Medical Investigator; University of New Mexico: Albuquerque, NM, USA, 2020.
43. Montoye, H.J. Energy costs of exercise and sport. In *Nutrition in Sport*; Maughan, R.J., Ed.; Blackwell Science Ltd: Oxford, UK, 2000; pp. 53–72.
44. Ruff, C.B. Long bone articular and diaphyseal structure in old world monkeys and apes. I: Locomotor effects. *Am. J. Phys. Anthr.* **2002**, *119*, 305–342. [CrossRef]
45. Borcard, D.; Legendre, P.; Drapeau, P. Partialling out the spatial component of ecological variation. *Ecology* **1992**, *73*, 1045–1055. [CrossRef]
46. Mardia, K.; Bookstein, F.; Moreton, I. Statistical assessment of bilateral symmetry of shapes. *Biometrika* **2000**, *87*, 285–300. [CrossRef]
47. Perelle, I.B.; Ehrman, L. An international study of human handedness: The data. *Behav. Genet.* **1994**, *24*, 217–227. [CrossRef]
48. McManus, I.C. The history and geography of human handedness. In *Language Lateralization and Psychosis*; Sommer, I.E.C., Kahn, R.S., Eds.; Cambridge University Press: Cambridge, UK, 2009; pp. 37–58.
49. Shaw, C.N. Is 'hand preference' coded in the hominin skeleton? An in-vivo study of bilateral morphological variation. *J. Hum. Evol.* **2011**, *61*, 480–487. [CrossRef] [PubMed]

 *symmetry*

Article

Biomechanical Evaluation on the Bilateral Asymmetry of Complete Humeral Diaphysis in Chinese Archaeological Populations

Yuhao Zhao [1,2,3], Mi Zhou [4], Haijun Li [5], Jianing He [6], Pianpian Wei [7,8] and Song Xing [1,2,*]

1   Key Laboratory of Vertebrate Evolution and Human Origins, Institute of Vertebrate Paleontology and Paleoanthropology, Chinese Academy of Sciences, Beijing 100044, China; zhaoyuhao@ivpp.ac.cn
2   CAS Center for Excellence in Life and Paleoenvironment, Beijing 100044, China
3   College of Earth and Planetary Sciences, University of Chinese Academy of Sciences, Beijing 100049, China
4   Institute of Archeology and Cultural Relics of Hubei Province, Wuhan 430077, China; zhoumi@hbww.org
5   School of Ethnology and Sociology, Minzu University of China, Beijing 100081, China; lindavy@163.com
6   Center for the Study of Chinese Archaeology, Peking University, Beijing 100871, China; hejianing@pku.edu.cn
7   Ministry of Education Key Laboratory of Contemporary Anthropology, Department of Anthropology and Human Genetics, School of Life Sciences, Fudan University, Shanghai 200433, China; weipianpian@fudan.edu.cn
8   Centre for the Exploration of the Deep Human Journey, Faculty of Science, University of the Witwatersrand, 2050 Johannesburg, South Africa
*   Correspondence: xingsong@ivpp.ac.cn

Abstract: Diaphyseal cross-sectional geometry (CSG) is an effective indicator of humeral bilateral asymmetry. However, previous studies primarily focused on CSG properties from limited locations to represent the overall bilateral biomechanical performance of humeral diaphysis. In this study, the complete humeral diaphyses of 40 pairs of humeri from three Chinese archaeological populations were scanned using high-resolution micro-CT, and their biomechanical asymmetries were quantified by morphometric mapping. Patterns of humeral asymmetry were compared between sub-groups defined by sex and population, and the representativeness of torsional rigidity asymmetry at the 35% and 50% cross-sections (J_{35} and J_{50} asymmetry) was testified. Inter-group differences were observed on the mean morphometric maps, but were not statistically significant. Analogous distribution patterns of highly asymmetrical regions, which correspond to major muscle attachments, were observed across nearly all the sexes and populations. The diaphyseal regions with high variability of bilateral asymmetry tended to present a low asymmetrical level. The J_{35} and J_{50} asymmetry were related to the overall humeral asymmetry, but the correlation was moderate and they could not reflect localized asymmetrical features across the diaphysis. This study suggests that the overall asymmetry pattern of humeral diaphysis is more complicated than previously revealed by individual sections.

Keywords: contralateral asymmetry; limb bone; biomechanical analysis; rigidity

Citation: Zhao, Y.; Zhou, M.; Li, H.; He, J.; Wei, P.; Xing, S. Biomechanical Evaluation on the Bilateral Asymmetry of Complete Humeral Diaphysis in Chinese Archaeological Populations. *Symmetry* **2021**, *13*, 1843. https://doi.org/10.3390/sym13101843

Academic Editor: Antoine Balzeau

Received: 28 August 2021
Accepted: 28 September 2021
Published: 2 October 2021

Publisher's Note: MDPI stays neutral with regard to jurisdictional claims in published maps and institutional affiliations.

1. Introduction

Humeral bilateral asymmetry has been extensively studied in orthopedics, forensics, and paleo/archaeological anthropology [1–3]. Handedness can be inferred from the bilateral asymmetry of the upper limb [4–6]. Evidence from living athletes of unilaterally dominated sports (such as tennis and cricket) suggests a close relationship between humeral bilateral asymmetry and behavioral laterality [7–9]. A combined study of endocranial and humeral asymmetry can shed light on how the human body responds to dependent asymmetrical stimuli across biologically independent anatomical regions [10]. These applications make humeral bilateral asymmetry an effective approach for reconstructing the behaviors of past human populations [11–16].

Long bone diaphyses show great plasticity to remodel in response to mechanical loadings across a lifetime, especially prior to sexual maturity [17–21]. This remodeling makes diaphyseal cross-sectional geometry (CSG) a more effective indicator of bilateral upper-limb use and asymmetry compared to other linear measurements, such as articular breath or bone length [13,20,22–24].

Polar moment of area (J) and second moment of area (SMA) are two commonly adopted CSG parameters in humeral biomechanical analysis. J indicates the cross-section's torsional and average bending rigidity, whereas SMA denotes the exact bending rigidity along a certain axis of a cross-section [3,25]. Owing to the difficulties of obtaining sequential histological cross-sections, most earlier studies focused on the CSG properties of cross-sections placed at 35% or 50% of the humeral biomechanical length (see the definition made by Ruff [26]). J at the 35% cross-section (J_{35}) can reasonably estimate the minimum rigidity of humeral diaphysis and avoids the interference of other anatomical features, as it is situated below the distal edge of deltoid tuberosity and above the supracondylar crest [13,14,16,27–29]. J at the 50% cross-section (J_{50}) provides reasonable estimates of midshaft rigidity [9,14,30–34], and is known to be a reliable indicator of hand preference [5]. When evaluating the directional biomechanical performance of a cross-section, most previous studies only calculated the maximum/minimum SMA or SMA along the standard anatomical axis (anteroposterior or mediolateral) to avoid the complexity of acquiring CSG values in multiple directions [35–37].

However, CSG properties of limited cross-sections and directions are insufficient to estimate the overall biomechanical performance of long bone diaphysis, especially in studies about humeral bilateral asymmetry. According to experimental data from professional baseball players, tensile and shear strains vary among different diaphyseal sections during throwing activities [38], and the degree of bilateral asymmetry evaluated by J was variable along the humeral shaft [16,38]. The shape asymmetry of different cross-sections also indicates that the asymmetry patterns vary in different anatomical directions [24,39].

Morphometric mapping is a 2D visualizing method that is commonly used for displaying the distribution patterns of morphometric and biomechanical properties across the entire diaphysis of a long bone [40,41]; for example, the distribution patterns of cortical bone thickness along the femoral diaphysis, visualized by morphometric maps, differentiate in Neanderthals and *Homo erectus* from modern humans [42,43]. Additionally, morphometric maps, quantifying the external radius across the entire femoral diaphysis, reveal the ontogenetic disparities between wild and captive chimpanzees [44]. The cortical structure of hallucal metatarsals, represented by morphometric maps of cortical bone thickness and bending rigidity, reflects locomotor adaptations of humans, chimpanzees, and gorillas [45]. Finally, morphometric mapping has been established to be an effective approach for quantifying the humeral biomechanical asymmetry across the complete diaphysis [16].

Factors such as geographic location, chronological age, subsistence pattern, and sex are known to influence the pattern of humeral asymmetry in human populations. Varying degrees of humeral asymmetry have been detected among Upper Paleolithic populations from Europe, Africa, and Asia [11]. European samples show a general decrease in humeral asymmetry from the early Upper Paleolithic populations through to the 20th century [13,22]. Foragers and farmers from the pre-Hispanic American Southwest present different humeral asymmetry patterns [36]. Due to the existence of the sexual division of labor, modern human populations with various geographic locations, chronological ages, and subsistence patterns tend to exhibit diverse sexual dimorphism patterns in humeral asymmetry [31,36,37,46].

In the present study, we aim to (1) generate a more comprehensive understanding of humeral asymmetry by evaluating the biomechanical performance across complete diaphysis compared to previous studies, which only used individual cross-sections; and (2) check the reliability of using J_{35} and J_{50} to represent the overall humeral biomechanical performance in bilateral asymmetry analysis. To fulfil these targets, specimens were scanned using high-resolution micro-computed tomography (micro-CT), and morphometric mapping was applied to quantify the overall biomechanical asymmetry of humeral diaphysis for its effectiveness in visualization and statistical analysis. To cover as wide a variety of specimens as possible, 40 pairs of humeri from three Chinese archaeological populations, which differ in geographic location, chronological age, and subsistence strategy, were selected to represent East Asian modern humans in the present study.

2. Materials and Methods

2.1. Materials

Forty pairs of modern human humeri are included in this study. All paired humeri were collected from archaeological sites with populations that varied in geographic location, chronological age, and subsistence pattern. Agricultural and nomadic/gathering populations are included because these lifestyles were the dominant subsistence patterns in pre-industrial East Asia. The subsistence patterns of these populations were determined by associated burial assemblages and relevant historical records. The populations will be referred to by their geographic locations, which are as follows: ①Hubei population (HB): 9 males and 4 females collected from agricultural sites from Hubei Province, Central China spanning Qin-Han-Tang dynasties (221 BC ~ 907 AD). For some sites of this population, analyses of charred plant remains indicate that *Setaria italica* and *Panicum miliaceum* were the main food crops [47]. ②Henan population (HN): 6 males and 5 females collected from an agricultural population from Junzicun cemetery, Henan Province, North China, which dates to Qing dynasty (1636 AD ~ 1912 AD). Historical records indicate that an agricultural economy was the dominant lifestyle of this population [48]. ③Xinjiang population (XJ): 10 males and 6 females collected from nomadic populations attributed to Subeixi culture (1000 BC ~ 200 BC) from the Turpan Basin, Xinjiang Province, Northwest China. Burial assemblages such as bows, arrows, and stone artifacts for males and spinning wheels and potteries for females indicate a subsistence pattern of nomadism and gathering [49–51]. All individuals were adults. Their age and sex were determined according to cranial and pelvic osteological indicators. All humeral specimens were intact, well preserved, and showed no symptoms of osteoporosis or other pathologies.

2.2. Data Collecting and Processing

All humeri were scanned by a 450 kV micro-CT scanner (designed by Institute of High Energy Physics, Chinese Academy of Sciences) located in Key Laboratory of Vertebrate Evolution and Human Origins, Institute of Vertebrate Paleontology and Paleoanthropology, Chinese Academy of Sciences. The scanning was performed under a voltage of 380 kV, current of 1.5 mA, 360° rotation with a step of 0.5°, and an isometric voxel size of 160 μm. Raw data were virtually reconstructed and segmented in VGStudio Max 3.0. Volume renderings of all humeri were aligned to anatomical position using the standard protocol defined by Ruff [26]. To ensure that the humeri were consistently aligned and to avoid inter-observer error, all alignments were made by one author (Y.Z.). Paired humeri were always aligned synchronously. Three-dimensional meshes of each aligned humerus were generated and saved as PLY files in Avizo 8.1 for the following analyses.

2.3. Cross-Sectional Geometric Parameters Calculation

Customized in-house scripts, mainly sourced from R package 'morphomap', were applied to calculate the CSG parameters [52]. For each humerus, the single-layer periosteum and endosteum surface meshes were firstly detached from the original humeral mesh. Second, 61 equidistant cross-sections were extracted from the surface meshes along the proximodistal diaphysis (between 20 and 80% of the biomechanical length). Third, 360 equiangular landmarks were placed along both the inner and outer contours on each cross-section. Finally, J values of the cross-sections at 35% and 50% of biomechanical length (J_{35} and J_{50}), and SMA values of 360 directions on 61 cross-sections were calculated based on the landmark coordinates.

2.4. Bilateral Asymmetry Quantification

Commonly used practices for assessing bilateral asymmetry are absolute asymmetry ($[(max - min)/((max + min)/2)] \times 100\%$) and directional asymmetry ($[(right - left)/((right + left)/2)] \times 100\%$). However, absolute asymmetry is not appropriate in this study, as the magnitude relationship between the left and right side is not consistent among different landmarks at humeral diaphysis. However, our study still focuses on absolute information of overall bilateral asymmetry, so directional asymmetry is also not suitable, because it does not eliminate the impact of handedness as well as behavioral laterality, which is not the issue this study attempts to investigate and may bring about bias to the conclusion. Therefore, bilateral asymmetry was quantified using dominant asymmetry ($[(dominant - non\text{-}dominant)/((dominant + non\text{-}dominant)/2)] \times 100\%$). The dominant side was decided according to the magnitude of J_{50}, given that it is a valid indicator of handedness [5].

2.5. Morphometric Mapping

The SMA asymmetry values were obtained using the dominant asymmetry equation for all 21,960 (360×61) landmarks, and the results for each paired humeri were deposited in a matrix with 61 rows (sorted by the order of cross-sections) and 360 columns (sorted by the order of directions). These matrices were then visualized as morphometric maps to display the distribution characteristics of bending rigidity asymmetry along the proximodistal humeral diaphysis (Figures 1 and A1). The asymmetry values of J_{35} and J_{50} for all individuals were also calculated using the same equation.

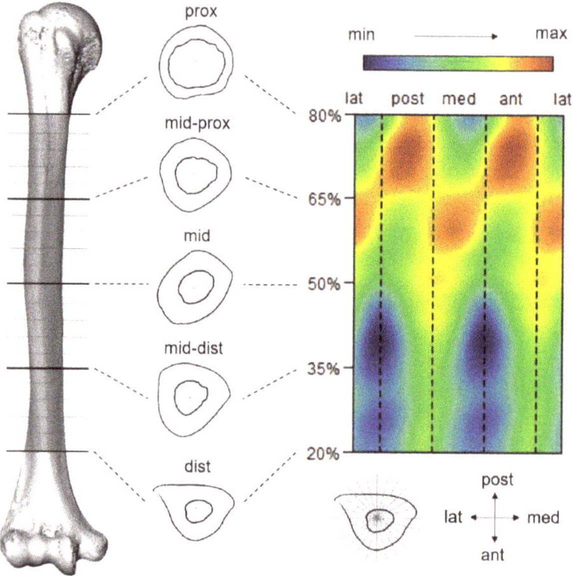

Figure 1. The positional and directional correspondence between humeral external structure, diaphyseal cross-sections, and morphometric map exhibiting bending rigidity asymmetry. Abbreviations for anatomical terms are as follows: prox: proximal; mid: middle; dist: distal; lat: lateral; post: posterior; med: medial; ant: anterior.

2.6. Methods to Estimate the Variation of Humeral Biomechanical Asymmetry

To explore the variation in humeral asymmetry patterns in modern humans, 40 individuals were divided into sub-groups defined by sex and population. The three populations, which varied in geographic location, chronological age, and subsistence pattern, were supposed to vary in their habitual behaviors, so population was set as one variable. Sexual division of labor is an important issue when discussing historical populations, and the sexual dimorphism of humeral asymmetry can be affected by non-behavioral factors such as genetics or hormones [27]. Therefore, sex was set as another variable. Mean morphometric maps exhibiting SMA asymmetry values for each sub-group were qualitatively compared. Additionally, a two-way multivariate analysis of variance (MANOVA) was conducted to quantitatively test whether sex and/or population were significant sources of variation. When fitting the regression model for MANOVA, SMA asymmetry values at all landmarks were set as the dependent variables, while sex and population were set as the independent variables with interaction. Customized in-house scripts, mainly sourced from R package 'geomorph' and 'RRPP', were utilized to conduct MANOVA [53,54]. In addition, the coefficients of variation (CV) for SMA asymmetry values at all landmarks were calculated in sub-groups and visualized by morphometric maps to exhibit intra-group variation characteristics. Only sub-groups defined by sex or by population were included in this analysis to reduce the impact of outliers.

2.7. Methods to Test the Representativeness of J_{35} or J_{50} Asymmetry

The reliability of using J_{35} or J_{50} asymmetry to represent the overall humeral asymmetry was tested using several statistical methods. First, a multivariate regression model was built on all specimens to statistically test the degree of correlation between overall SMA asymmetry and J asymmetry. When fitting the model, the SMA asymmetry values at all landmarks were set as the dependent variables, and the J_{35} or J_{50} asymmetry value as the independent variable. Customized in-house scripts, mainly sourced from R package 'geomorph' and 'RRPP', were utilized to carry out this fitting [53,54]. Second, to investigate the association of every SMA asymmetry value and J asymmetry value across the entire humeral diaphysis, the correlation coefficients between each SMA asymmetry value and J_{35} or J_{50} asymmetry value (CC_{35} and CC_{50}) were calculated within sub-groups. The same protocols for visualizing SMA asymmetry values were applied to CC results to generate morphometric maps. The CC morphometric maps of sub-groups were qualitatively compared to reveal inter-group variations.

3. Results

3.1. Pattern of Humeral Biomechanical Asymmetry in Modern Humans

The mean morphometric maps exhibiting SMA asymmetry values for each sub-group and pooled samples are presented in Figure 2. Hubei females and males are more asymmetrical in the near-anterolateral posteromedial aspect along the entire proximodistal diaphysis. The degree of asymmetry is transversely uniform around the mid-distal diaphysis for Hubei females, and around the midshaft for Hubei males. Hubei males have higher anteroposterior asymmetry from the proximal to mid-proximal diaphysis. Henan females have a restricted area of relatively higher anteroposterior asymmetry around the mid-proximal diaphysis, while Henan males are more asymmetrical in the near-anterolateral posteromedial aspect spanning the mid-proximal to distal diaphysis. Both Xinjiang females and males have reinforced anteroposterior asymmetrical areas around the proximal diaphysis, as well as the region between the proximal to mid-proximal diaphysis, mediolaterally. The region with a relatively higher asymmetry of Xinjiang males extends from the midshaft to the distal diaphysis in the near-anterolateral posteromedial aspect.

For the mean morphometric maps that are defined only by population, Hubei is more asymmetrical across the entire proximodistal diaphysis in the near-anterolateral posteromedial aspect, with a reinforcement of anteroposterior asymmetry along the proximal to mid-proximal diaphysis. The region with high asymmetry for Henan is located in the anterolateral posteromedial aspect between the mid-proximal to distal diaphysis. Xinjiang has higher anteroposterior asymmetry around the proximal diaphysis, connecting with another area with high mediolateral asymmetry around the mid-proximal diaphysis, which continuously extends to the midshaft in the anterolateral posteromedial aspect. Hubei and Xinjiang are more asymmetrical than Henan, according to their overall magnitude of SMA asymmetry values. For the mean morphometric maps that are defined only by sex, females are more anteroposteriorly asymmetrical between the proximal and mid-distal diaphysis, with a reinforcement of asymmetry near the mid-proximal section. The distribution patterns of males resemble that of Xinjiang, but the regions with highest asymmetry at the proximal and mid-proximal diaphysis are not so prominent, and the region with relatively higher asymmetry along the distal half of the diaphysis in the anterolateral posteromedial aspect is more developed. Males are more asymmetrical than females in general. The mean morphometric map for pooled samples shows uniform areas of asymmetry spanning from the proximal diaphysis, anteroposteriorly, to the mid-proximal diaphysis, mediolaterally, and continuing distally in the anterolateral posteromedial aspect.

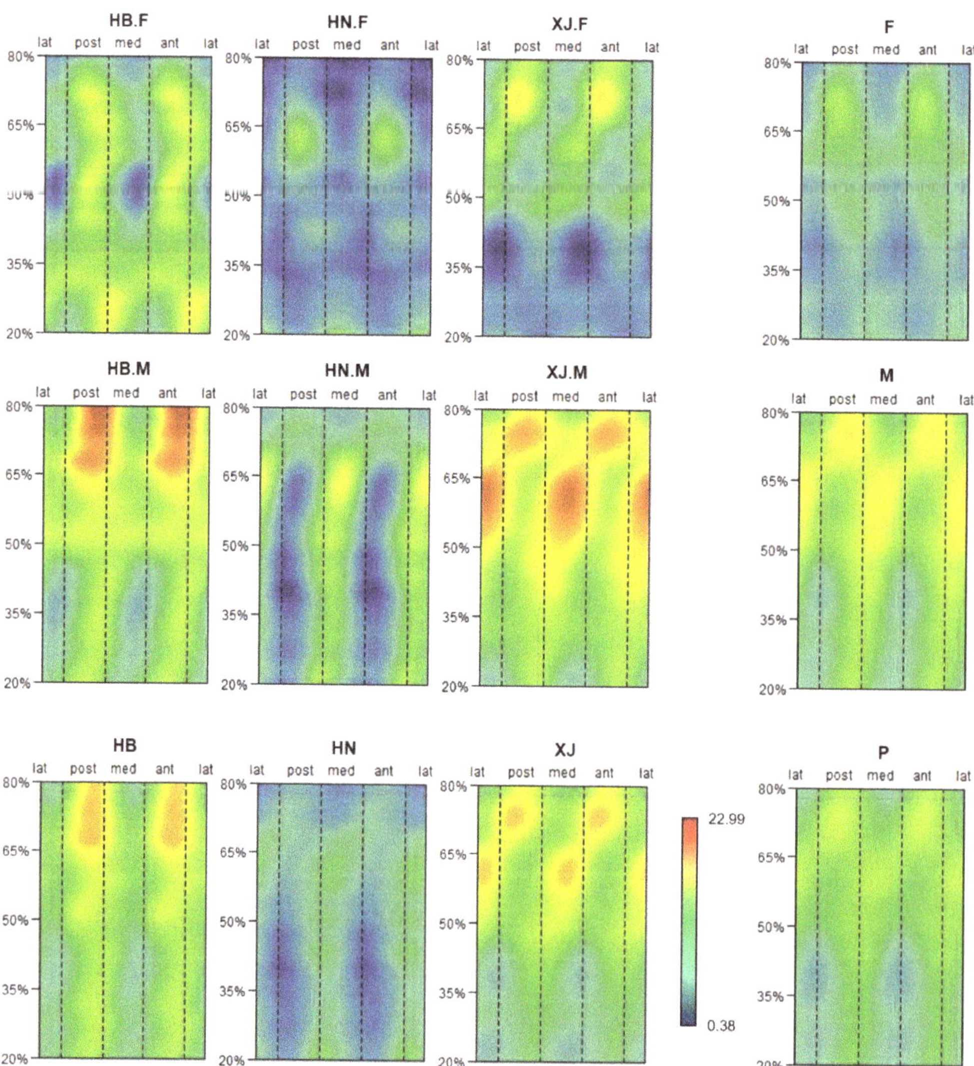

Figure 2. Mean morphometric maps exhibiting SMA asymmetry values for sub-groups and pooled samples (**P**). Sub-groups are defined by population, sex, and the pairwise combination of these two factors. Populations include Hubei (**HB**), Henan (**HN**), and Xinjiang (**XJ**); sexes include female (**F**) and male (**M**). All mean morphometric maps are under the same chromatic scale.

According to the results of MANOVA (Table 1), the differences sourced from sex ($P = 0.11$), population ($P = 0.296$), and the interaction of sex and population ($P = 0.783$) are not statistically significant. The R-squared values reveal that sex, population, and the interaction of sex and population accounted for 5.49%, 5.99%, and 2.74% of the total variation, respectively. Residuals accounted for 85.77% of the total variation.

Table 1. MANOVA results interpreting the differences between sexes and among populations.

	Df	SS	MS	Rsq	F	Z	P (>F)
Sex	1	5,309,415	5,309,415	0.05494	2.1778	1.31140	0.110
Population	2	5,792,218	2,896,109	0.05993	1.1879	0.58014	0.296
Sex:Population	2	2,652,077	1,326,039	0.02744	0.5439	−0.78875	0.783
Residuals	34	82,891,403	2,437,982	0.85769			
Total	39	96,645,113					

Df: degree of freedom; SS: sums of squares; MS: mean squares; Rsq: R-squared values.

The CV morphometric maps show nearly identical distribution patterns across all the sub-groups and pooled samples (Figure 3). Relatively high CV values are concentrated in the region between the middle and mid-distal diaphysis, and at the distal extreme in the anteromedial posterolateral aspect. Similarly high CV values appear at the proximal section, mediolaterally, but to a smaller extent compared to the distal section. Henan has localized regions of higher CV values at the proximal extreme, mediolaterally, and at the mid-distal diaphysis in the anteromedial posterolateral aspect, but displays no other differences compared to Hebei and Xinjiang. Females present higher overall CV values than males.

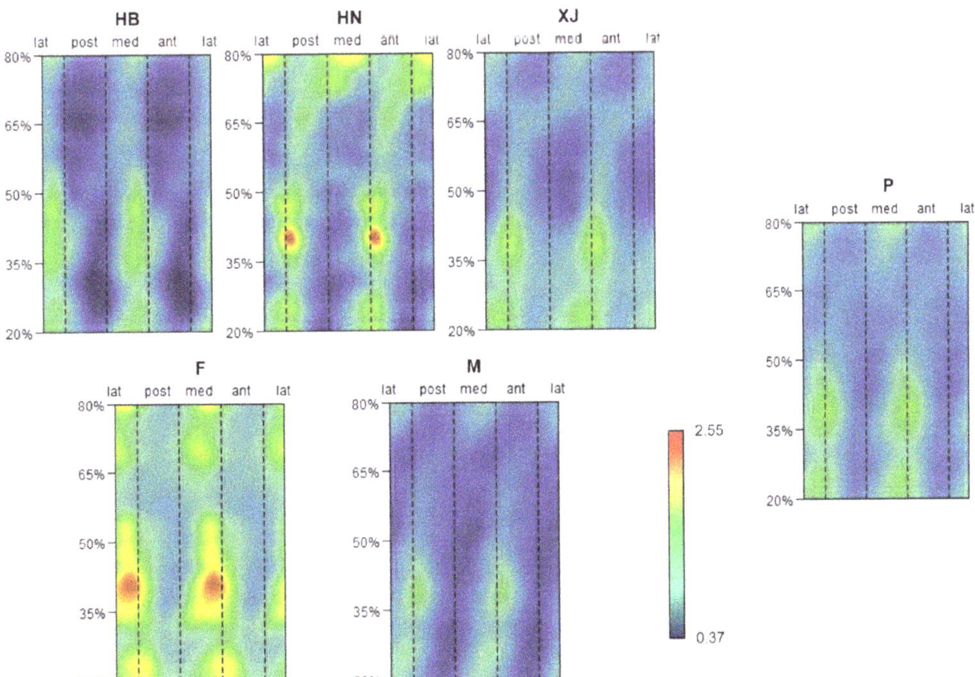

Figure 3. Morphometric maps exhibiting the coefficient of variation (CV) for SMA asymmetry values in sub-groups and pooled samples (**P**). Sub-groups are defined by population and sex. Populations include Hubei (**HB**), Henan (**HN**), and Xinjiang (**XJ**); sexes include female (**F**) and male (**M**). All CV morphometric maps are under the same chromatic scale.

3.2. Representativeness of J_{35} and J_{50} Bilateral Asymmetry

Table 2 and Figure 4 show the result of a multivariate regression fitting all the SMA asymmetry values on the J_{35} or J_{50} asymmetry value using pooled samples. The results of J_{35} and J_{50} asymmetry are highly significant ($P < 0.001$), indicating that the multivariate regression model is effective. According to the R-squared values, J_{35} asymmetry accounts for 48.66% of the total variation, whereas J_{50} asymmetry accounts for 50.93%. The remaining variations are explained by the residuals, which is 51.34% in the J_{35} asymmetry model and 49.07% in the J_{50} asymmetry model.

Table 2. Multivariate regression of all SMA asymmetry values on J asymmetry value.

	Df	SS	MS	Rsq	F	Z	P (>F)
J_{35} asymmetry	1	47,026,558	47,026,558	0.48659	36.015	3.5479	0.001 **
Residuals	38	49,618,555	1,305,751	0.51341			
Total	39	96,645,113					
J_{50} asymmetry	1	49,218,068	49,218,068	0.50927	39.435	3.7473	0.001 **
Residuals	38	47,427,046	1,248,080	0.49073			
Total	39	96,645,113					

Df: degree of freedom; SS: sums of squares; MS: mean squares; Rsq: R-squared values; **: statistically highly significant.

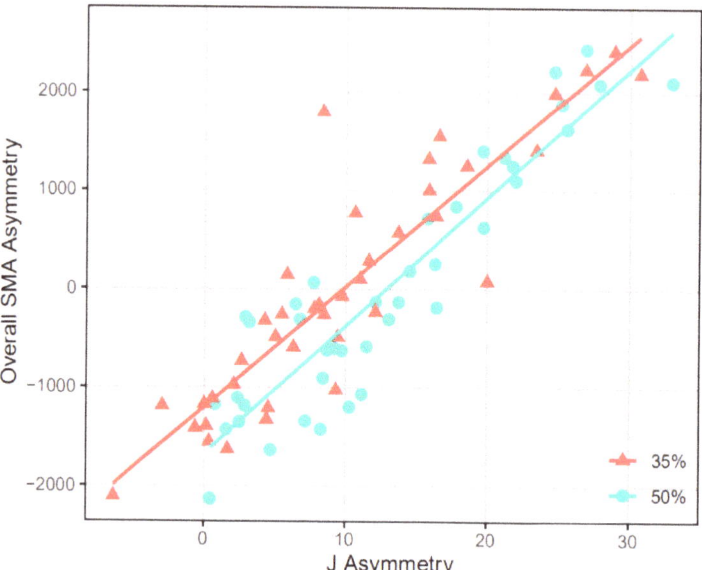

Figure 4. Multivariate regression of all SMA asymmetry values on J asymmetry value at the 35% or 50% cross-section using pooled samples.

The CC morphometric maps of the sub-groups and pooled samples are shown in Figure 5. Across all the CC morphometric maps, the SMA asymmetry values and the J_{35} or J_{50} asymmetry value are positively correlated among the entire humeral diaphysis, except for some areas of Henan. When specific to the morphometric maps of CC_{35}, high CC_{35} values are detected primarily among the distal half of the diaphysis, particularly around the mid-distal to distal section, while lower CC_{35} values are more inclined to distribute anteroposteriorly over the proximal half of the diaphysis. Henan differs from the other sub-groups in that its SMA asymmetry values are negatively correlated with the J_{35} asymmetry value in the region between the mid-proximal and middle diaphysis, anteroposteriorly. For the morphometric maps of CC_{50}, high CC_{50} values are found between the proximal and middle diaphysis, anteroposteriorly, which gradually shift in the anterolateral posteromedial aspect, from the middle to distal diaphysis. Comparatively, low CC_{50} values tend to follow the approximately anterolateral posteromedial aspect between the mid-distal and distal diaphysis. In comparison to other sub-groups, Henan exhibits a distinct distribution pattern of CC_{50} values at the distal humeral section, mediolaterally, with the SMA asymmetry values being negatively correlated with the J_{50} asymmetry value.

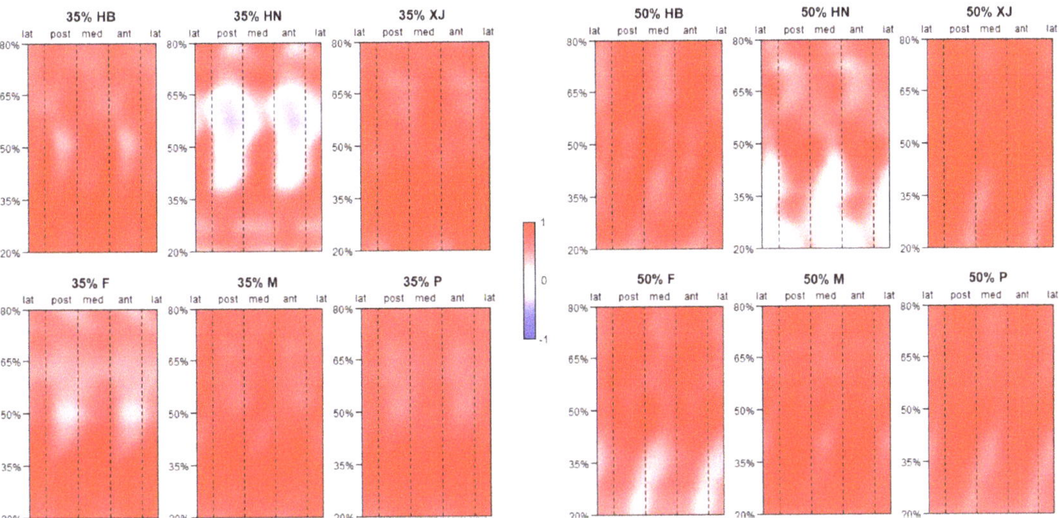

Figure 5. Morphometric maps exhibiting the correlation coefficient (CC) between SMA asymmetry values and J asymmetry value at the 35% or 50% cross-section in sub-groups and pooled samples (**P**). Sub-groups are defined by population and sex. Populations include Hubei (**HB**), Henan (**HN**), and Xinjiang (**XJ**); sexes include female (**F**) and male (**M**). All CC morphometric maps are under the same chromatic scale.

4. Discussion

The objective of this study was to reveal the humeral asymmetry patterns of East Asian modern humans with diverse backgrounds, by evaluating the biomechanical performance across complete humeral diaphysis rather than individual cross-sections only, as well as to identify the reliability of torsional rigidity at the 35% and 50% cross-sections (J_{35} and J_{50}) in bilateral asymmetry analysis.

By quantifying the overall bending rigidity asymmetry of humeral proximodistal diaphysis using morphometric mapping of SMA asymmetry values, the variation range and pattern of humeral asymmetry in East Asian modern humans represented by our samples were investigated. In all the sub-groups, male humeri are more asymmetrical than female humeri. The Henan population has lower humeral asymmetry overall compared to the Hubei and Xinjiang populations. Although three populations show unique distributions of bending rigidity asymmetry, the inter-group differences are not significant in MANOVA. This suggests that, at least for the samples used in this study, the behavioral differences among different populations and between different sexes are not significant enough to generate discernable differences in bilateral asymmetry. The relatively small sample size of the present study might be a factor in this result. Future studies with larger sample sizes and populations from more varied backgrounds may reveal significant differences.

Overall, the mean morphometric maps of most the sub-groups and pooled samples show the following common distribution pattern: the asymmetry of the proximal section is reinforced anteroposteriorly, connecting it to another relatively asymmetrical area between the mid-proximal and middle diaphysis, mediolaterally, and finally extending to the distal end in the anterolateral posteromedial aspect. Previous research found that humeral asymmetry was most prominent at the midshaft and decreased towards both the proximal and distal diaphyseal ends, and this pattern can be attributed to the general mechanical model that bending loads should be the greatest at mid-diaphyseal regions [55]. However, as revealed in the present study, the proximal to middle diaphysis tends to have a higher asymmetrical level than the distal half, and the differences tend to be more prominent among different anatomical directions than between different sections along the humeral diaphysis. This asymmetry pattern emphasizes the necessity of examining multiple anatomical directions when analyzing bilateral asymmetry, and suggests that the mechanism regulating the response of the long bone to external stimuli might be more complicated than previously understood.

As some highly asymmetrical regions correspond with the positions of major muscle attachments, such as deltoid tuberosity and the crest of the greater tubercle [56], the distribution of areas with reinforced asymmetry might reflect adaptions to muscle loadings, which were proved to be an important determinant of upper-limb strength [57–59]. In our study, factors such as genetic regulation and health condition can be excluded from the elements influencing the bilateral asymmetry because the analysis was based on paired humeri from the same individual. However, more experimental evidences are needed to verify this hypothesis in future studies.

According to the results of the CV morphometric maps, the variability in bilateral asymmetry is not consistent across the humeral diaphysis. Highly variable regions are restricted to the distal half of humeral diaphysis in the anteromedial posterolateral aspect, corresponding to the medial/lateral border and medial/lateral supracondylar. Since this feature is shared by all the sub-groups as well as the pooled data, it may represent a generality of East Asian modern humans. It is noteworthy that highly variable regions on the humeral diaphysis tend to overlap with areas presenting a low asymmetrical level, which may be a signal of relative insensitivity to lateralized mechanical stimuli (see previous paragraph). Previous studies found that humeral distal articular properties, such as articular surface area, did not just respond to mechanical loadings, but were also ontogenetically constrained and genetically canalized [60]. As the structure of the medial/lateral border and medial/lateral supracondylar are closely related to the distal articular morphology, according to their anatomical adjacency [56], one possible interpretation for the high variability of asymmetry is that these regions might present fluctuating asymmetry that is attributable to genetic, nutrient, and health factors instead of the mechanical environment alone [60–62].

This study supports the previous perspective that torsional rigidity at a specific cross-section (35% or 50% of the humeral biomechanical length) can be used to indicate the overall biomechanical asymmetry of humeral diaphysis, because the multivariate regression model built on all the specimens is effective, and a positive correlation exists between the SMA asymmetry and J asymmetry at most diaphyseal locations. However, we should also note that a single J asymmetry value cannot convey the complexity of the entire humerus' asymmetry. The correlation between overall SMA asymmetry and J asymmetry is moderate, because J_{35} and J_{50} asymmetry can only explain about half of the total variation in humeral bilateral asymmetry. In addition, the degree of correlation between SMA asymmetry and J asymmetry varies across the humeral diaphysis, and is only strong in specific regions.

5. Conclusions

This study evaluated humeral biomechanical asymmetry across complete humeral diaphysis based on high-resolution micro-CT, and by quantifiable visualization and statistical methods. Using specimens from three Chinese archaeological populations that varied in geographic location, chronological age, and subsistence pattern, the pattern of humeral asymmetry in East Asian modern humans was investigated. Distinct humeral asymmetry patterns are observed on the mean morphometric map, but are not statistically significant. Analogous distributions of highly asymmetrical regions and CV are observed across nearly all the sexes and populations, indicating possible universality of the humeral asymmetry pattern in East Asian modern humans. These highly asymmetrical regions correspond with major muscle attachments. The diaphyseal regions that are highly varied in bilateral asymmetry tend to present a low asymmetrical level. Although J_{35} and J_{50} asymmetry are related to the overall humeral asymmetry, it can only explain about half of the total variation. These findings suggest that the overall biomechanical asymmetry of humeral diaphysis is more complicated than previously assumed. This study complements previous findings on humeral asymmetry, and accumulate data and knowledge for future works in this area.

Author Contributions: Conceptualization, Y.Z.; methodology, Y.Z.; software, Y.Z.; validation, Y.Z., S.X.; formal analysis, Y.Z.; investigation, Y.Z.; resources, M.Z., H.L., J.H.; data curation, Y.Z., S.X.; writing—original draft preparation, Y.Z.; writing—review and editing, P.W., S.X.; visualization, Y.Z.; supervision, S.X.; project administration, S.X.; funding acquisition, H.L., S.X. All authors have read and agreed to the published version of the manuscript.

Funding: This research was funded by the Strategic Priority Research Program of the Chinese Academy of Sciences (XDB26000000), the National Natural Science Foundation of China (41872030), and the National Social Science Fund of China (19BKG039, 19VJX066).

Institutional Review Board Statement: Not applicable.

Informed Consent Statement: Not applicable.

Data Availability Statement: The data supporting this study are available from the corresponding author on reasonable request.

Acknowledgments: The authors thank Yemao Hou, Pengfei Yin, and Jiawei Ma for their help in scanning and image processing. The authors also thank Xiujie Wu and Mackie O'Hara Ali for optimizing the manuscript.

Conflicts of Interest: The authors declare no conflict of interest. The funders had no role in the design of the study; in the collection, analyses, or interpretation of data; in the writing of the manuscript, or in the decision to publish the results.

Appendix A

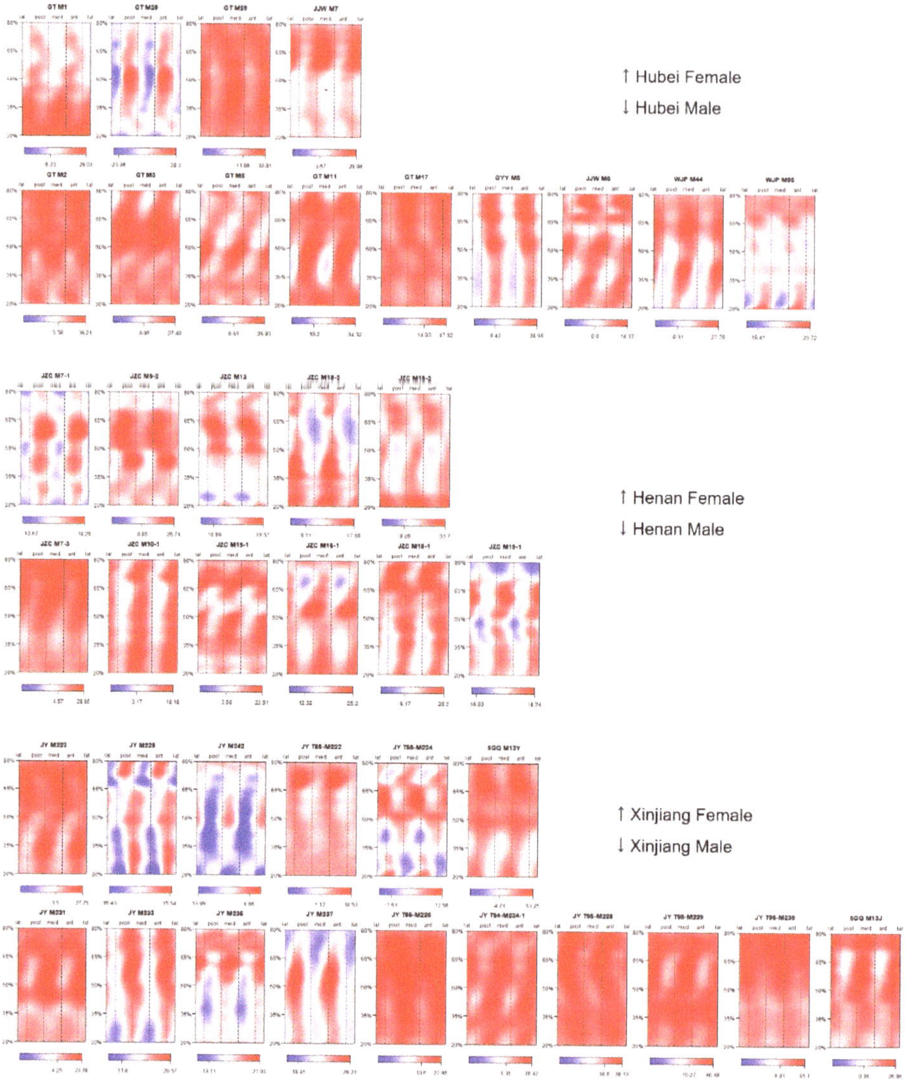

Figure A1. Morphometric maps exhibiting SMA asymmetry values of all specimens.

References

1. DeLude, J.A.; Bicknell, R.T.; MacKenzie, G.A.; Ferreira, L.M.; Dunning, C.E.; King, G.J.W.; Johnson, J.A.; Drosdowech, D.S. An anthropometric study of the bilateral anatomy of the humerus. *J. Shoulder Elbow Surg.* **2007**, *16*, 477–483. [CrossRef] [PubMed]
2. Ubelaker, D.H.; Zarenko, K.M. Can handedness be determined from skeletal remains? A chronological review of the literature. *J. Forensic Sci.* **2012**, *57*, 1421–1426. [CrossRef] [PubMed]
3. Ruff, C.B. Biomechanical analyses of archaeological human skeletons. In *Biological Anthropology of the Human Skeleton*, 3rd ed.; Katzenberg, M.A., Grauer, A.L., Eds.; Wiley-Blackwell: Hoboken, NJ, USA, 2018; pp. 189–224.
4. Faurie, C.; Raymond, M. Handedness, homicide and negative frequency-dependent selection. *Proc. R. Soc. B* **2005**, *272*, 25–28. [CrossRef] [PubMed]

5. Shaw, C.N. Is 'hand preference' coded in the hominin skeleton? An in-vivo study of bilateral morphological variation. *J. Hum. Evol.* **2011**, *61*, 480–487. [CrossRef] [PubMed]
6. Ozener, B. Extreme behavioral lateralization and the remodeling of the distal humerus. *Am. J. Hum. Biol.* **2012**, *24*, 436–440. [CrossRef] [PubMed]
7. Haapasalo, H.; Kannus, P.; Sievänen, H.; Pasanen, M.; Uusi-Rasi, K.; Heinonen, A.; Oja, P.; Vuori, I. Effect of long-term unilateral activity on bone mineral density of female junior tennis players. *J. Bone Miner. Res.* **1998**, *13*, 310–319. [CrossRef] [PubMed]
8. Bass, S.L.; Saxon, L.; Daly, R.M.; Turner, C.H.; Robling, A.G.; Seeman, E.; Stuckey, S. The effect of mechanical loading on the size and shape of bone in pre-, peri-, and postpubertal girls: A study in tennis players. *J. Bone Miner. Res.* **2002**, *17*, 2274–2280. [CrossRef] [PubMed]
9. Shaw, C.N.; Stock, J.T. Habitual throwing and swimming correspond with upper limb diaphyseal strength and shape in modern human athletes. *Am. J. Phys. Anthropol.* **2009**, *140*, 160–172. [CrossRef]
10. Balzeau, A.; Ball-Albessard, L.; Kubicka, A.M. Variation and correlations in departures from symmetry of brain torque, humeral morphology and handedness in an archaeological sample of Homo sapiens. *Symmetry* **2020**, *12*, 432. [CrossRef]
11. Shackelford, L.L. Regional variation in the postcranial robusticity of late upper paleolithic humans. *Am. J. Phys. Anthropol.* **2007**, *133*, 655–668. [CrossRef] [PubMed]
12. Sparacello, V.S.; d'Ercole, V.; Coppa, A. A bioarchaeological approach to the reconstruction of changes in military organization among Iron Age Samnites (Vestini) From Abruzzo, Central Italy. *Am. J. Phys. Anthropol.* **2015**, *156*, 305–316. [CrossRef] [PubMed]
13. Sládek, V.; Ruff, C.B.; Berner, M.; Holt, B.; Niskanen, M.; Schuplerová, E.; Hora, M. The impact of subsistence changes on humeral bilateral asymmetry in Terminal Pleistocene and Holocene Europe. *J. Hum. Evol.* **2016**, *92*, 37–49. [CrossRef] [PubMed]
14. Sparacello, V.S.; Villotte, S.; Shackelford, L.L.; Trinkaus, E. Patterns of humeral asymmetry among Late Pleistocene humans. *Comptes Rendus Palevol* **2017**, *16*, 680–689. [CrossRef]
15. Wei, P.; Lu, H.; Carlson, K.J.; Zhang, H.; Hui, J.; Zhu, M.; He, K.; Jashashvili, T.; Zhang, X.; Yuan, H.; et al. The upper limb skeleton and behavioral lateralization of modern humans from Zhaoguo Cave, southwestern China. *Am. J. Phys. Anthropol.* **2020**, *173*, 671–696. [CrossRef] [PubMed]
16. Wei, P.; Zhao, Y.; Walker, C.S.; He, J.; Lu, X.; Hui, J.; Shui, W.; Jin, L.; Liu, W. Internal structural properties of the humeral diaphyses in an early modern human from Tianyuan Cave, China. *Quat. Int.* **2021**, *591*, 107–118. [CrossRef]
17. Hsieh, Y.-F.; Robling, A.G.; Ambrosius, W.T.; Burr, D.B.; Turner, C.H. Mechanical loading of diaphyseal bone in vivo: The strain threshold for an osteogenic response varies with location. *J. Bone Miner. Res.* **2001**, *16*, 2291–2297. [CrossRef] [PubMed]
18. Burr, D.B.; Robling, A.G.; Turner, C.H. Effects of biomechanical stress on bones in animals. *Bone* **2002**, *30*, 781–786. [CrossRef]
19. Pearson, O.M.; Lieberman, D.E. The aging of Wolff's "law": Ontogeny and responses to mechanical loading in cortical bone. *Am. J. Phys. Anthropol.* **2004**, *125*, 63–99. [CrossRef] [PubMed]
20. Ruff, C.; Holt, B.; Trinkaus, E. Who's afraid of the big bad wolff? "Wolff is law" and bone functional adaptation. *Am. J. Phys. Anthropol.* **2006**, *129*, 484–498. [CrossRef] [PubMed]
21. Gosman, J.H.; Hubbell, Z.R.; Shaw, C.N.; Ryan, T.M. Development of cortical bone geometry in the human femoral and tibial diaphysis. *Anat. Rec.* **2013**, *296*, 774–787. [CrossRef]
22. Sládek, V.; Berner, M.; Holt, B.; Niskanen, M.; Ruff, C.B. Past human manipulative behavior in the European Holocene as assessed through upper limb asymmetry. In *Skeletal Variation and Adaptation in Europeans*; Wiley-Blackwell: Hoboken, NJ, USA, 2018; pp. 163–208.
23. Perchalski, B.; Placke, A.; Sukhdeo, S.M.; Shaw, C.N.; Gosman, J.H.; Raichlen, D.A.; Ryan, T.M. Asymmetry in the cortical and trabecular bone of the human humerus during development. *Anat. Rec.* **2018**, *301*, 1012–1025. [CrossRef] [PubMed]
24. Hong, E.; Kwak, D.-S.; Kim, I.-B. Morphometric evaluation of detailed asymmetry for the proximal humerus in Korean population. *Symmetry* **2021**, *13*, 862. [CrossRef]
25. Ruff, C.B.; Burgess, M.L.; Bromage, T.G.; Mudakikwa, A.; McFarlin, S.C. Ontogenetic changes in limb bone structural proportions in mountain gorillas (Gorilla beringei beringei). *J. Hum. Evol.* **2013**, *65*, 693–703. [CrossRef] [PubMed]
26. Ruff, C.B. Long bone articular and diaphyseal structure in old world monkeys and apes. I: Locomotor effects. *Am. J. Phys. Anthropol.* **2002**, *119*, 305–342. [CrossRef] [PubMed]
27. Weiss, E. Humeral cross-sectional morphology from 18th century Quebec prisoners of war: Limits to activity reconstruction. *Am. J. Phys. Anthropol.* **2005**, *126*, 311–317. [CrossRef] [PubMed]
28. Wescott, D.J.; Cunningham, D.L. Temporal changes in Arikara humeral and femoral cross-sectional geometry associated with horticultural intensification. *J. Archaeol. Sci.* **2006**, *33*, 1022–1036. [CrossRef]
29. Sládek, V.; Berner, M.; Sosna, D.; Sailer, R. Human manipulative behavior in the Central European Late Eneolithic and Early Bronze Age: Humeral bilateral asymmetry. *Am. J. Phys. Anthropol.* **2007**, *133*, 669–681. [CrossRef] [PubMed]
30. Rhodes, J.A.; Knüsel, C.J. Activity-related skeletal change in medieval humeri: Cross-sectional and architectural alterations. *Am. J. Phys. Anthropol.* **2005**, *128*, 536–546. [CrossRef] [PubMed]
31. Hill, E.C.; Pearson, O.M.; Durband, A.C.; Walshe, K.; Carlson, K.J.; Grine, F.E. An examination of the cross-sectional geometrical properties of the long bone diaphyses of Holocene foragers from Roonka, South Australia. *Am. J. Phys. Anthropol.* **2020**, *172*, 682–697. [CrossRef] [PubMed]

32. Cowgill, L.W.; Mednikova, M.B.; Buzhilova, A.P.; Trinkaus, E. The Sunghir 3 Upper Paleolithic juvenile: Pathology versus persistence in the Paleolithic. *Int. J. Osteoarchaeol.* **2015**, *25*, 176–187. [CrossRef]

33. Shaw, C.N.; Stock, J.T. Extreme mobility in the Late Pleistocene? Comparing limb biomechanics among fossil Homo, varsity athletes and Holocene foragers. *J. Hum. Evol.* **2013**, *64*, 242–249. [CrossRef] [PubMed]

34. Trinkaus, E. Epipaleolithic human appendicular remains from Ein Gev I, Israel. *Comptes Rendus Palevol* **2018**, *17*, 616–627. [CrossRef]

35. Kubicka, A.M.; Nowaczewska, W.; Balzeau, A.; Piontek, J. Bilateral asymmetry of the humerus in Neandertals, Australian aborigines and medieval humans. *Am. J. Phys. Anthropol.* **2018**, *167*, 46–60. [CrossRef]

36. Ogilvie, M.D.; Hilton, C.E. Cross-sectional geometry in the humeri of foragers and farmers from the prehispanic American Southwest: Exploring patterns in the sexual division of labor. *Am. J. Phys. Anthropol.* **2011**, *144*, 11–21. [CrossRef] [PubMed]

37. Sparacello, V.S.; Pearson, O.M.; Coppa, A.; Marchi, D. Changes in skeletal robusticity in an iron age agropastoral group: The samnites from the Alfedena necropolis (Abruzzo, Central Italy). *Am. J. Phys. Anthropol.* **2011**, *144*, 119–130. [CrossRef] [PubMed]

38. Warden, S.J.; Mantila Roosa, S.M.; Kersh, M.E.; Hurd, A.L.; Fleisig, G.S.; Pandy, M.G.; Fuchs, R.K. Physical activity when young provides lifelong benefits to cortical bone size and strength in men. *Proc. Natl. Acad. Sci. USA* **2014**, *111*, 5337–5342. [CrossRef] [PubMed]

39. Wilson, L.A.B.; Humphrey, L.T. A Virtual geometric morphometric approach to the quantification of long bone bilateral asymmetry and cross-sectional shape. *Am. J. Phys. Anthropol.* **2015**, *158*, 541–556. [CrossRef]

40. Bondioli, L.; Bayle, P.; Dean, C.; Mazurier, A.; Puymerail, L.; Ruff, C.; Stock, J.T.; Volpato, V.; Zanolli, C.; Macchiarelli, R. Technical note: Morphometric maps of long bone shafts and dental roots for imaging topographic thickness variation. *Am. J. Phys. Anthropol.* **2010**, *142*, 328–334. [CrossRef]

41. Zhao, Y.; Zhou, M.; Wei, P.; Xing, S. 2D visualization and quantitative analysis of the humeral diaphysis cortical thickness. *Acta Anthropol. Sin.* **2020**, *39*, 632–647. [CrossRef]

42. Puymerail, L.; Ruff, C.B.; Bondioli, L.; Widianto, H.; Trinkaus, E.; Macchiarelli, R. Structural analysis of the Kresna 11 *Homo erectus* femoral shaft (Sangiran, Java). *J. Hum. Evol.* **2012**, *63*, 741–749. [CrossRef] [PubMed]

43. Puymerail, L.; Volpato, V.; Debénath, A.; Mazurier, A.; Tournepiche, J.-F.; Macchiarelli, R. A Neanderthal partial femoral diaphysis from the "grotte de la Tour", La Chaise-de-Vouthon (Charente, France): Outer morphology and endostructural organization. *Comptes Rendus Palevol* **2012**, *11*, 581–593. [CrossRef]

44. Morimoto, N.; Ponce de León, M.S.; Zollikofer, C.P.E. Exploring femoral diaphyseal shape variation in wild and captive chimpanzees by means of morphometric mapping: A test of Wolff's law. *Anat. Rec.* **2011**, *294*, 589–609. [CrossRef] [PubMed]

45. Jashashvili, T.; Dowdeswell, M.R.; Lebrun, R.; Carlson, K.J. Cortical structure of hallucal metatarsals and locomotor adaptations in hominoids. *PLoS ONE* **2015**, *10*, e0117905. [CrossRef]

46. Cameron, M.E.; Pfeiffer, S. Long bone cross-sectional geometric properties of Later Stone Age foragers and herder-foragers. *S. Afr. J. Sci.* **2014**, *110*, 01–11. [CrossRef]

47. Tian, J.; Tang, L.; Shi, D.; Luo, Y.; Zhao, Z. Research on the charred plant remains unearthed from the Jijiawan site in Fang County, Hubei Province. *Cult. Relics South. China* **2019**, *5*, 180–188. (In Chinese) [CrossRef]

48. Xing, S.; Carlson, K.J.; Wei, P.; He, J.; Liu, W. Morphology and structure of Homo erectus humeri from Zhoukoudian, Locality 1. *PeerJ* **2018**, *6*, e4279. [CrossRef] [PubMed]

49. Wang, L.; Xiao, G.; Liu, Z.; Lv, E.; Wu, Y. Brief report on excavation of Jiayi cemeteries in Turpan, Xinjiang. *Turfanological Res.* **2014**, *1*, 1–19. [CrossRef]

50. Xing, K.; Zhang, Y.; Li, Z. 新疆鄯普三个桥墓葬发掘简报. *Cult. Relics* **2002**, *6*, 46–56. (In Chinese). Available online: http://www.cqvip.com/qk/97337x/2002006/11400528.html (accessed on 27 August 2021).

51. Xiao, G. The Review and Study of Jia Yi Cemetery in Turfan, Xinjiang. Master's Thesis, Northwest University, Xi'an, China, 2018.

52. Profico, A.; Bondioli, L.; Raia, P.; O'Higgins, P.; Marchi, D. morphomap: An R package for long bone landmarking, cortical thickness, and cross-sectional geometry mapping. *Am. J. Phys. Anthropol.* **2021**, *174*, 129–139. [CrossRef]

53. Collyer, M.L.; Adams, D.C. RRPP: An r package for fitting linear models to high-dimensional data using residual randomization. *Methods Ecol. Evol.* **2018**, *9*, 1772–1779. [CrossRef]

54. Adams, D.C.; Otárola-Castillo, E. geomorph: An r package for the collection and analysis of geometric morphometric shape data. *Methods Ecol. Evol.* **2013**, *4*, 393–399. [CrossRef]

55. Zelazny, K.G.; Sylvester, A.D.; Ruff, C.B. Bilateral asymmetry and developmental plasticity of the humerus in modern humans. *Am. J. Phys. Anthropol.* **2021**, *174*, 418–433. [CrossRef] [PubMed]

56. White, T.D.; Black, M.T.; Folkens, P.A. Arm: Humerus, Radius, and Ulna. In *Human Osteology*, 3rd ed.; Academic Press: San Diego, CA, USA, 2012; pp. 175–198.

57. Ruff, C. Growth in bone strength, body size, and muscle size in a juvenile longitudinal sample. *Bone* **2003**, *33*, 317–329. [CrossRef]

58. Schoenau, E.; Neu, C.M.; Mokov, E.; Wassmer, G.; Manz, F. Influence of puberty on muscle area and cortical bone area of the forearm in boys and girls. *J. Clin. Endocrinol. Metab.* **2000**, *85*, 1095–1098. [CrossRef] [PubMed]

59. Niinimäki, S. The relationship between musculoskeletal stress markers and biomechanical properties of the humeral diaphysis. *Am. J. Phys. Anthropol.* **2012**, *147*, 618–628. [CrossRef] [PubMed]

60. Lieberman, D.E.; Devlin, M.J.; Pearson, O.M. Articular area responses to mechanical loading: Effects of exercise, age, and skeletal location. *Am. J. Phys. Anthropol.* **2001**, *116*, 266–277. [CrossRef] [PubMed]

61. Plochocki, J.H. Bilateral variation in limb articular surface dimensions. *Am. J. Hum. Biol.* **2004**, *16*, 328–333. [CrossRef] [PubMed]
62. Hallgrímsson, B.; Willmore, K.; Hall, B.K. Canalization, developmental stability, and morphological integration in primate limbs. *Am. J. Phys. Anthropol.* **2002**, *119*, 131–158. [CrossRef]

 *symmetry*

Article

Get a Grip: Variation in Human Hand Grip Strength and Implications for Human Evolution

Ameline Bardo [1,*], Tracy L. Kivell [1,2], Katie Town [1], Georgina Donati [3], Haiko Ballieux [4], Cosmin Stamate [5], Trudi Edginton [6] and Gillian S. Forrester [7]

[1] Skeletal Biology Research Centre, School of Anthropology and Conservation, University of Kent, Canterbury CT2 7NR, UK; t.l.kivell@kent.ac.uk (T.L.K.); kt299@kent.ac.uk (K.T.)
[2] Department of Human Evolution, Max Planck Institute for Evolutionary Anthropology, D-04103 Leipzig, Germany
[3] Department of Psychiatry, Warneford Hospital, University of Oxford, Oxford OX3 7JX, UK; georgina.donati@psych.ox.ac.uk
[4] Psychology, School of Social Sciences, University of Westminster, London W1W 6UW, UK; H.Ballieux@westminster.ac.uk
[5] Department of Computer Science, Birkbeck, University of London, London WC1E 7HX, UK; cosmin@dcs.bbk.ac.uk
[6] Department of Psychology, City University of London, London EC1V 0HB, UK; Trudi.Edginton@city.ac.uk
[7] Department of Psychological Sciences, Birkbeck, University of London, Malet Street, London W1CE 7HX, UK; g.forrester@bbk.ac.uk
* Correspondence: ameline.bardo@gmail.com

Citation: Bardo, A.; Kivell, T.L.; Town, K.; Donati, G.; Ballieux, H.; Stamate, C.; Edginton, T.; Forrester, G.S. Get a Grip: Variation in Human Hand Grip Strength and Implications for Human Evolution. *Symmetry* **2021**, *13*, 1142. https://doi.org/10.3390/sym13071142

Academic Editor: Antoine Balzeau

Received: 14 May 2021
Accepted: 8 June 2021
Published: 26 June 2021

Publisher's Note: MDPI stays neutral with regard to jurisdictional claims in published maps and institutional affiliations.

Abstract: Although hand grip strength is critical to the daily lives of humans and our arboreal great ape relatives, the human hand has changed in form and function throughout our evolution due to terrestrial bipedalism, tool use, and directional asymmetry (DA) such as handedness. Here we investigate how hand form and function interact in modern humans to gain an insight into our evolutionary past. We measured grip strength in a heterogeneous, cross-sectional sample of human participants (n = 662, 17 to 83 years old) to test the potential effects of age, sex, asymmetry (hand dominance and handedness), hand shape, occupation, and practice of sports and musical instruments that involve the hand(s). We found a significant effect of sex and hand dominance on grip strength, but not of handedness, while hand shape and age had a greater influence on female grip strength. Females were significantly weaker with age, but grip strength in females with large hands was less affected than those with long hands. Frequent engagement in hand sports significantly increased grip strength in the non-dominant hand in both sexes, while only males showed a significant effect of occupation, indicating different patterns of hand dominance asymmetries and hand function. These results improve our understanding of the link between form and function in both hands and offer an insight into the evolution of human laterality and dexterity.

Keywords: power grip strength; directional asymmetry; hand dominance; hand shape; manual activities; human evolution; functional morphology

1. Introduction

The hand is essential to how modern humans interact with their environment, as it was for our extinct relatives. The enhanced dexterity of the human hand is unique among living primates and is generally considered to have evolved through both (1) adaptation to bipedalism and a relaxation of locomotor selective pressures on the hands and (2) increasingly more complex tool production and use in hominins (i.e., group consisting of modern humans and our closely related extinct relatives) [1,2]. The use of stone tools would have allowed early hominins to access different and potentially higher-quality foods (e.g., marrow) [3,4]. The manufacture and use of even relatively simple stone tools, such as Oldowan technology (2.6–1.7 million years ago) [5,6], would have required both increased cognitive function (e.g., learning, working memory/future thinking, planning and decision-making etc.) [7,8] and particular biomechanical demands on the anatomy of the hands [9–11]. Thus, it is likely that tool production and use played a critical role in shaping both cognitive development (e.g., with the crucial role of social learning) [12] and hand morphology. For example, a long, powerful thumb and relatively short fingers facilitate the forceful precision and power-squeeze gripping that are considered to be unique human abilities [1,13]. Although modern humans are also adept as using their hands for locomotion [14,15], the upper limbs are predominantly used for manipulation.

Humans are unique among primates in the strength of population-level hand directional asymmetry (DA) or laterality (i.e., preference for one hand, called the dominant hand, over the other, non-dominant hand), with 85–90% of humans being right-handed regardless of geographical region and ethnicity [16–19]. Non-human primates also show population laterality for object manipulation, but not with the same strength as that found in humans ([20–22], see [23]) and their laterality can also vary depending on the complexity of the manual task (e.g., bimanual manipulative action versus tool use) [24]. Moreover, motor skill biases for tool use in chimpanzees may be supported by anatomically asymmetric, left-biased brain regions analogous to Broca's and Wernicke's area in humans [25], brain regions that are both implicated in the perception and production of speech. Handedness (i.e., side preference for the right or the left hand) in humans is thought to have played an important role in the lateralisation of the human brain for language [26] and the emergence of other complex cognitive functions, including tool use [27,28], manual gestures [29,30], and throwing [31]. Greenfield [32] proposed that it was the motor sequencing for tool use—requiring dexterous hierarchical motor activities—that paved a way for the emergence of language that likely emerged first in the form of hand gestures [33,34]. Thus, more dexterous hands may have increased object manipulation capabilities that, in turn, increased hemispheric specialisation and DA, suggesting that the capacities for tool use and language evolved together [32]. However, when population-level handedness first evolved within the hominin clade remains unclear [35,36].

Hand size, shape [37–40], and bone morphology are also highly variable among recent human populations [41,42]. How this variability potentially affects hand function may provide insights into the evolution of the human hand. For example, ergonomic studies have shown that handle design is important for hand grip performance (e.g., time to complete the task and strength use) [43] and that hand size strongly affects performance [44], indicating the importance of designing tools in accordance with current anthropometric data. Moreover, individuals with relatively longer fingers and therefore larger joint surfaces require less force during stone tool production than those with shorter fingers [45]. Key and Lycett [46] found that through experimental stone tool use, grip strength was the primary contributing biometric factor for stone cutting efficiency. Therefore, both hand shape and hand strength were likely important factors in the efficient stone tool production and use, and would have played an important role in the evolution of hominin cultural technology.

Hand grip strength is commonly measured in a clinical or sports medicine context as an indicator of overall muscle strength [47–51]. Grip strength reflects the gross power of the hand and has been found to be strongly associated with physical activity [52–55], as well as anthropometric traits, such as age and sex [56–58], hand length and shape [59,60], handedness [61], and body mass index [62–64]. For example, males typically have a stronger average grip strength than females [56,65]. In both sexes, hand asymmetry in grip strength was found, with the dominant hand (defined as the hand used most within the context of object manipulation) is approximatively 10% stronger than the non-dominant hand [61], although this difference is more pronounced, and is therefore more of a DA, for right-handed individuals than left-handed individuals [66]. Furthermore, hand size has been shown to be positively correlated with grip strength for both sexes [67–70]. It was also found that hand shape influences grip strength [59,71], such that, for both sexes, people with bigger hands (i.e., large hand length and width) were significantly stronger than people with smaller hands. Moreover, Carlson [72] proposed that, although variation in hand grip strength primarily reflected differences in soft tissue and skeletal morphology, changes in grip strength across the lifespan were also significantly influenced by neural mechanisms (e.g., central nervous system recruiting motoneurons to mediate the control of coordinated movement). Thus, grip strength can be used as a marker of brain health [72]. Indeed, maximum grip strength provides a discriminating measure of cognitive function, such as how central nervous system disorders (e.g., vascular disorders, structural disorders or degeneration) affect the quality of motor coordination [72]. The rate of decline in cognitive function (e.g., motor and perceptual speed, memory, and spatial functioning) has also been shown to correlate with a decline in grip strength, especially towards the end of life [73].

However, most previous studies of grip strength have focused on specific populations [57,74–76], occupations and activities [51,52,54], or sex and age [77–79] with the aim to better understand health, but these same methods may also be useful for understanding the broader scope of form and function from an evolutionary perspective. Although informative, these studies do not fully capture the potential variability in the key factors that can affect grip strength, particularly hand size, shape, and daily use. To broaden our understanding of the link between hand form and function, this study aims to evaluate the variation of grip strength in a heterogeneous and international group of human adults across the lifespan. We test the potential influence of age, sex, asymmetry in hand dominance and preference (i.e., right- vs. left-hand), hand shape, and lifestyle factors (i.e., occupation, practice of sport and music) on grip strength. In this context, we explore hand asymmetry in grip strength as an indicator of brain/manual lateralisation, with hand dominance (i.e., significant difference between the dominant and the non-dominant hand, without taking into account the left-right direction) and DA (i.e., pattern of bilateral variation observed when one side, right or left, is significantly stronger than the other). Based on previous studies, we predict that (1) males will be significantly stronger than females; (2) younger participants will be stronger than the oldest participants; (3) hand asymmetry will be found, with the dominant hand significantly stronger than the non-dominant hand and that this effect will be stronger for right-handed compared with left-handed individuals indicating DA; (4) participants with wider hands (i.e., a hand wider than it is long) will be significantly stronger than those with smaller or longer hands (a hand longer than it is wide); and (5) participants that regularly practice sport, music, or occupational activities that engage their hands will be stronger than those who do not.

2. Materials and Methods

2.1. Sample

The participants were visitors to an interactive, three-month public engagement and citizen science collaboration project called Me, Human (www.mehuman.io/, accessed on 10 February 2021) hosted by Live Science at the London Science Museum between 02/07/2019 and 31/09/2019. Me, Human consisted of a series of experiments exploring motor-sensory behavioural biases and cognitive ability. Experiments included measurements of grip strength, dexterity, cognitive puzzles, and functional brain laterality, in which participants could engage in as many of the experiments as they wanted. Volunteer participants first completed a baseline demographic questionnaire, including date of birth, sex, and handedness for writing (our hand asymmetry indicator), before engaging in the experiment(s). More than 1600 individuals participated in the Me, Human experiments, of which n = 1286 took part in the 'Get a grip' experiment that measured grip strength and hand size, and collected further information about lifestyle and daily activities using the hands. Within this sample, 719 were classified as 'adults' between the ages 17–83 years old because the hand is fully developed (i.e., complete fusion of hand bones) by 17 years of age [80,81]. The remainder of the sample (n = 567) were children and adolescents (6–16 year-olds), and were excluded from this study. Of the total adult sample, 57 participants were removed from the data set due to incomplete data or because they had a recorded hand and/or arm injury within one year of the testing date, making the final sample size n = 662. Of the

662 participants, 89.6% (N = 593) self-reported as right-handed, 9.0% (N = 60) left-handed, and 1.4% (N = 9) as ambidextrous. Participants self-reported as ambidextrous were excluded from subsequent analyses, with the remaining sample containing 653 participants (Table 1; Figure S1). Our sample was divided into 10 age categories (Table 1) of five-year intervals, excluding the first (17–19 years) and last (60 years and older) age categories.

Table 1. Sample used in the analyses with details on the number of participants for each sex by age range and self-reported handedness, type of job, and according to thus practicing a musical instrument and sport.

Age (Years)	Dominance to Write	Total Participants		Forceful Manual Labour		Office Job/Work		Precision Manual Work		Playing a Musical Instrument		Practicing Sport	
		M	F	M	F	M	F	M	F	M	F	M	F
17–19	R	25	42	-	-	25	42	-	-	12	19	20	29
	L	3	5	-	-	3	5	-	-	1	3	2	5
20–24	R	43	56	6	4	32	46	5	6	15	27	36	33
	L	3	2	-	1	3	1	-	-	2	1	2	1
25–29	R	31	30	6	5	17	22	1	6	7	9	20	16
	L	3	4	-	-	1	3	2	1	2	1	1	-
30–34	R	21	34	2	-	17	28	2	6	7	10	17	16
	L	3	4	1	1	2	3	-	-	1	-	2	2
35–39	R	22	33	5	5	17	24	-	4	8	8	18	17
	L	4	4	-	-	4	3	-	1	1	-	1	1
40–44	R	28	47	5	2	21	42	2	3	9	13	17	21
	L	3	4	1	1	2	2	-	1	1	2	2	1
45–49	R	30	51	5	3	22	42	4	6	4	12	20	24
	L	-	6	1	-	-	6	-	-	-	1	-	2
50–54	R	23	31	2	-	19	29	2	2	6	11	15	19
	L	4	4	1	-	3	4	-	-	2	1	-	2
55–59	R	2	12	-	-	2	11	-	1	2	3	2	8
	L	1	1	-	-	1	1	-	-	-	1	1	-
60 and +	R	16	23	1	-	14	20	1	3	4	4	6	10
	L	2	-	-	-	2	-	-	-	2	-	2	-
Total		257	396	36	22	207	334	19	40	86	126	184	207

Males (M), Females (F), right-handed (R), left-handed (L).

All participants gave their written informed consent before participating in the study. The study was conducted in accordance with the Declaration of Helsinki, and the protocol was approved by the Department of Psychological Sciences Ethics Committee at Birkbeck (ref: 181996), University of London.

2.2. Data Collection Procedure

2.2.1. Questionnaire

Participants were asked first if they had a hand or arm injury in the 12 months prior to the test date, and only those participants who answered "no" were allowed to continue the experiment. Participants were then asked several multiple-choice binary questions about their type of occupation, if they regularly played musical instruments (e.g., violin, guitar, piano, saxophone, flute, drums), or engaged in sport using their hands (e.g., rock climbing, bouldering, gymnastics, acrobatics, archery, racket sports, lifting, cricket, golf, hand ball games and bike riding (including commuting to work)). Regarding occupation, participants could choose between (1) office job or work that requires limited manual strength (e.g., typing, shop teller); (2) precision manual work (e.g., jeweller, dressmaker, artist, lab technician); or (3) forceful manual labour (e.g., builder, carpenter, farmer). We considered as "office job/work" students and stay-at-home parents, who would use their hands for a variety of tasks, but none that were specialised enough to be considered "precision" or "forceful" manual labourer. From this questionnaire we created three measurements: music with two levels (yes/no), sport with two levels (yes/no), and occupation with three levels (office/precision/forceful).

2.2.2. Grip Strength

Grip strength was measured using a Jamar dynamometer (Sammons Preston, USA) while the participant was in a seated position. The size of the grip span was adjusted according to the hand size of the participant, visually evaluated by the experimenter, and tested by the participant before they did the grip test. The experimenter first demonstrated the appropriate sitting position and how to hold and use the dynamometer with the elbow bent at 90 degrees, as recommended by the American Society of Hand Therapy (ASHT) [82]. A poster demonstrating the appropriate posture and arm/hand position was also visible to the participant (Figure 1A). The participant was asked to squeeze the dynamometer to their maximum ability for two seconds. Grip strength was measured (lbs) for the left and right hand and after a rest of approximately one minute, each hand was measured again. An average of both measures for each hand was used in the analysis. All staff and volunteers of the Me, Human project were trained to measure grip strength with the same protocol to limit measurement error.

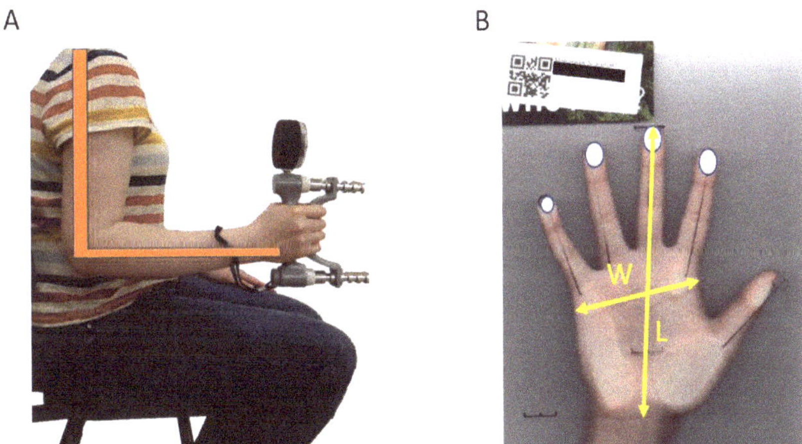

Figure 1. Appropriate posture and arm/hand position for the grip strength on the poster showed to participant (**A**) and the measurements taken on hand scans (**B**). W = hand width, L = hand length.

2.2.3. Hand Size and Shape

We measured hand size and shape from scans using a flatbed scanner (Epson Perfection V39). Participants were asked to place each hand palm side down, lining up their fingers to fit within an outline drawn on a transparent plastic sheet and keeping the fingers and palm flat (Figure 1B). Two differently sized outline transparencies for intermediate and large hands were available to allow a participant to best align their hand in a standardised manner (Figure 2). A 2 cm scale was included in each transparent plastic sheet to facilitate the accurate measurement of hand size from the scans. Hand size and shape were measured from each scan using freeware tpsDig2 software version 2.31 [83]. The hand width (W) was measured from the radial side of the second metacarpal joint to the ulnar side of the fifth metacarpal joint and hand length (L) from midline of the distal wrist crease to the tip of the middle finger (Figure 1B), following [59]. A ratio of hand width to length (W/L) was used as an indicator of hand shape, such as in [59]. We denoted hands with a ratio >0.5 as 'wide' hands and hands with a ratio <0.5 as 'long' hands. To correct the potential effect of hand size on grip strength, W*L was used as an estimate of hand area to quantify the relative grip strength (i.e., grip strength/hand area). All measurements were taken by one researcher (KT) and to test interobserver error, 20% of both right and left hands (n = 266) were measured by a second researcher (AB). An intraclass correlation coefficient (ICC) was used to test for interobserver reliability in R (R Core Team, 2020) with the "irr" package [84]. Both measurements of right and left hands were consistent between the two observers (ICC = 0.981, $p < 0.0001$), indicating an excellent reproducibility and repeatability of the measurements. Therefore, we only used the measurements of the first researcher for all subsequent analyses.

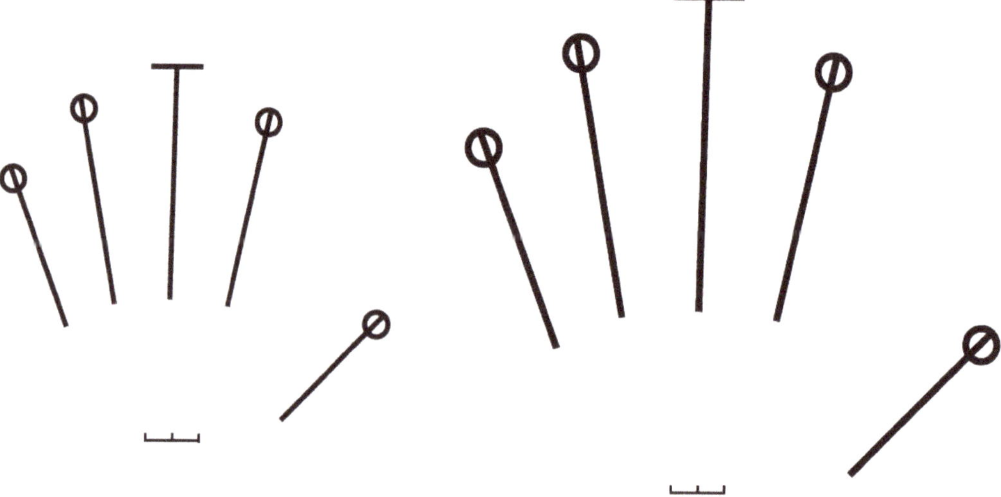

Figure 2. The two different sizes of outline transparencies (intermediate to the left and for large hands to the right) used on the flatbed scanner to allow the participant to best align their hand in a standardised manner. The scale of 2 cm was placed in the middle of the palm.

2.3. Statistical Analyses

Shapiro–Wilk Normality tests ($p > 0.05$) revealed that all data were normally distributed. We used an ANOVA to test our prediction that males would be stronger than females and that the dominant hand (defined here as the hand used to write) would be significantly stronger than the non-dominant hand (dominant hand asymmetry) using, first, absolute grip strength and, second, relative grip strength (i.e., grip strength/hand area). A Levene's test was performed to test the homogeneity of variance between males and females and for both hands. An ANOVA was also used to assess the difference in absolute grip strength between both hands across age categories within (1) males, (2) females, and (3) right and left-handers (sexes pooled) (DA).

Next, we fitted four linear multiple regressions to predict the four outcome measures of: (1) male dominant hand, (2) male non-dominant hand, (3) female dominant hand, (4) female non-dominant hand. Our predictor variables were age, occupation, hand shape, hand preference, playing music, and playing sport. These six predictor variables were considered as fixed effects and grip strength was considered a random effect. The function "predictorEffects" from the package "effect" [85] was used to graphically represent the model effects. An ANOVA was performed for each model to statistically test the effect of the predictor variables on grip strength. Tukey corrections were used for post-hoc analyses. All tests were performed with R 3.6.3 [86] with level of significance set at $p \leq 0.05$.

3. Results

First, we tested whether there were differences in grip strength between males and females and, for each sex, between both hands to test the effect of hand dominance asymmetry using ANOVAs. We investigated potential differences in both absolute grip strength and relative grip strength, in which hand area was used as a proxy for size. The results of the ANOVA showed that males were significantly stronger than females for both absolute grip strength (F(1, 1303) = 1782.72, $p < 0.0001$, η2 = 0.58) (Table 2 and Figure 3) and relative grip strength (F(1, 1303) = 820.36, $p < 0.0001$, η2 = 0.39), and both males and females were significantly stronger with their dominant hand compared with their non-dominant hand (grip strength, F(1, 1303) = 16.16, $p < 0.0001$, η2 = 0.005; relative grip strength, F(1, 1303) = 16.88, $p < 0.0001$, η2 = 0.007), indicating dominant hand asymmetry. When looking at the strength of this asymmetry, we found a mean difference in grip strength between the two hands of 5.5% for males, ranging between 1.9% (35–39 years old) and 11.4% (55–59 years old), and 4.2% for females, ranging between 0.4% (40–44 years old) and 8.8% (60 years and older) (Table S1). As a result, we searched for possible differences in the strength of hand dominance asymmetry across ages, and we found no significant interaction between age category and grip strength difference between dominant and non-dominant hands for either males (F(9, 494) = 0.134, $p > 0.05$) or females (F(9, 772) = 0.207, $p > 0.05$) (Table S1). We tested for the homogeneity of variance between males and females, and for both hands, males showed significantly more variation than females across all age categories (F(3, 1302) = 41.822, $p < 0.0001$, η2 = 0.09).

Table 2. Summary statistics for grip strength (Ibs) of the dominant and non-dominant hand according to the sex and the age categories.

Age (Years)	Sex	Dominant Hand		Non-Dominant Hand	
		Mean	SD	Mean	SD
17–19	M	87.9	17.7	81.1	17.8
	F	55.5	13	51.8	12.5
20–24	M	93.1	18.9	88.9	18.1
	F	56.3	11.7	54.2	11.7
25–29	M	101	22.3	96	19.2
	F	59.4	11.6	55.6	11.6
30–34	M	106	19	96.7	20.5
	F	56.8	15.3	54.2	13.6
35–39	M	99.8	18.1	97.9	17.9
	F	59.8	9.36	57.7	10.3
40–44	M	102	21.1	96.7	19.1
	F	57.7	11.5	57.5	12.1
45–49	M	93.1	21	87.9	21.1
	F	54.7	12.1	52.3	12.4
50–54	M	92.3	17.7	87.8	17
	F	55.4	12.3	54.2	11.2
55–59	M	102	25	90.8	22.7
	F	56.9	10.6	54.5	8.95
60 and +	M	81.4	22.2	75.1	17.1
	F	48.4	9.31	44.1	11

Males (M), Females (F), Standard deviation of the mean (SD).

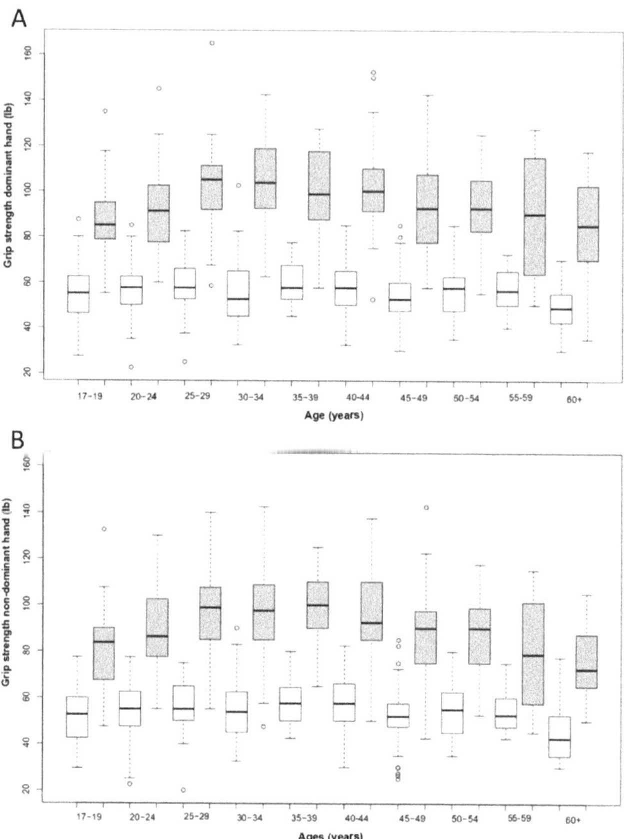

Figure 3. Grip strength (lbs) performance in males (grey) and females (white) for the dominant hand (**A**) and the non-dominant hand (**B**). The boxplots show medians (solid line) and interquartile ranges of grip strength according to age groups. The dotted lines indicate variability outside the upper and lower quartiles, except for "outliers" (dots).

We then tested differences in handedness, as an indicator of DA, in absolute grip strength (sexes pooled). There were no significant differences in grip strength between right- and left-handed participants for either hand (F(1, 1302) = 0.180, $p > 0.05$), although left-handed individuals had, on average, a higher hand dominance asymmetry with a stronger difference in grip strength between their dominant and non-dominant hands (males = 8.2%, females = 6.4%) compared with right-handed individuals (males = 5.2%, females = 4.0%). No significant interaction was found between age category, handedness and grip strength difference between dominant and non-dominant hands (F(9, 1266) = 0.028, $p > 0.05$) (Table S1).

We tested for the effect of the predictor variables (age, occupation, hand shape, hand preference, playing music, and playing sport) on grip strength using Fitting linear models. Results revealed that male grip strength of the dominant hand showed a trend towards being affected by hand shape (Figure 4) but not significantly so (F(1, 249) = 3.562, $p = 0.06$). However, for the non-dominant hand, male grip strength was significantly affected by hand shape (F(1, 249) = 9.489, $p < 0.01$, $\eta2 = 0.034$), such that males with wider hands were stronger than males with longer hands (Figure 4). Grip strength for the non-dominant hand was also significantly affected by occupation (F(2, 249) = 5.278, $p < 0.01$, $\eta2 = 0.038$), such that males doing forceful manual labour were significantly stronger than males doing an office job (post-hoc analyses, $p < 0.05$), and males who practiced sports (F(1, 249) = 4.125, $p < 0.05$, η2 = 0.015) were stronger than males who did not (Figure 4).

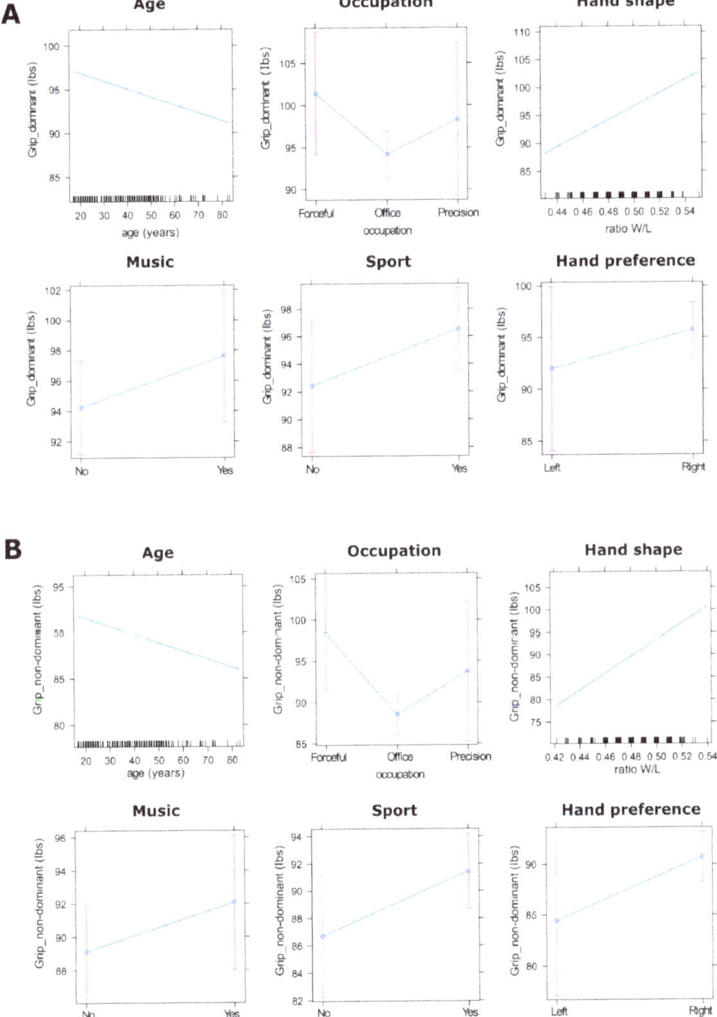

Figure 4. Predictor effect plots for the fitting linear models in males for the dominant hand (**A**) and non-dominant hand (**B**). On the age and hand shape graphics, the blue shaded area is a pointwise confidence band for the fitted values at a level of confidence of 95%. The rug plots at the bottom of the graphs shows the location of the age values and the ratio W/L values. On the other graphics, the pink bars represent the confidence intervals at a level of 95%.

For females, linear modelling revealed that grip strength of the dominant hand was significantly affected by hand shape (F(1, 388) = 4.733, $p < 0.05$, η2 = 0.017), with females with wider hands being stronger than females with longer hands, and by age (F(1, 388) = 5.369, $p < 0.05$, η2 = 0.013), such that younger females (~<30 age) were significantly stronger than older females (~>50 age; Figure 5). For the non-dominant hand, female grip strength was also significantly affected by hand shape (F(1, 388) = 5.891, $p < 0.02$, η2 = 0.014) and by age (F(1, 388) = 4.463, $p < 0.05$, η2 = 0.011), following the same pattern as the dominant hand.

However, females practicing hand sports also had a significantly stronger non-dominant hand ($F(1, 388) = 4.858$, $p < 0.05$, $\eta 2 = 0.012$) than females who did not (Figure 5). Given the effect of age on female grip strength, we tested which factors potentially interacted with age. Linear modelling revealed that for both hands, a significant interaction was found for grip strength between age (continuous) and hand shape (dominant hand, $F(1, 388) = 4.123$, $p < 0.05$, $\eta 2 = 0.010$; non-dominant hand, $F(1, 388) = 6.092$, $p < 0.05$, $\eta 2 = 0.015$). Moreover, while younger females (~<30 age) showed a similar grip strength regardless of differences in hand shape, older females (~>50 age) with wider hands were stronger than older females with longer hands (Figure S2).

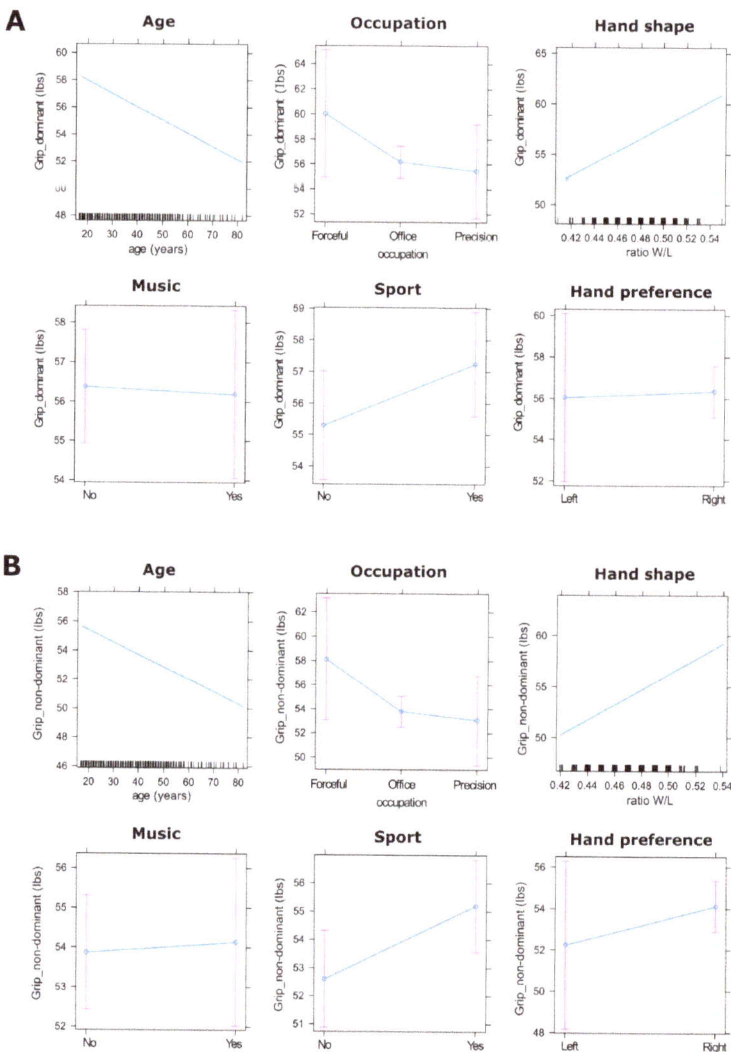

Figure 5. Predictor effect plots for the fitting linear models in females for the dominant hand (**A**) and non-dominant hand (**B**). On the age and hand shape graphics, the blue shaded area is a pointwise confidence band for the fitted values at a level of confidence of 95%. The rug plots at the bottom of the graphs shows the location of the age values and the ratio W/L values. On the other graphics, the pink bars represent the confidence intervals at a level of 95%.

4. Discussion

This study investigated different factors that predict hand grip strength in a large, heterogenous adult human sample. While some of the results support the findings of previous studies, our study sheds new light on the variability in grip strength relative to sex, age, hand shape, hand dominance asymmetry (i.e., laterality) and daily hand use. We discuss these results below and their implications for understanding the evolution of the human hand.

Consistent with findings from previous studies [78,79] and our predictions, we found that males were stronger than females. However, our study also investigated relative grip strength and found that males remained significantly stronger than females even when accounting for variation in hand size. This result is consistent with Leyk et al. [62], who showed that untrained males were, on average, stronger than highly trained female athletes. We also found a significant effect of hand dominance asymmetry, in which the dominant hand was significantly stronger than the non-dominant hand in both sexes, with an average difference being slightly higher for males (5.5%) than for females (4.2%) across all age categories. These mean differences were lower than the reported average of 10% higher grip strength for the dominant hand compared with the non-dominant hand for both sexes reported in previous studies [61,74,87]. Interestingly, we found that the difference in hand strength between the hands seemed to vary with age in both sexes (Table S1), but the differences were not significant. This result could be due to the uneven distributions of the sample across the age categories (e.g., more participants in the 20–24 age category than the 60 years and older), but also requires further investigation through a larger study to examine this general premise of 10% difference in grip strength between the dominant and nondominant hands by examining different age categories.

We also found that the grip strength of both hands decreased significantly with age in females but not in males, which partially supported our expectations. Previous studies have shown an effect of age on grip strength for both sexes [77–79]. The non-significant effect of age for both hands in males in our study may reflect the greater variability in grip strength (Figure 3). Middle-aged participants (ages 35–39 for males and 40–44 for females) showed more hand symmetry with limited differences in grip strength between both hands, while younger and older individuals showed greater asymmetry in hand dominance. Although previous studies have shown that laterality decreased with age [88], this was not the case for grip strength in our male sample. The dominant hand was always significantly stronger than the non-dominant hand for both sexes, even for participants in older age categories (Table S1). One reason why this might be the case is that previous studies have typically only assessed differences in the average grip strength across all ages [61,89,90], and thus we demonstrate for the first time, to our knowledge, important grip strength variation at specific life stages for both sexes.

Our results showed that grip strength in our sample was not an indicator of handedness. We did not find a significant effect of self-reported hand preference, although left-handed participants tended to show a larger difference of grip strength between the dominant and non-dominant hand (males = 8.2%, females = 6.4%) compared with right-handed participants (males = 5.2%, females = 4.0%; Table S1). This result did not support our expectation or previous research showing more symmetry in grip strength between the hands in left-handed compared to right-handed people showing more DA [66,75,91]. However, previous work has yielded mixed results, with some studies finding that left-handed individuals had a relatively stronger non-dominant hand [92,93]. The results of our study (and previous research) may be biased by differences in sample size (n = 60 left-handed vs. n = 594 right-handed participants), given the much lower proportion of left-handed individuals across human populations [16–19]. Left-handed individuals may also be expected to show more symmetry in grip strength between both hands (i.e., a relatively stronger right hand), because the world is adapted for right-handed individuals. Thus, our results are somewhat unexpected and require further investigation through a larger study of left-handed people across their lifespan to better understand the potential differences in grip strength between right- and left-handed individuals.

We found that grip strength of the non-dominant hand was also significantly influenced by hand shape in both males and females, and in both hands for females only. Participants with wider hands were stronger than participants with longer hands, which is consistent with previous studies that reported that people with wider hands tend to have greater muscular strength (when controlling for height) [67–69]. The fact that hand shape did not significantly influence grip strength of the dominant hand in males may also reflect the greater variability in grip strength for males compared with females. Interestingly, we found that the effect of hand shape is stronger for older females (~>50 age) than younger females (~<30 age), with older females with wider hands being stronger than older females with longer hands. This variation by age may reflect younger females being, on average, more active than older females and potentially using both hands more frequently and/or with more muscular force during a variety of daily activities. In contrast, older females are more likely to develop osteoarthritis within the hand [94,95], and patients with this disease show weaker grip strength in the affected hand than healthy individuals [96]. As humans with longer digits appear to have relatively larger articular areas [45], females with longer digits could be less susceptible to osteoarthritis, and thus could show less reduction in hand strength than females with shorter fingers. Additional research is needed to investigate the potential effect of hand shape on grip strength in older females and the potential clinical implications of this.

Variation in hand dominance asymmetry and hand function was observed according to the lifestyle factors (i.e., occupation, practice of sport and music). We found that the type of occupation had a significant effect on grip strength for males but not for females, which is consistent with the findings of Hossain et al. [76]. The female result could be explained by the relatively fewer number of female participants doing, for example, forceful manual labour (22 females against 36 males), which potentially could have affected the analysis. In particular, we found that males engaging in forceful manual work were significantly stronger than those doing 'office work'. This result supports that of previous studies [52,54] (but see [76]). However, we found an effect only for the non-dominant hand. This result likely reflects the fact that manual labour occupations often involve using both hands more forcefully and frequently than office work does, thus increasing muscle strength [54]. Indeed, middle-aged males doing forceful manual work showed greater similarities in grip strength between the dominant and non-dominant hand compared to office workers (Table S1). We found similar results for the practice of manual sports, which significantly affected grip strength in the non-dominant hand for both sexes, while there was no effect from practicing a musical instrument. Together, these results suggest that middle-aged individuals practicing regular manual activities that require the forceful use of both hands have less strength difference between the two hands (i.e., greater symmetry), while office and precision workers, who are doing more fine motor manipulation and using general tools more often with their dominant hand, may have a greater asymmetry in grip strength between the two hands.

Our findings have interesting implications for the study of human evolution. Both hands are important for modern human daily activities; however, experimental studies have demonstrated the importance of having two strong upper limbs, and hands in particular, for prehistoric activities, such as tool production/use behaviours [1,9–11,97–99], carrying [100], hunting, picking fruit, or dismembering an animal carcass [101]. For some hominins, powerful grip strength in both hands would be critical for climbing as well [102–104]. Thus, there would likely be negative selection for having weak hand grip strength throughout hominin evolution [105–107]. We found a significant influence of hand shape on grip strength such that individuals with wider hands were significantly stronger than those with longer hands. If this relationship held true in the past, there may have been increased selection for relatively shorter fingers and proportionally wider hands. Indeed, fossil evidence demonstrates that hand proportions have changed throughout human evolution (e.g., [103,108]) and that these changes likely improved dexterity [109–111] and potentially grip strength [112,113].

In modern hunter-gatherer populations, greater grip strength in the dominant hand is associated with better hunting outcomes among Hadza males, but not for Yali males [114]. It would be interesting to also measure the strength of the non-dominant hand in hunter-gatherer populations to test the hypothesis of the importance of high grip strength in both hands for hunting and other manual activities. An increase in sedentism among recent (non-foraging) humans correlates with a decrease in cortical bone strength (reviewed in [115]) and a reduction in trabecular bone density [116–118] throughout the skeleton, including the hands [116,117]. Given bone's ability to reflect variation in loading throughout life via (re-)modelling, this research suggests that recent, more sedentary humans have a reduced level of forceful manual activity compared with that of hunter-gatherers and/or that increased sedentism has resulted in systemic changes to bone structure throughout the skeleton.

It is possible that the population-level hand asymmetry or bias well-documented in modern human populations, inferred in Neandertals [28,36] and potentially earlier *Homo* species [119], is related to advances in technological and cultural innovations [28,120] or, more generally, to task complexity [24,119,121]. In turn, more frequent use of a dominant hand, rather than both hands, for diverse activities could have favoured an increase in hemispheric specialisation and vice versa. However, early hominins (e.g., *Australopithecus*, *Homo habilis*), and particularly those that likely still used their hands for locomotion, may have been under stronger selection for bimanual manipulative ability and grip strength, such as in extant great apes [119]. Thus, it is important to investigate both hands in studies of grip strength and laterality to provide a broader evolutionary understanding. In their research on lateralisation through prehistorical tools, Steele and Uomini [122] also highlighted the importance of studying the roles of both hands during bimanual activities (i.e., what they call a "Complementary Role Differentiation" model). We also require a greater appreciation of the effect of lateralisation of specific behaviours on upper limb and hand bone morphology [123–127] to better understand the evolution of human dexterity related to strength of laterality.

There are some limitations to this study which should be considered. As these data were collected as part of the larger Me, Human project and within the rules of the Live Science scheme of the London Science Museum, we were limited to a specific amount of time in which we could keep participants at any one experimental station. As a result, we were not able to collect more detailed data on daily hand use (e.g., the number of musical instruments or specific sports played and for how long). Moreover, although participants were not selected and came voluntarily to the Live Science experiment, potential selection bias could be present and may have affected our results. First, we had more young participants and parents accompanying their children compared to participants from 55 years and older. Moreover, the experiments took place during the summer months when the museum was more likely to attract international visitors. We did not record the ethnicity of the participants while previous, more targeted, studies have shown variation in grip strength across different populations [57,74–76]. Future research on large and more diverse groups of humans providing greater detail on ethnicity and specific hand use activities may provide a more nuanced understanding of the links between performance (e.g., grip strength), hand asymmetry, and hand shape or how variation in hand shape and size impacts dexterity.

5. Conclusions

To conclude, we found that adult human grip strength was influenced by a variety of factors, including age, sex, hand shape, and hand dominance asymmetry (i.e., laterality), consistent with previous studies. We also demonstrated for the first time that (1) grip strength varies throughout the lifespan, with more pronounced differences at specific life stages and (2) that the practice of different manual activities through occupation and sport also influence grip strength, particularly in males. These results emphasise the importance of physical manual activities for the attenuation of age-related grip strength loss in a clinical context. These findings may also inform ergonomic research on modern anthropotechnical systems that rely on grip strength data [128,129]. Our results highlight the importance of studying the grip strength of both hands, rather than just the dominant hand, in relation to the above factors, to better understand the link between form and function of the hand, in both modern populations and in our evolutionary past.

Supplementary Materials: The following are available online at https://www.mdpi.com/2073-8994/13/7/1142/s1, Figure S1: Distribution of participants used in the analyses according to ages groups and with right-handed (R) and left-handed (L) participants, Figure S2: Predictor effect plots for the fitting linear models of the interaction between age and hand shape for grip strength in female, for the dominant hand (A) and the non-dominant hand (B), Table S1: Details of the difference in percentage between hand grip strength in dominant and non-dominant hands for each sex (M = males; F = females) and age groups, according to right and left-handers, practicing of sport and the different occupations.

Author Contributions: Conception and Design of the research, G.S.F., G.D., T.E., T.L.K., A.B., H.B., C.S.; Data Acquisition, G.S.F., G.D., T.E., T.L.K., A.B., H.B., C.S., K.T.; Analysis, A.B., K.T.; Writing—Original Draft Preparation, A.B.; Writing—Review and Editing, A.B., T.L.K., G.S.F., G.D., T.E., H.B., C.S.; Project Administration, G.S.F.; Funding Acquisition, G.S.F. All authors have read and agreed to the published version of the manuscript.

Funding: This research was funded by The Waterloo Foundation, grant number 917-3756.

Institutional Review Board Statement: The study was conducted according to the guidelines of the Declaration of Helsinki, and approved by the Department of Psychological Sciences Ethics Committee at Birkbeck (ref: 181996, 14/05/2019), University of London.

Informed Consent Statement: Informed consent was obtained from all subjects involved in the study.

Acknowledgments: We thank the London Science Museum and the Live Science scheme for allowing us to conduct this study, and the visitors who agreed to be part of the citizen science collaboration project called Me, Human. We thank the volunteers and research assistants: M. Ali, M. Diletta, R. Filippi, L. Fish, E. Forrester, S. Forrester, A. Ghanemi, C. Grevel, J. Hall, E. Jackson, C. Marsh, B. Meyer, S. Pem, E. Periche, E. Ranzato, Y. Slaveva, B. Stewart, and B. Todd for their help in collecting participant data. The authors thank the reviewers for their helpful comments, and Antoine Balzeau for the invitation to submit our work in the special issues "Symmetry in Human Evolution, From Biology to Behaviours".

Conflicts of Interest: The authors declare no conflict of interest.

References

1. Marzke, M.W.; Wullstein, K.L.; Viegas, S.F. Evolution of the power (squeeze) grip and its morphological correlates in hominids. *Am. J. Phys. Anthropol.* **1992**, *89*, 283–298. [CrossRef]
2. Marzke, M.W. Tool making, hand morphology and fossil hominins. *Philos. Trans. R. Soc. Lond. B. Biol. Sci.* **2013**, *368*, 20120414. [CrossRef]
3. Toth, N.; Schick, K. Early stone industries and inferences regarding language and cognition. In *Tools, Language and Cognition in Human Evolution*; Gibson, K.R., Ingold, T., Eds.; Cambridge University Press: Cambridge, UK, 1993; pp. 346–362.
4. McPherron, S.P.; Alemseged, Z.; Marean, C.W.; Wynn, J.G.; Reed, D.; Geraads, D.; Bobe, R.; Béarat, H.A. Evidence for stone-tool-assisted consumption of animal tissues before 3.39 million years ago at Dikika, Ethiopia. *Nature* **2010**, *466*, 857–860. [CrossRef] [PubMed]
5. Toth, N.; Schick, K. The Oldowan: The tool making of early hominins and chimpanzees compared. *Annu. Rev. Anthropol.* **2009**, *38*, 289–305. [CrossRef]
6. Braun, D.R.; Aldeias, V.; Archer, W.; Arrowsmith, J.R.; Baraki, N.; Campisano, C.J.; Deino, A.L.; DiMaggio, E.N.; Dupont-Nivet, G.; Engda, B.; et al. Earliest known Oldowan artifacts at> 2.58 Ma from Ledi-Geraru, Ethiopia, highlight early technological diversity. *Proc. Natl. Acad. Sci. USA* **2019**, *116*, 11712–11717. [CrossRef]
7. Stout, D.; Toth, N.; Schick, K.; Chaminade, T. Neural correlates of Early Stone Age toolmaking: Technology, language and cognition in human evolution. *Philos. Trans. R. Soc. Lond. B Biol. Sci.* **2008**, *363*, 1939–1949. [CrossRef]
8. Toth, N.; Schick, K. An overview of the cognitive implications of the Oldowan Industrial Complex. *Azania Archaeol. Res. Afr.* **2018**, *53*, 3–39. [CrossRef]

9.   Williams, E.M.S.; Gordon, A.D.; Richmond, B.G. Upper limb kinematics and the role of the wrist during stone tool production. *Am. J. Phys. Anthropol.* **2010**, *43*, 134–145. [CrossRef] [PubMed]
10.   Williams, E.M.S.; Gordon, A.D.; Richmond, B.G. Hand pressure distribution during Oldowan stone tool production. *J. Hum. Evol.* **2012**, *62*, 520–532. [CrossRef]
11.   Williams-Hatala, E.M.; Hatala, K.G.; Gordon, M.; Key, A.; Kasper, M.; Kivell, T.L. The manual pressures of stone tool behaviors and their implications for the evolution of the human hand. *J. Hum. Evol.* **2018**, *119*, 14–26. [CrossRef] [PubMed]
12.   Stout, D.; Hecht, E.E. Evolutionary neuroscience of cumulative culture. *Proc. Natl. Acad. Sci. USA* **2017**, *114*, 7861–7868. [CrossRef] [PubMed]
13.   Feix, T.; Romero, J.; Schmiedmayer, H.B.; Dollar, A.M.; Kragic, D. The grasp taxonomy of human grasp types. *IEEE Trans. Hum. Mach. Syst.* **2015**, *46*, 66–77. [CrossRef]
14.   Venkataraman, V.V.; Kraft, T.S.; Dominy, N.J. Tree climbing and human evolution. *Proc. Natl. Acad. Sci. USA* **2013**, *110*, 1237–1242. [CrossRef] [PubMed]
15.   Wieczorek, B.; Kukla, M.; Warguła, Ł. The symmetric nature of the position distribution of the human body center of gravity during propelling manual wheelchairs with innovative propulsion systems. *Symmetry* **2021**, *13*, 154. [CrossRef]
16.   Annett, M. *Left, Right, Hand and Brain: The Right Shift Theory*; LEA Publishers: London, UK, 1985; 416p.
17.   Porac, C.; Coren, S. *Lateral Preferences and Human Behavior*; Springer: New York, NY, USA, 1981; 283p.
18.   McManus, I.C. The history and geography of human handedness. In *Language Lateralization and Psychosis*; Sommer, I.E.C., Kahn, R.S., Eds.; Cambridge University Press: Cambridge, UK, 2009; pp. 37–57.
19.   Llaurens, V.; Raymond, M.; Faurie, C. Why are some people left-handed? An evolutionary perspective. *Philos. Trans. R. Soc. Lond. B Biol. Sci.* **2009**, *364*, 881–894. [CrossRef] [PubMed]
20.   Forrester, G.S.; Quaresmini, C.; Leavens, D.A.; Mareschal, D.; Thomas, M.S.C. Human handedness: An inherited evolutionary trait. *Behav. Brain Res.* **2013**, *237*, 200–206. [CrossRef] [PubMed]
21.   Forrester, G.S. Hand, limb and other motor preferences: Methodological Considerations. In *Lateralized Brain Functions*; Vallortigara, G., Rogers, L., Eds.; Humana Press: New York, NY, USA, 2017; pp. 121–152.
22.   Hopkins, W.D.; Phillips, K.A.; Bania, A.; Calcutt, S.E.; Gardner, M.; Russell, J.; Schaeffer, J.; Lonsdorf, E.V.; Ross, S.R.; Schapiro, S.J. Hand preferences for coordinated bimanual actions in 777 great apes: Implications for the evolution of handedness in hominins. *J. Hum. Evol.* **2011**, *60*, 605–611. [CrossRef]
23.   Papademetriou, E.; Sheu, C.F.; Michel, G.F. A meta-analysis of primate hand preferences, particularly for reaching. *J. Comp. Psychol.* **2005**, *119*, 33–48. [CrossRef]
24.   Bardo, A.; Pouydebat, E.; Meunier, H. Do bimanual coordination, tool use, and body posture contribute equally to hand preferences in bonobos? *J. Hum. Evol.* **2015**, *82*, 159–169. [CrossRef]
25.   Hopkins, W.D.; Russell, J.L.; Cantalupo, C. Neuroanatomical correlates of handedness for tool use in chimpanzees (*Pan troglodytes*): Implication for theories on the evolution of language. *Psychol. Sci.* **2007**, *18*, 971–977. [CrossRef]
26.   Corballis, M.C. Cerebral asymmetry: Motoring on. *Trends Cogn. Sci.* **1998**, *2*, 152–157. [CrossRef]
27.   Gibson, K.R.; Ingold, T. *Tools, Language and Cognition in Human Evolution*; Cambridge University Press: Cambridge, UK, 1993; 496p.
28.   Uomini, N.T. The prehistory of handedness: Archaeological data and comparative ethology. *J. Hum. Evol.* **2009**, *57*, 411–419. [CrossRef]
29.   Corballis, M.C. From mouth to hand: Gesture, speech and the evolution of right- handedness. *Behav. Brain Sci.* **2003**, *26*, 199–260. [CrossRef]
30.   Meguerditchian, A.; Vauclair, J.; Hopkins, W.D. On the origins of human handedness and language: A comparative review of hand preferences for bimanual coordinated actions and gestural communication in nonhuman primates. *Dev. Psychobiol.* **2013**, *55*, 637–650. [CrossRef]
31.   Calvin, W.H. A stone's throw and its launch window: Timing precision and its implication for language and hominids brain. *J. Theor. Biol.* **1983**, *104*, 121–135. [CrossRef]
32.   Greenfield, P.M. Language, tools and brain: The ontogeny and phylogeny of hierarchically organized sequential behavior. *Behav. Brain Sci.* **1991**, *14*, 531–595. [CrossRef]
33.   Bradshaw, J.L.; Nettleton, N.C. Language lateralization to the dominant hemisphere: Tool use, gesture and language in hominid evolution. *Curr. Psychol.* **1982**, *2*, 171–192. [CrossRef]
34.   Armstrong, D.F.; Stokoe, W.C.; Wilcox, S.E. *Gesture and the Nature of Language*; Cambridge University Press: Cambridge, UK, 1995; 270p.
35.   Uomini, N.T. Handedness in Neandertals. In *Neanderthal Lifeways, Subsistence and Technology*; Conard, N.J., Richter, J., Eds.; Springer: Dordrecht, Netherlands, 2011; pp. 139–154.
36.   Estalrrich, A.; Rosas, A. Handedness in Neandertals from the El Sidrón (Asturias, Spain): Evidence from instrumental striations with ontogenetic inferences. *PLoS ONE* **2013**, *8*, e62797. [CrossRef] [PubMed]
37.   Courtney, A.J. Hand anthropometry of Hong Kong Chinese females compared to other ethnic groups. *Ergonomics* **1984**, *27*, 1169–1180. [CrossRef] [PubMed]
38.   Okunribido, O.O. A survey of hand anthropometry of female rural farm workers in Ibadan, Western Nigeria. *Ergonomics* **2000**, *43*, 282–292. [CrossRef]
39.   Kar, S.K.; Ghosh, S.; Manna, I.; Banerjee, S.; Dhara, P. An investigation of hand anthropometry of agricultural workers. *J. Hum. Ecol.* **2003**, *14*, 57–62. [CrossRef]

40. Imrhan, S.N.; Sarder, M.D.; Mandahawi, N. Hand anthropometry in Bangladeshis living in America and comparisons with other populations. *Ergonomics* **2009**, *52*, 987–998. [CrossRef] [PubMed]
41. Lazenby, R.; Smashnuk, A. Osteometric variation in the Inuit second metacarpal: A test of Allen's Rule. *Int. J. Osteoarchaeol.* **1999**, *9*, 182–188. [CrossRef]
42. Betti, L.; Lycett, S.J.; von Cramon-Taubadel, N.; Pearson, O.M. Are human hands and feet affected by climate? A test of Allen's rule. *Am. J. Phys. Anthropol.* **2015**, *158*, 132–140. [CrossRef]
43. Dianat, I.; Nedaei, M.; Nezami, M.A.M. The effects of tool handle shape on hand performance, usability and discomfort using masons' trowels. *Int. J. Ind. Ergon.* **2015**, *45*, 13–20. [CrossRef]
44. Kong, Y.K.; Lowe, B.D. Optimal cylindrical handle diameter for grip force tasks. *Int. J. Ind. Ergon.* **2005**, *35*, 495–507. [CrossRef]
45. Rolian, C.; Lieberman, D.E.; Zermeno, J.P. Hand biomechanics during simulated stone tool use. *J. Hum. Evol.* **2011**, *61*, 26–41. [CrossRef]
46. Key, A.J.; Lycett, S.J. Investigating interrelationships between Lower Palaeolithic stone tool effectiveness and tool user biometric variation: Implications for technological and evolutionary changes. *Archaeol. Anthrop. Sci.* **2018**, *10*, 989–1006. [CrossRef]
47. Mathiowetz, V.; Kashman, N.; Volland, G.; Weber, K.; Dowe, M.; Rogers, S. Grip and pinch strength: Normative data for adults. *Arch. Phys. Med. Rehabil.* **1985**, *66*, 69–74.
48. Bohannon, R.W. Dynamometer measurements of hand-grip strength predict multiple outcomes. *Percept. Mot. Skills* **2001**, *93*, 323–328. [CrossRef]
49. Bohannon, R.W. Hand-grip dynamometry predicts future outcomes in aging adults. *J. Geriat. Phys. Ther.* **2008**, *31*, 3–10. [CrossRef]
50. Cuesta-Vargas, A.; Hilgenkamp, T. Reference values of grip strength measured with a Jamar dynamometer in 1526 adults with intellectual disabilities and compared to adults without intellectual disability. *PLoS ONE* **2015**, *10*, e0129585. [CrossRef] [PubMed]
51. Zaggelidis, G. Maximal isometric handgrip strength (HGS) in Greek elite male judo and karate athletes. *Sport Sci. Rev.* **2016**, *25*, 320. [CrossRef]
52. Josty, I.C.; Tyler, M.P.H.; Shewell, P.C.; Roberts, A.H.N. Grip and pinch strength variations in different types of workers. *J. Hand Surg.* **1997**, *22*, 266–269. [CrossRef]
53. Nevill, A.M.; Holder, R.L. Modelling handgrip strength in the presence of confounding variables: Results from the Allied Dunbar National Fitness Survey. *Ergonomics* **2000**, *43*, 1547–1558. [CrossRef]
54. Saremi, M.; Rostamzadeh, S. Hand Dimensions and Grip Strength: A Comparison of Manual and Non-manual Workers. In Proceedings of the 20th Congress of the International Ergonomics Association, Florence, Italy, 26–30 August 2018; Bagnara, S., Tartaglia, R., Albolino, S., Alexander, T., Fujita, Y., Eds.; Springer International Publishing: Cham, Switzerland, 2019; pp. 520–529.
55. Burdukiewicz, A.; Pietraszewska, J.; Andrzejewska, J.; Chromik, K.; Stachoń, A. Asymmetry of Musculature and Hand Grip Strength in Bodybuilders and Martial Artists. *Int. J. Environ. Res. Public Health* **2020**, *17*, 4695. [CrossRef]
56. Desrosiers, J.; Bravo, G.; Hébert, R.; Dutil, E. Normative data for grip strength of elderly men and women. *Am. J. Occup. Ther.* **1995**, *49*, 637–644. [CrossRef]
57. Vaz, M.; Hunsberger, S.; Diffey, B. Prediction equations for hand grip strength in healthy Indian male and female subjects encompassing a wide age range. *Ann. Hum. Biol.* **2002**, *29*, 131–141. [CrossRef]
58. Bohannon, R.W.; Bear-Lehman, J.; Desrosiers, J.; Massy-Westropp, N.; Mathiowetz, V. Average grip strength: A meta-analysis of data obtained with a Jamar dynamometer from individuals 75 years or more of age. *J. Geriatr. Phys. Ther.* **2007**, *30*, 28–30. [CrossRef] [PubMed]
59. Clerke, A.M.; Clerke, J.P.; Adams, R.D. Effects of hand shape on maximal isometric grip strength and its reliability in teenagers. *J. Hand Ther.* **2005**, *18*, 19–29. [CrossRef]
60. Hone, L.S.; McCullough, M.E. 2D: 4D ratios predict hand grip strength (but not hand grip endurance) in men (but not in women). *Evol. Hum. Behav.* **2012**, *33*, 780–789. [CrossRef]
61. Petersen, P.; Petrick, M.; Connor, H.; Conklin, D. Grip strength and hand dominance: Challenging the 10% rule. *Am. J. Occup. Ther.* **1989**, *43*, 444–447. [CrossRef] [PubMed]
62. Leyk, D.; Gorges, W.; Ridder, D.; Wunderlich, M.; Rüther, T.; Sievert, A.; Essfeld, D. Hand-grip strength of young men, women and highly trained female athletes. *Eur. J. Appl. Physiol.* **2007**, *99*, 415–421. [CrossRef] [PubMed]
63. Angst, F.; Drerup, S.; Werle, S.; Herren, D.B.; Simmen, B.R.; Goldhahn, J. Prediction of grip and key pinch strength in 978 healthy subjects. *BMC Musculoskelet. Disord.* **2010**, *11*, 1–6. [CrossRef] [PubMed]
64. Massy-Westropp, N.M.; Gill, T.K.; Taylor, A.W.; Bohannon, R.W.; Hill, C.L. Hand Grip Strength: Age and gender stratified normative data in a population-based study. *BMC Res. Notes* **2011**, *4*, 1–5. [CrossRef]
65. Hanten, W.P.; Chen, W.-Y.; Austin, A.A.; Brooks, R.E.; Carter, H.C.; Law, C.A.; Morgan, M.K.; Sanders, D.J.; Swan, C.A.; Vanderslice, A.L. Maximum grip strength in normal subjects from 20 to 64 years of age. *J. Hand. Ther.* **1999**, *12*, 193–200. [CrossRef]
66. Bohannon, R.W. Grip strength: A summary of studies comparing dominant and nondominant limb measurements. *Percept. Mot. Ski.* **2003**, *96*, 728–730. [CrossRef] [PubMed]
67. Aghazadeh, F.; Waikar, A.M.; Lee, K.S.; Backhouse, T.; Davis, P. Impact of anthropometric variables and sex on grip strength. In *Advances in Industrial Ergonomics and Safety*; Mital, A., Ed.; Taylor & Francis: London, UK, 1989; Volume I, pp. 501–505.
68. Yakou, T.; Yamamoto, K.; Koyama, M.; Hyodo, K. Sensory evaluation of grip using cylindrical objects. *JSME Int. J. Ser. C Mech. Syst. Mach. Elem. Manuf.* **1997**, *40*, 730–735. [CrossRef]

69. Crawford, J.O.; Wabine, E.; Nayak, L. The interaction between lid diameter, height and shape on wrist torque exertion in younger and older adults. *Ergonomics* **2002**, *45*, 922–933. [CrossRef]
70. Kong, Y.K.; Kim, D.M. The relationship between hand anthropometrics, total grip strength and individual finger force for various handle shapes. *Int. J. Occup. Saf. Ergon.* **2015**, *21*, 187–192. [CrossRef]
71. Barut, Ç.; Demirel, P.; Kiran, S. Evaluation of hand anthropometric measurements and grip strength in basketball, volleyball and handball players. *Anatomy* **2008**, *2*, 55–59. [CrossRef]
72. Carson, R.G. Get a grip: Individual variations in grip strength are a marker of brain health. *Neurobiol. Aging* **2018**, *71*, 189–222. [CrossRef] [PubMed]
73. Praetorius Björk, M.; Johansson, B.; Hassing, L.B. I forgot when I lost my grip-strong associations between cognition and grip strength in level of performance and change across time in relation to impending death. *Neurobiol. Aging* **2016**, *38*, 68–72. [CrossRef] [PubMed]
74. Lau, V.W.-S.; Ip, W.-Y. Comparison of power grip and lateral pinch strengths between the dominant and non-dominant hands for normal Chinese male subjects of different occupational demand. *Hong Kong Physiother. J.* **2006**, *24*, 16–22. [CrossRef]
75. Mitsionis, G.; Pakos, E.E.; Stafilas, K.S.; Paschos, N.; Papakostas, T.; Beris, A.E. Normative data on hand grip strength in a Greek adult population. *Int. Orthop.* **2009**, *33*, 713–717. [CrossRef]
76. Hossain, M.G.; Zyroul, R.; Pereira, B.P.; Kamarul, T. Multiple regression analysis of factors influencing dominant hand grip strength in an adult Malaysian population. *J. Hand Surg.* **2012**, *37*, 65–70. [CrossRef] [PubMed]
77. Bohannon, R.W.; Peolsson, A.; Massy-Westropp, N.; Desrosiers, J.; Bear-Lehman, J. Reference values for adult grip strength measured with a Jamar dynamometer: A descriptive meta-analysis. *Physiotherapy* **2006**, *92*, 11–15. [CrossRef]
78. Nicolay, C.W.; Walker, A.L. Grip strength and endurance: Influences of anthropometric variation, hand dominance, and gender. *Int. J. Ind. Ergon.* **2005**, *35*, 605–618. [CrossRef]
79. Wang, Y.C.; Bohannon, R.W.; Li, X.; Sindhu, B.; Kapellusch, J. Hand-grip strength: Normative reference values and equations for individuals 18 to 85 years of age residing in the United States. *J. Orthop. Sports Phys. Ther.* **2018**, *48*, 685–693. [CrossRef]
80. Cardoso, H.F.V.; Severino, R.S.S. The chronology of epiphyseal union in the hand and foot from dry bone observations. *Int. J. Osteoarchaeol.* **2010**, *20*, 737–746. [CrossRef]
81. Cunningham, C.; Scheuer, L.; Black, S. *Developmental Juvenile Osteology*, 2nd ed.; Academic Press: Cambridge, MA, USA, 2016; pp. 283–350.
82. Fess, E.E. Grip strength. In *Clinical Assessment Recommendations*, 2nd ed.; Casanova, J.S., Ed.; American Society of Hand Therapists: Chicago, IL, USA, 1992; pp. 41–45.
83. Rohlf, F.J. *tpsDig2 Software*; Version 2.31; The State University of New York at Stony Brook. Stony Brook, NY, USA, 2017; Available online: http://www.sbmorphometrics.org/soft-dataacq.html (accessed on 12 October 2020).
84. Gamer, M.; Lemon, J.; Fellows, I.; Singh, P. *irr: Various Coefficients of Interrater Reliability and Agreement*; R Package Version 0.84.1.; The R Foundation for Statistical Computing: Vienna, Austria, 2019; Available online: https://CRAN.R-project.org/package=irr (accessed on 3 March 2021).
85. Fox, J.; Weisberg, S. Visualizing Fit and Lack of Fit in Complex Regression Models with Predictor Effect Plots and Partial Residuals. *J. Stat. Softw.* **2018**, *87*, 1–27. [CrossRef]
86. R Core Team. *R: A Language and Environment for Statistical Computing*; R Foundation for Statistical Computing: Vienna, Austria, 2020; Available online: https://www.R-project.org/ (accessed on 6 June 2020).
87. Golden, C.J.; Espe-Pfeifer, P.; Wachsler-Felder, J. *Neuropsychological Interpretation of Objective Psychological Tests. Motor and Sensory Tests*; Springer: New York, NY, USA, 2002; 243p.
88. Porac, C. Are age trends in adult hand preference best explained by developmental shifts or generational differences? *Can. J. Exp. Psychol.* **1993**, *47*, 697–713. [CrossRef]
89. Armstrong, C.A.; Oldham, J.A. A comparison of dominant and non-dominant hand strengths. *J. Hand Surg. Eur. Vol.* **1999**, *24*, 421–425. [CrossRef] [PubMed]
90. Crosby, C.A.; Wehbé, M.A. Hand strength: Normative values. *J. Hand Surg.* **1994**, *19*, 665–670. [CrossRef]
91. Incel, N.A.; Ceceli, E.; Durukan, P.B.; Erdem, H.R.; Yorgancioglu, Z.R. Grip strength: Effect of hand dominance. *Singap. Med. J.* **2002**, *43*, 234–237.
92. Lewandowski, L.; Kobus, D.A.; Church, K.L.; Van Orden, K. Neuropsychological implications of hand preference versus hand grip performance. *Percep. Mot. Skills* **1982**, *55*, 311–314. [CrossRef]
93. Ziyagil, M.A.; Gürsoy, R.; Dane, Ş.; Türkmen, M.; Çebi, M. Effects of handedness on the hand grip strength asymmetry in Turkish athletes. *Compr. Psychol.* **2015**, *4*, 20. [CrossRef]
94. Bagis, S.; Sahin, G.; Yapici, Y.; Cimen, Ö.B.; Erdogan, C. The effect of hand osteoarthritis on grip and pinch strength and hand function in postmenopausal women. *Clin. Rheumatol.* **2003**, *22*, 420–424. [CrossRef]
95. Fontana, L.; Neel, S.; Claise, J.M.; Uqhetto, S.; Catilina, P. Osteoarthritis of the thumb carpometacarpal joint in women and occupational risk factors: A case-control study. *J. Hand. Surg. Am.* **2007**, *32*, 459–465. [CrossRef]
96. Ceceli, E.; Gül, S.; Borman, P.N.; Uysal, S.R.; Okumuş, M. Hand function in female patients with hand osteoarthritis: Relation with radiological progression. *Hand* **2012**, *7*, 335–340. [CrossRef]
97. Marzke, M.W.; Toth, N.; Schick, K.; Reece, S.; Steinberg, B.; Hunt, K.; Linscheid, R.L.; An, K.N. EMG study of hand muscle recruitment during hard hammer percussion manufacture of Oldowan tools. *Am. J. Phys. Anthropol.* **1998**, *105*, 315–332. [CrossRef]

98. Key, A.J.; Dunmore, C.J. The evolution of the hominin thumb and the influence exerted by the non-dominant hand during stone tool production. *J. Hum. Evol.* **2015**, *78*, 60–69. [CrossRef]
99. Shaw, C.N.; Hofmann, C.L.; Petraglia, M.D.; Stock, J.T.; Gottschall, J.S. Neandertal humeri may reflect adaptation to scraping tasks, but not spear thrusting. *PLoS ONE* **2012**, *7*, e40349. [CrossRef]
100. Key, A.J. Manual loading distribution during carrying behaviors: Implications for the evolution of the hominin hand. *PLoS ONE* **2016**, *11*, e0163801. [CrossRef]
101. Eshed, V.; Gopher, A.; Galili, E.; Hershkovitz, I. Musculoskeletal stress markers in Natufian hunter-gatherers and Neolithic farmers in the Levant: The upper limb. *Amer. J. Phys. Anth.* **2004**, *123*, 303–315. [CrossRef] [PubMed]
102. Ward, C.V. Interpreting the posture and locomotion of *Australopithecus afarensis*: Where do we stand? *Am. J. Phys. Anthropol.* **2002**, *119* (Suppl. 35), 185–215. [CrossRef]
103. Kivell, T.L. Evidence in hand: Recent discoveries and the early evolution of human manual manipulation. *Philos. Trans. R. Soc. Lond. B. Biol. Sci.* **2015**, *370*, 20150105. [CrossRef] [PubMed]
104. Carlson, K.J.; Green, D.J.; Jashashvili, T.; Pickering, T.R.; Heaton, J.L.; Beaudet, A.; Stratford, D.; Crompton, R.; Kuman, K.; Bruxelles, L.; et al. The pectoral girdle of StW 573 ('Little Foot') and its implications for shoulder evolution in the Hominina. *J. Hum. Evol.* **2021**, 102983. [CrossRef]
105. Cartmill, M. Climbing. In *Functional Vertebrate Morphology*; Hildebrand, M., Bramble, D.M., Liem, K.F., Wake, D.B., Eds.; Harvard University Press: Cambridge, MA, USA, 1985; pp. 73–88.
106. Preuschoft, H. What does "arboreal locomotion" mean exactly and what are the relationships between "climbing", environment and morphology? *Z. Morphol. Anthropol.* **2002**, *83*, 171–188.
107. Kappelman, J.; Ketcham, R.A.; Pearce, S.; Todd, L.; Akins, W.; Colbert, M.W.; Feseha, M.; Maisano, J.A.; Witzel, A. Perimortem fractures in Lucy suggest mortality from fall out of tall tree. *Nature* **2016**, *537*, 503–507. [CrossRef]
108. Rolian, C.; Gordon, A.D. Reassessing manual proportions in *Australopithecus afarensis*. *Am. J. Phys. Anthropol.* **2013**, *152*, 393–406. [CrossRef]
109. Niewoehner, W.A.; Bergstrom, A.; Eichele, D.; Zuroff, M.; Clark, J.T. Manual dexterity in Neanderthals. *Nature* **2003**, *422*, 395. [CrossRef]
110. Feix, T.; Kivell, T.L.; Pouydebat, E.; Dollar, A.M. Estimating thumb–index finger precision grip and manipulation potential in extant and fossil primates. *J. R. Soc. Interface* **2015**, *12*, 20150176. [CrossRef]
111. Almécija, S.; Alba, D.M. On manual proportions and pad-to-pad precision grasping in *Australopithecus afarensis*. *J. Hum. Evol.* **2014**, *73*, 88–92. [CrossRef]
112. Niewoehner, W.A. Neanderthal hands in their proper perspective. In *Neanderthals Revisited: New Approaches and Perspectives*; Harvati, K., Harrison, T., Eds.; Springer: Berlin, Germany, 2008; pp. 157–190.
113. Bardo, A.; Moncel, M.H.; Dunmore, C.J.; Kivell, T.L.; Pouydebat, E.; Cornette, R. The implications of thumb movements for Neanderthal and modern human manipulation. *Sci. Rep.* **2020**, *10*, 1–12. [CrossRef]
114. Misiak, M.; Butovskaya, M.; Oleszkiewicz, A.; Sorokowski, P. Digit ratio and hand grip strength are associated with male competition outcomes: A study among traditional populations of the Yali and Hadza. *Am. J. Hum. Biol.* **2020**, *32*, e23321. [CrossRef]
115. Ruff, C.B.; Larsen, C.S. Long bone structural analyses and the reconstruction of past mobility: A historical review. In *Reconstructing Mobility*; Carlson, K.J., Marchi, D., Eds.; Springer: Boston, MA, USA, 2014; pp. 13–29.
116. Chirchir, H.; Kivell, T.L.; Ruff, C.B.; Hublin, J.J.; Carlson, K.J.; Zipfel, B.; Richmond, B.G. Recent origin of low trabecular bone density in modern humans. *Proc. Natl. Acad. Sci. USA* **2015**, *112*, 366–371. [CrossRef] [PubMed]
117. Chirchir, H.; Ruff, C.B.; Junno, J.A.; Potts, R. Low trabecular bone density in recent sedentary modern humans. *Am. J. Phys. Anthropol.* **2017**, *162*, 550–560. [CrossRef] [PubMed]
118. Ryan, T.M.; Shaw, C.N. Gracility of the modern *Homo sapiens* skeleton is the result of decreased biomechanical loading. *Proc. Natl. Acad. Sci. USA* **2015**, *112*, 372–377. [CrossRef] [PubMed]
119. Cashmore, L.; Uomini, N.; Chapelain, A. The evolution of handedness in humans and great apes: A review and current issues. *J. Anthropol. Sci.* **2008**, *86*, 7–35.
120. Provins, K.A. The specificity of motor skill and manual asymmetry: A review of the evidence and its implications. *J. Mot. Behav.* **1997**, *29*, 183–192. [CrossRef]
121. Fagot, J.; Vauclair, J. Manual laterality in nonhuman primates: A distinction between handedness and manual specialization. *Psychol. Bull.* **1991**, *109*, 76. [CrossRef]
122. Steele, J.; Uomini, N. Can the archaeology of manual specialization tell us anything about language evolution? A survey of the state of play. *Cambr. Arch. J.* **2009**, *19*, 97–110. [CrossRef]
123. Lazenby, R.A. Identification of sex from metacarpals: Effect of side asymmetry. *J. Forensic Sci.* **1994**, *39*, 1188–1194. [CrossRef] [PubMed]
124. Lazenby, R.A.; Cooper, D.M.L.; Angus, S.; Hallgrmsson, B. Articular constraint, handedness, and directional asymmetry in the human second metacarpal. *J. Hum. Evol.* **2008**, *54*, 857–885. [CrossRef] [PubMed]
125. Stephens, N.B.; Kivell, T.L.; Gross, T.; Pahr, D.H.; Lazenby, R.A.; Hublin, J.J.; Hershkovitz, I.; Skinner, M.M. Trabecular architecture in the thumb of *Pan* and *Homo*: Implications for investigating hand use, loading, and hand preference in the fossil record. *Am. J. Phys. Anthropol.* **2016**, *161*, 603–619. [CrossRef] [PubMed]

126. Boonsaner, K.; Louthrenoo, W.; Meyer, S.; Schumacher, H.R., Jr. Effect of dominancy on severity in rheumatoid arthritis. *Br. J. Rheumatol.* **1992**, *31*, 77–80. [CrossRef]
127. Sarringhaus, L.A.; Stock, J.T.; Marchant, L.F.; McGrew, W.C. Bilateral asymmetry in the limb bones of the chimpanzee (*Pan troglodytes*). *Am. J. Phys. Anthropol.* **2005**, *128*, 840–845. [CrossRef]
128. Warguła, Ł.; Kukla, M.; Krawiec, P.; Wieczorek, B. Impact of Number of Operators and Distance to Branch Piles on Woodchipper Operation. *Forests* **2020**, *11*, 598. [CrossRef]
129. Wieczorek, B.; Kukla, M. Biomechanical Relationships Between Manual Wheelchair Steering and the Position of the Human Body's Center of Gravity. *J. Biomech. Eng.* **2020**, *142*. [CrossRef]

Review

Limb Preference in Animals: New Insights into the Evolution of Manual Laterality in Hominids

Grégoire Boulinguez-Ambroise [1,*,†], Juliette Aychet [2,*,†] and Emmanuelle Pouydebat [3]

[1]   Department of Anatomy and Neurobiology, Northeast Ohio Medical University (NEOMED), Rootstown, OH 44272, USA
[2]   Laboratoire Ethologie Cognition Développement (UR LECD), Université Paris Nanterre, 92000 Nanterre, France
[3]   Département Adaptations du Vivant, UMR 7179 C.N.R.S/M.N.H.N., 57 Rue Cuvier, 75231 Paris, France; emmanuelle.pouydebat@mnhn.fr
*   Correspondence: gregoire.boulinguez-ambroise@cri-paris.org (G.B.-A.); juliette.aychet@gmail.com (J.A.)
†   These authors contributed equally to this work.

Abstract: Until the 1990s, the notion of brain lateralization—the division of labor between the two hemispheres—and its more visible behavioral manifestation, handedness, remained fiercely defined as a human specific trait. Since then, many studies have evidenced lateralized functions in a wide range of species, including both vertebrates and invertebrates. In this review, we highlight the great contribution of comparative research to the understanding of human handedness' evolutionary and developmental pathways, by distinguishing animal forelimb asymmetries for functionally different actions—i.e., potentially depending on different hemispheric specializations. Firstly, lateralization for the manipulation of inanimate objects has been associated with genetic and ontogenetic factors, with specific brain regions' activity, and with morphological limb specializations. These could have emerged under selective pressures notably related to the animal locomotion and social styles. Secondly, lateralization for actions directed to living targets (to self or conspecifics) seems to be in relationship with the brain lateralization for emotion processing. Thirdly, findings on primates' hand preferences for communicative gestures accounts for a link between gestural laterality and a left-hemispheric specialization for intentional communication and language. Throughout this review, we highlight the value of functional neuroimaging and developmental approaches to shed light on the mechanisms underlying human handedness.

Keywords: handedness; grasping; gesture; brain asymmetry; vertebrates; invertebrates; primates; ontogeny; evolution

Citation: Boulinguez-Ambroise, G.; Aychet, J.; Pouydebat, E. Limb Preference in Animals: New Insights into the Evolution of Manual Laterality in Hominids. *Symmetry* **2022**, *14*, 96. https://doi.org/10.3390/sym14010096

Academic Editor: Antoine Balzeau

Received: 29 November 2021
Accepted: 1 January 2022
Published: 7 January 2022

Publisher's Note: MDPI stays neutral with regard to jurisdictional claims in published maps and institutional affiliations.

1. Introduction

Humans exhibit a strong right hand preference for manual actions, which is consistently observed across tasks at the population-level and is so referred as "handedness" [1,2]. Recent meta-analyses assessed more precisely the strong manual bias observed for manipulating items and for different manual tasks. Handedness is usually assessed using questionnaires (e.g., Annett's Hand Preference Questionnaire, Edinburgh Handedness Inventory, Rennes Laterality Questionnaire [3–5]) asking for the preferred hand (right vs. left) when performing a variety of manual tasks: the hand used for writing, for using different tools (e.g., hammer, scissors, toothbrush . . .), performing tasks like unscrewing a lid or threading a needle, or to communicate through iconic, symbolic, or deictic gestures (i.e., physically representing a shape or movement, having arbitrary meanings, or directing other's attention, respectively). While 10.6% of the human population shows a left hand preference, the right-handedness prevalence lies thus around 90% [6]. It is to be noted that values may vary according to the way handedness is measured and exclusion criteria for certain categories of the population (e.g., elite athletes), as highlighted in the five meta-analyses run by Papadatou-Pastou et al. [6].

However, even if this manual bias exhibited for reaching, grasping, and manipulating objects or even interacting and communicating with conspecifics is being better assessed, the mechanisms underlying human handedness are still widely debated on both theoretical and empirical grounds. The large corpus of studies on this topic suggests that, besides genetic factors, non-genetic environmental factors play a significant role and need further considerations [7–9].

The presence of such a population-level right-side bias (i.e., similar proportions) has been demonstrated in hominin species prior to *Homo sapiens*, namely in *Homo neanderthalensis* [10]; as evidence of this, previous studies investigated asymmetric morphological traits of the fossil record like asymmetries in the humeral shape or dental striations, but also asymmetrical retouch patterns on Paleolithic artifacts (i.e., when producing stone or bone tools [11]). In *Homo habilis*, on a maxilla dated to ~1.8 mya (OH-65, found in Olduvai Bed), Frayer et al. [12] documented the earliest evidence for right-handedness (i.e., oblique labial striations) in the hominin fossil record. Another category of evidence from fossils are the asymmetries of the endocast (i.e., cranial vault) as some authors suggested specific patterns of the petalias (i.e., one of the brain hemispheres protruding towards the other, causing an impression on the inner surface of the skull-that can still be visible in fossil skulls) to be associated with right- or left-handedness [10].

Indeed, the evolution of the human brain led to a cerebral lateralization: while some organs are duplicated (i.e., kidneys, lungs), the two hemispheres of the human brain display a functional specialization associated with structural asymmetries [13–15]. This dissociation of specialized processes of left and right hemispheres permits to optimize the associated functions, for instance the language for the left hemisphere, and emotional signals' processing for the right one [16,17]. As the nerve fibers of the motor cortices are contralaterally innervated, the dominant hemisphere processes can manifest as contralateral motor behaviors [18], such as handedness.

Cerebral lateralization is not specific to humans and has been well established in many other vertebrates such as birds, fishes, and amphibians (see [19] for a review; [20,21]), and forelimb preferences (at the individual or population level, for one specific task or across tasks) have been widely demonstrated among non-human animals [9,22]. Cerebral and associated behavioral lateralizations may be beneficial for animals in terms of cognitive and motor performance, notably permitting spatial gain within the brain [21,23] or the processing of several simultaneous tasks [24–26].

Because of the hemispheric specialization, hand use may be mainly processed by different brain hemispheres according to the action performed, and shows differences in lateralization (right- versus left-hand dominance). In this review, we present the results of studies conducted in a wide variety of species (including both vertebrates and invertebrates) that allow us to discuss the potential mechanisms underlying human handedness by identifying three categories of "manual" actions: (1) towards inanimate targets, (2) towards animate targets (i.e., self, conspecific)—that may involve emotion processing, and (3) communicative gestures—involving language-related functions. Comparative research done in the past years has been a real opportunity to better understand the different functions in which limb use is lateralized, and thus better assess the adaptive explanations for the evolution of limb lateralization by better understanding the different selective pressures that may have driven this evolution. Recent studies have also further considered that—besides adaptive explanations—the acquisition of handedness may be related to variations in developmental trajectories in other traits across ontogeny.

2. Current Developmental and Evolutionary Hypotheses on Object Manipulation Laterality

2.1. Is Handedness Genetically Determined?

In humans, as hand preferences run in families, many studies in the past worked on a genetic model [27–30], but no gene has been linked to the expression of handedness. Running genome-wide association analyses (GWAS) with large sample sizes, recent studies investigated more precisely how many loci are involved in determining handedness: the results of the GWAS showed only a handful of significant associations [31–33]. For instance, Cuellar-Partida et al. [31] analyzed data collected on a considerable sample of 1,766,671 individuals (right-handed: 86.88%; left-handed: 10.99%; ambidextrous: 2.13%) and GWAS revealed only 48 statistically significant variants. Furthermore, a meta-analysis on handedness in twins showed that the rate of handedness concordance was higher in monozygotic twins compared to dizygotic twins [34], supporting the idea that genetic factors do play a role in the determination of handedness. However, the heritability of handedness in humans has been evaluated around 24% [30], which is a relatively modest value showing that genetic factors explain less than one quarter of the variance in human handedness, thus contributing only partially to handedness. Also in mice, if the degree of paw preferences is under the influence of genetic effects, these effects only drive the direction of the preference a little; it is to be noticed, however, that some studies brought out that different strains of mice differ in strength and direction (for a meta-analysis in mice, *Mus musculus*, *Apodemus agrarius*, and rats, *Rattus rattus*, see [35]).

These results suggest that other nongenetic factors may also play a significant role in the development of handedness, explaining the remaining variance. In this regard, investigating limb preference in animals brings further elements to better picture all the factors that may affect the development of this trait.

2.2. The Insights of Ontogeny

As nongenetic factors that may influence the acquisition of handedness, a growing number of studies investigate early life parameters. For instance, still in the mice animal model, it has been reported that prenatal stress can affect paw preference pattern, even transgenerationally via epigenetic mechanisms [36,37]. As the mother acts on the immediate developmental environment of the fetus and then of the infant [38], some works focused on potential effects of mother-infant interactions during ontogeny. Since the maternal intrauterine environment is asymmetric [39], it has been suggested that the position of the fetus may play a role in the development of lateralization in the motor system [40,41], such as handedness. In humans, according to Ververs et al. [42], there is a clear lateral bias at 38 weeks of pregnancy for a rightward head turning. A majority of human fetuses seem to place their back on the left side of the mother because of asymmetries of the pregnant uterus shape. For cephalic fetuses (i.e., positioned head-down, which is the majority case) lying on the left maternal side, a rightward head turning could be explained by the fact that it allows them to face maternal movements when walking, or also because the maternal front part is clearer, exposing the fetus to more light but also to more tactile and auditory stimulations [43]. Ultrasound scans made it possible to demonstrate that limb movements emerge during fetal life: young human fetuses already grab the umbilical cord, push the uterine wall, and even repeat hand-mouth contacts [44,45]; while unimanual movements are visible between 8 and 10 weeks of gestation, hand face contacts are recorded from 12 weeks [46,47]. A consequence of the asymmetry in head position for hand use may be that in case of a rightward head turning, the fetus' right hand is more likely to touch the mouth than the left one, which may consequently "encourage movements of the right arm more than the left (...) as the fetus becomes sensitive to sensorimotor contingencies" [43]. Thumb sucking, with a rightward bias, is a very early demonstration of manual asymmetry, observed in utero [48].

Just like in humans, some non-human primate species such as chimpanzees (*Pan troglodytes*), gorillas (*Gorilla gorilla*), and olive baboons (*Papio anubis*) show an interesting asymmetry in maternal cradling behavior. Not only do mothers show a bias at individual-level but also a left-side bias at population-level, which means the use of left arm is favored over the right arm to cradle the infant in a majority of individuals [49,50]. In chimpanzees, Hopkins et al. [51] have found an inverse relationship between this maternal ventro-ventral cradling bias and the offspring hand preference for simple reaching at the age of 3 years. In olive baboons, if cradled on the left, the infant embraces and holds onto the left side of the mother with its right arm, the left hand being free, and vice versa. The hand that is not recruited for clinging on the fur is free for reaching and fine manipulative grasping actions, involving potentially greater motor and neurological stimulation than the other hand. In fact, in this species, early postnatal individual hand preference for unimanual grasping within the first months of age has been positively correlated with the maternal cradling lateralization (infants cradled on the left side of the mother are left-handed, and vice-versa; [52]). Hand preferences assessed later in the development, from 9 to 10 months of age, are less dependent on maternal cradling bias and less consistent with the earlier developmental stages, especially in infants initially cradled on the right maternal side. These findings suggest that maternal cradling behavior might be the first environmental factor that affects the development of early handedness in infant monkeys before being weaned from the mother and letting other mother-independent factors change its onto-genetic trajectory. As maternal left-cradling bias likely reflects brain right hemisphere specialization for emotion [53–55], the early emergence of handedness in baboons might be indirectly related to emotion processing. In Barbary macaques (*Macaca sylvanus*) as well, early life asymmetries in mothers' and infants' behavior seems to affect the development of hand preference: while maternal cradling is lateralized at individual-level, the infants' nipple preference is correlated with their hand preference [56]. However, early postnatal infant lateralization remains poorly investigated in non-human primates. In human and chimpanzee neonates, the only few data available so far reported manual performance asymmetries in the strength of grasping responses [57,58]. In a few other primate species, data on the development of manual grasping and its early lateralization are available at juvenile stages: in capuchins (in *Sapajus apella*, an individual hand preference by 5–6 months of age [59,60], marmosets (in *Callithrix jacchus*, an individual hand preference for unimanual reaching by 5–8 months [61], and rhesus macaques (in *Macaca mulatta*, a population-level bias for both unimanual reaching and bimanual tasks by 4–11 months [62]). Regarding the ontogeny of limb preferences in non-primate species, Wells and Millsopp [63] investigated the development of paw preferences in the domestic cat (*Felis silvestris catus*) and reported a significant effect of age: while individuals were more ambilateral at 12 weeks of age than at later developmental stages, paw preferences at 6 months and at 1 year of age were positively correlated. In marsupial species, the red-necked wallaby (*Macropus rufogriseus*) and the eastern gray kangaroo (*Macropus giganteus*) show a left-forelimb preference (for manipulating food) at population-level as soon as the pouch young stage (approximately 6–7 and 7–9 months old, respectively) [64]. In the eastern gray kangaroo, the authors compared limb-preferences in manipulative behavior at different developmental stages, namely before and shortly after individuals display the bipedal posture (young-at-foot, approximately 11–15 months old): as they observed no difference between these two juvenile stages and the adult stage, the authors concluded that "manual lateralization in bipedal marsupials is not determined by the acquisition of habitual bipedality" but precedes it in the course of ontogenesis [64] (p. 1). Interestingly, in the American lobster crustacean species (*Homarus americanus*), while normal differential claw use during on-togeny induces one claw to transform into the specialized crusher claw, induced insufficient stimulation in laboratory conditions during a specific developmental stage leads to no specialization [65,66], highlighting the strong role that behavioral asymmetry may have "inducing and orienting morphological and subsequently functional asymmetry" [9].

2.3. Brain Correlates of Lateralized Manual Actions

At adult stage, the asymmetric use of the hands for manipulative manual tasks in humans has been correlated to contralateral brain structural asymmetries within a section of the central sulcus related to the motor hand area [67]. Outside the human species, cerebral lateralization has been well established in many other vertebrates such as birds, fishes, and amphibians (see [19] for a review; [20,21,68]. Direction and degree of hand preference for a bimanual task (i.e., tube task, see [69] in nonhuman primates such as baboons, capuchin monkeys, or chimpanzees have been found to be associated—just like in humans—with contralateral neuro-structural asymmetries in the primary motor cortex including the surface of the motor hand area surface, its neuronal densities, or its adjacent central sulcus depth [70–75]). Furthermore, as in humans, the hand preferences tested in a large population of adult olive baboons, for both unimanual and bimanual tasks, are consistent over time [76].

Regarding non-primate species, Australian parrots' footedness is correlated with eye lateralization for discriminating food items, supporting—according to the authors—a functional explanation for the evolution of handedness in vertebrates (Figure 1) [77]. A recent study investigated the association between brain size and parrots' (psittacine) foot preference [78]. It has been shown that cerebral lateralization enhances the brain capacity by allowing parallel processing of sensory information (e.g., to forage efficiently while remaining vigilant for predators) [26]. As the Australian parrot species known for having foot preferences also has a better ability to perform certain manipulative and cognitive tasks compared to species with no foot preference [79], Kaplan and Rogers [78] asked the following question: «Do species with footedness have larger brains, or is footedness a way of compensating for having a smaller brain?» (p. 2). The authors found in several Australian parrots that species with larger brains (i.e., absolute brain mass) have stronger foot preference and that left-footedness is stronger in species with a larger brain. Moreover, the authors found foot preference to be associated with the size of a brain area (i.e., the nidopallium) recruited for higher cognitive tasks, so that species with stronger left-foot preferences have larger brains, with a larger volume of the nidopallium (compared to the whole brain) [78].

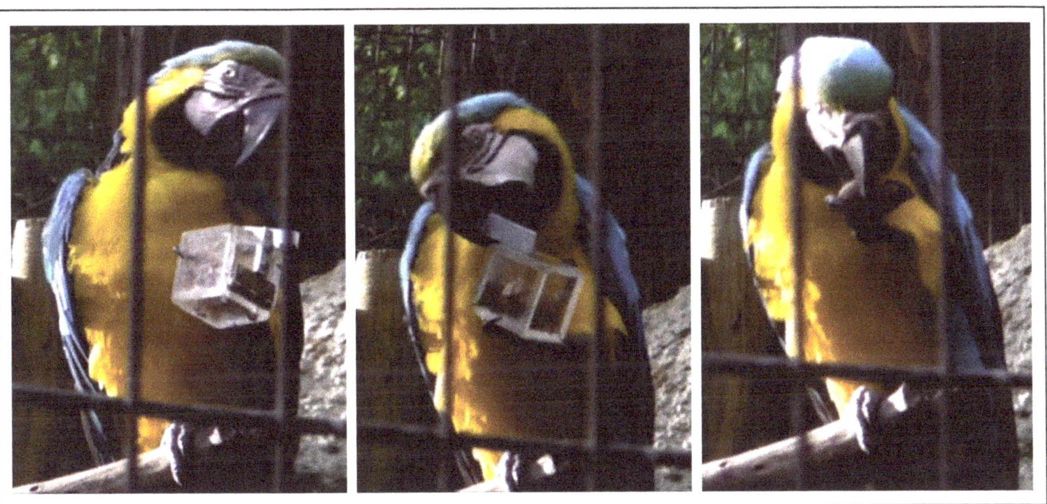

Figure 1. A blue-and-yellow macaw (*Ara ararauna*) opening a box to retrieve a food reward (i.e., a nut) using its left foot and its mouth (at the Ménagerie du Jardin des Plantes, MNHN, Paris). See Brunon et al. [80] for a description of the manipulative repertoire of blue and yellow macaws. Photograph credit: Emmanuelle Pouydebat.

Interestingly, even if the eight arms of octopuses (*Octopus vulgaris*) were traditionally thought to be equipotential, Byrne et al. [81] demonstrated a preference for frontal arms in reaching and exploring objects, as well as a preference for a specific arm to reach into a maze and retrieve a food item. Given the structure of the octopus neural network with each arm possessing its own network operating it (i.e., all arms being coordinated by a central hub in the head), it would be interesting to investigate whether cerebral asymmetries related to limb preference are shaped in these species.

Several structural asymmetries have been observed in the fetal brain, during human development [82–86]. Further studies, benefitting from the improvements in MRI technology, should help to determine whether contralateral hemispheric specialization of the brain within the central sulcus is present at early developmental stages, its potential change across ontogeny, and whether it predicts hand preference at later stages.

2.4. Morphological Limb Specialization

A recent study (published in this volume) assessed grip strength—a common indicator of overall muscle strength—in a large sample of humans (i.e., 662 individuals aged 17 to 83 years), testing the effects of hand dominance (i.e., asymmetric use of the dominant vs. the non-dominant hand, without considering the left-right direction) and handedness [87]. The authors found that both males and females are significantly stronger with their dominant hand compared with their non-dominant hand; however, they found no significant difference in grip strength between right- and left-handed individuals [87].

Whether limb preference is associated with asymmetric body traits in the limb morphology is especially visible in crustaceans, namely the brachyuran crustaceans (i.e., crabs) and lobsters. For instance, In American lobsters, if both the left and right claws are initially similar, they transform during a given developmental stage and become morphologically different: one being a large slow-acting (i.e., closing slowly, made of only slaw muscle fibers) "molar-toothed" crusher claw, and the other being a minor fast-acting (mainly fast fibers) "incisor-toothed" cutter claw; both being used when foraging [65,88]. The pattern of this claw asymmetry in American lobsters appears to be random, with half of the population having the major claw on the right and the other half having it on the left side [9]. Such a 1:1 ratio in limb asymmetry suggests no advantage to any of the two groups. Additionally, in fiddler crabs (genus *Uca*), while a few species have been reported to be predominantly right-clawed (i.e., major claw on the right side; with a population-level bias greater than 95%), most species show populations with equal numbers of right-clawed and left-clawed individuals (for a review, see [89]). These differences in limb (i.e., claw) asymmetry between crab species make it difficult to conclude whether this trait is under selective pressure or a bimodal trait [90]. Perhaps, next studies may further investigate the differences observed between these species by carefully taking into consideration the functional context in which asymmetric claws are used: whether there are more recruited in feeding behaviors or in interactions towards conspecifics (i.e., animate targets) like courtship or fights. Interestingly, in the males of the *Uca vocans dampieri* species—which fight using their enlarged major claw—it has been shown that only 1.4% of males are left-clawed, and that being left-clawed was a disadvantage for fighting (i.e., left-clawed males were both less likely to engage in a fight and less likely to win a fight; [91]).

2.5. The Effects of Posture and Locomotion Mode

If several primate species display a right-hand preference for bimanual manipulative tasks that is associated with a left-hemisphere specialization, some other primate species—namely lemur species—show a left-handedness for reaching food items, suggesting right hemisphere prehension specialization (for reviews, see [92,93]). Facing this paradoxical finding of left-handedness in strepsirrhine species (e.g., sifakas, black lemurs, indris), Mac-Neilage et al. developed in 1987 the "Postural Origins (PO) theory" as an evolutionary theory of handedness in primates [93–95]. The PO theory relies on the fact that several strepsirrhines species show a "vertical clinging and leaping" [96] locomotor style and display unimanual predation: one side of the body ensures anchor to the substrate while the other grabs the prey [93,97]. The authors thus suggested two complementary specializations: a "left hand-right hemisphere specialization for unimanual predatory prehension" and a specialization of the right side of the body for postural support that would be controlled by the left hemisphere [93,97]. Even if the PO theory focuses on primates, it is interesting to report that several parrot species—which are not primate, nor mammal species either—show a similar left-footedness predominance for grasping and holding food items while they use their right foot for perching or climbing (also using their beak to help them climb), (for a review, see [98]).

The PO theory further suggests that given "the greater physical strength of the right side of the body", the right hand would have become "the operative side", favored in object manipulation, in primate species abandoning the vertical clinging locomotor style-giving the forelimbs more freedom with regard to postural support—and showing a more omnivorous diet requiring more manipulative skills for foraging. An interaction between effects of postures and arborealism on the direction of grasping laterality is indeed observable in primate species, being biased in favor of the right hand in terrestrial and bipedal species and of the left hand in arboreal, quadrupedal ones [99–102]. For both a unimanual task (food grasping of grains) and a bimanual task (i.e., tube task: the two hands are used in an asymmetric but complementary matter, e.g., holding a tube with one hand and removing the food inside a tube with the other hand), adult olive baboons, which are mainly terrestrial monkeys, show a right-handedness predominance at population level [76]. Moreover, capuchin monkeys are well known as tool users [103], namely stones to crack nuts, and display a right hand preference for feeding [104] and coordinated-bimanual tasks [105]. Human right-handedness may have then derived from a selective pressure for tool use or coordinated bimanual manipulations [9]; as MacNeilage highlighted in his review, the conclusion of the Hook-Costigan and Rogers [104] study suggested that "tool use and right handedness may have evolved before bipedalism, and well before the apes and, indeed, humans evolved" (p. 195), although bipedalism seem to strengthen right-biased manual laterality consequently to the suppression of the locomotory function of hands [60].

If a large body of literature documents the lateralization of the primate limb motor systems at both individual and population levels [76,93,104,106–108], there is very little comparative research of manual lateralization in non-primate mammals [9,22]. Giljov et al. [109] assessed handedness in marsupial species, one of the other large mammalian lineages. The authors reported a population-level manual preference for multiple behaviors (e.g., unimanual feeding, grooming) in red (*Macropus rufus*) and grey (*Macropus giganteus*) kangaroos, which mainly display a bipedal gait, freeing the hands to perform other tasks. By comparing mainly bipedal and quadrupedal marsupial species, Giljov et al. [109] highlighted the crucial role that postural characteristics (i.e., bipedality), rather than phylogeny, may have play in the origin of handedness in mammals, beyond the order of primates.

These works on lateralization open many perspectives of comparison within tetrapods: the questions addressed and the new ones that arise can be applied to other species outside primates. However, it has to be noted that interspecies comparisons of handedness measured using different tasks has to be done carefully. A task effect related to variation in motor demand has been reported in several studies in both humans and non-human primates: namely, differences between unimanual and bimanual tasks when assessing the hand preference: handedness in unimanual grasping seems to be not as strong as in bimanual grasping [76,110–112], but also less sensitive than bimanual manipulations in detecting population-level bias [113,114]. About the literature focusing on humans, Fagard and Marks [110] highlighted that the use of different tasks to measure handedness (i.e., asymmetric bimanual actions vs. unimanual actions, reaching vs. manipulation) led to contradictory interpretations. To better assess the validity of the PO theory, for instance, handedness in strepsirrhine species should be additionally assessed for a bimanual coordinated manipulation such as in the tube task.

In addition to the above-cited observations suggesting that species locomotory style affects manual laterality, the PO theory is supported by findings on postural effects at the individual level in several mammal species. Indeed, an increased manual preference can be observed in human and non-human primates performing manual tasks in a bipedal compared to quadrupedal posture [60,115–118]. Similar observations have been made in other mammals, such as red-necked wallabies (*Macropus rufogriseus*) [115] (but see absence of postural effect in tree shrews, *Tupaia belangeri* [119], and in cats, *Felis silvestris catus* [120]). This corroborates the hypothesis that the need for postural support acts as a constraint on hand availability for manual actions, and so on manual laterality. The PO theory proposes this as a critical evolutionary mechanism which would have shaped handedness emergence.

2.6. Social Origins of Manual Laterality

The fact that social animals exhibit population-level forelimb preferences [22] also led to the hypothesis that the alignment of individual lateralization may be under specific social constraints. An evolutionary theory has been proposed regarding lateralities at the population level, which postulates that the alignment of individual lateralizations favors the coordination and cooperation between individuals of the same social group [21,121–123] First supported by observations of population-level behavioral asymmetries in social vertebrates (e.g., of flight behavior in fishes, [124]), this theory is also corroborated by group-level lateralizations of social invertebrates compared to solitary ones (see for review: [125]), suggesting a wide phenomenon in animal phylogeny.

In the case of manual actions, social constraints on laterality may arise from the need for inter-individual coordination to perform complex tasks and from social learning. The acquisition of tool use, for instance, may be facilitated if the learner uses the same hand as the teaching expert [123]. Parental hand preferences have been shown to affect the development of children's handedness, notably through social play involving object manipulation [7]. Diverse other social factors affect individuals' hand preferences [37], resulting in slight variations in the rate of left-handedness across different geographical and cultural regions [6,126,127]. One of the most striking examples of cultural pressures modulating handedness might be the constraints exerted against left-handedness for writing and eating, which directly affects the development of children hand preference [128,129].

In spite of these social constraints and of the advantages of the alignment of individual lateralization for intragroup coordination, one can note that in all species in which we observe a population-level laterality (including humans), there still exist a certain proportion of individuals lateralized in the opposite direction compared to the majority of others. Ghirlanda et al. [130] suggested that the frequency of minority laterality results from a costs-benefits balance of behavioral lateralization. Although the alignment of laterality is beneficial for cooperation, it disadvantages individuals in competition contexts, as their decisions become more predictable [24]. In humans, left-handers are indeed more frequently represented in competitive sports, and seem to benefit from strategic advantages (e.g., in tennis [131]). Recently, a large-sample study on professional boxers evidenced greater success for left-handed subjects [132], supporting the hypothesis that fighting interactions may have constituted an evolutionary constraint in favor of left-handedness in humans [133]. As individual- and population-level behavioral lateralizations amongst a species may result from the relative frequencies of cooperative and competitive social interactions [130,134], it may be hypothesized that the high proportion of right-handedness in humans results from a high need for inter-individual cooperation and coordination in manual tasks [135].

While the process of manipulating inanimate objects involves a left-hemisphere specialization (as shown by the contralateral left-brain asymmetries present in the predominant right-handed individuals), a growing number of studies in vertebrates support the idea that the grasping function when involved in interactions with animate targets (i.e., conspecifics, self) is processed differently, namely in relationship with the hemispheric specialization for emotion processing.

3. Manual Laterality for Living Object Manipulation and the Role of Emotional Lateralization

Along with the hypotheses previously mentioned regarding the origins of manual laterality, behavioral asymmetries in animals suggests that hemispheric specializations for specific cognitive mechanisms, such as emotion processing, might have driven the lateralization of associated manual tasks. In gorillas and chimpanzees, hand preference for manipulative actions has been shown to depend on the living nature of the target objects [136,137]. Although these apes exhibited a right-hand bias for inanimate object manipulation (i.e., objects and environment), they used the left as much as the right hand to act toward animate objects (i.e., self or conspecifics), supposedly because self- or socially-directed actions imply emotional processes in addition to manipulative ones. Recently, Baldachini et al. [138] reported concordant observations in Barbary macaques. Although the animacy of targets did not affect the direction and strength of manual laterality at the population level, individual lateralizations differed depending on whether actions were directed to an object or to a living being. These results are in favor of the hypothesis of (socially-driven) emotions affecting the laterality of manipulative actions.

3.1. Current Hypotheses on Emotional Lateralization

Different theories have been formulated and co-exist regarding emotional lateralization in vertebrates, i.e., to explain the differential involvement of the two brain hemispheres in the processing of emotions (see for reviews [139,140]). The "Right hemisphere theory", which postulates that the right hemisphere is specialized in both positive and negative emotion processing, is particularly relevant to explain the asymmetries which are observed in the expression and perception of emotional signals [16,141,142]. Notably, the facial and vocal expressions of emotions are associated with leftward oro-facial asymmetries in humans and other primates [143–147]. Moreover, behavioral leftward-biases have been evidenced for the perception of emotional stimuli (including intraspecific or interspecific signals) in numerous species ([148]; e.g., in humans [149,150]; chimpanzees [151]; olive baboons [152]; vervet monkeys, *Chlorocebus pygerythrus* [153]; and dogs, *Canis familiaris* [154,155]). Other theories suggest that both hemispheres are involved in emotional processing but that the left and right side of the brain are differently involved depending on the emotional valence or motivation elicited by the context. The "Valence theory" thus proposes that the right hemisphere is involved in the treatment of negative emotions, frequently associated with withdrawal behaviors, and that conversely, the left hemisphere is responsible of the processing of positive emotions, frequently associated with approach behaviors [156–158]. The theory has been supported by behavioral asymmetries expressed by numerous vertebrates in contexts with different emotional valences [148,155]. In the last decades, this assumption has been updated by differentiating affective hypotheses (i.e., the left and right hemispheres are respectively associated with positive and negative emotions) from motivational hypotheses (i.e., the left and right hemispheres are respectively associated with approach and withdrawal/flight motivations). This idea is strengthened by the fact that a positive relation was found between right-handedness and approach motivations in captive chimpanzees in an experimental context [159]. A similar association was observed in Geoffroy's marmosets (*Callithrix geoffroyi*), in which right-handed individuals presented with novel objects seemed less fearful and exhibited more frequent approach behaviors than left-handed subjects [160,161]. Although affective and motivational hypotheses have long been amalgamated, they can result in contrary predictions for some contexts such as aggressivity, which involve both negative emotional states and high approach motivations [139,162–164].

Some authors propose that the affective hypothesis may explain how the valence of a particular situation is experienced by individuals, whereas the motivational hypothesis may account for the decision-making process to approach or avoid an emotional stimulus [162,163,165]. It is to note that these different theories might all be compatible with one another, as they relate to different levels of cerebral processing of emotions [140,166].

As we may assume that self- or socially directed manual actions are more likely than others to be underlined by specific emotional states, the associated hand preferences may be representative of the role of emotional lateralization in the evolutionary and developmental history of handedness.

3.2. The Case of Self-Directed Manual Actions

Behaviors such as scratching or self-grooming may be identified as "displacement activities" in response of social or predatory stress in primates [167–172]. Therefore, several primatology studies have addressed the potential brachio-manual asymmetries in self-directed actions commonly considered as indicators of negative emotional states, but have led to discordant results. Some revealed a left-hand preference for self- touching and scratching in great apes, as expected according to both the "Right hemisphere hypothesis" and the "Valence theory" considering the negative emotional state associated with these behaviors, such as anxiety (in humans, gorillas, chimpanzees, and orangutans [173–175]). Human children exhibit similar leftward bias for actions directed to self [176], and left-handed face touch in fetuses has been evidenced to be associated with maternal stress [177]. Other studies showed a preference for the right hand for self-scratching in squirrel monkeys [178] and for self-rubbing but not scratching in chimpanzees under stressful situations [179,180], highlighting the effect of the type of self-directed on laterality. Bard et al. [181] also evidenced a right-hand preference for self-calming behaviors in young chimpanzees under human care (e.g., "hand-to-mouth" behaviors, such as thumb sucking, and "hand-to-hand" grasps, i.e., holding and pressing one hand with the other). Authors interpreted it as resulting from the left hemispheric specialization for anxiety regulation in mammals, and notably for dopaminergic reward circuits whose activity is affected by stressful stimuli [182–184]. Finally, other research works did not show any lateral bias for self-directed manipulation in primates [173,178,185–187], though it is to be noted that they were based on small samples of subjects, which may prevent the evidence of a population-level bias [188,189]. Very few studies in other animals as in primates reported forelimb laterality for self-directed actions (e.g., no lateral bias for autogrooming in rats and mice: Stieger et al., 2021), but noteworthy results on these topics arose from marsupial studies [109]. Comparably to the leftward lateralization of self-touching observed in great apes, red-necked wallabies (*Macropus rufogriseus*), Eastern grey kangaroos (*Macropus giganteus*), and red kangaroos (*Macropus rufus*) preferentially use their left limb for autogrooming in bipedal position [64,109,115], which suggests similar hemispheric specialization for emotional control in these marsupials than in primates. Interestingly, such lateralization for self-touching has not been reported for the Goodfellow's tree-kangaroos, *Dandrolagus goodfellowi*, which is mainly arboreal, suggesting some effect of species characteristics [109].

3.3. Laterality of Conspecifics-Directed Manipulative Actions

A difference in hand preference between unimanual interactions with inanimate targets (i.e., food, objects) and physical contacts made toward a conspecific has been reported in ape species-chimpanzees and gorillas: in the two studies conducted by Forrester et al. [136,137], while a group-level right hand preference for interaction with inanimate targets was confirmed in these species, no right-handed bias was reported toward conspecifics, further suggesting that manual lateralization reflects right- or left hemisphere processing according to the emotive or functional characteristics of the target. The right hemisphere of vertebrates seems specialized for the processing of social information, notably for the purpose of emotional signal perception [16,141,142] or individual recognition [190]. This may result in a higher involvement of this hemisphere for performing manual actions directed towards conspecifics compared to manual actions directed to inanimate objects, resulting in a higher use of the left side of the body.

Interestingly, forelimb actions directed to conspecifics may also be lateralized in invertebrate species. A greater involvement of the left body part in interactions towards conspecifics has been observed in insects: for instance, in the Mediterranean fruit flies

(*Ceratitis capitata*, "medflies"), Benelli et al. [191] observed a left-biased population level lateralization of aggressive displays executed with their forelegs. In fact, during a fight, when boxing with their forelegs (i.e., the boxing attacker raises a foreleg, hitting the opponent on the head or thorax), a majority of medflies (almost 70%) were "left-handed" (i.e., performing with their left foreleg significantly more than with their right one); moreover, the authors reported that performing aggressive displays with the left body parts (including foreleg and wing) enhanced fighting success compared to those performed with right body parts [191]. This lateralization in insects may not be homologous (i.e., be inherited from a common ancestor) to the left-hand/right-hemispheric preference observed in vertebrates for socially-directed actions, hence it reflects the possibly ubiquitous nature of the constraints that social interactions represent on the lateralization of social animals' behaviors [192].

3.4. The Case of Maternal Cradling

A phenomenon that raised lots of questions is the maternal left-cradling bias that has been demonstrated in humans (66–72% [193]) but also in great apes (chimpanzees and gorillas [50]) and more recently in olive baboons in the same proportions [49]. At the human population-level, inanimate objects (i.e., bags) are carried on the right side for the greatest part [194]. However, just a pillow adorned with a proto-face is enough to elicit a left-cradling bias in children [53]. A study even asked adult humans (including both women and men) to imagine themselves holding in their arms an object (i.e., either an expensive vase or an old shoebox) and then an infant (i.e., about 3 months of age): while a right-cradling bias was reported for both imagined objects, a left-cradling bias (i.e., 66%) was reported for holding the imagined infant [194]. When cradling their baby, mothers hold their infant in their arms close to their body, positioning the infant's face in one of their peri-personal hemispace (e.g., left side of their body) and supporting the weight with the corresponding arm (e.g., left arm), see Figure 2. The maternal left-cradling bias seems not to be related with the mother handedness [53]. Next to the manual preference for manipulating items, the heart position (i.e., soothing sound of heartbeats [195–197]) and cultural considerations [198] do not affect the left-cradling bias. The theory reaching a consensus combines visual field and cerebral hemispheric specialization. The maternal cradling bias would reflect the right-hemispheric dominance for emotional processing [55]. In fact, the brain right hemisphere is specialized in the perception of emotional facial expressions [16,199]. Since left-side cradling exposes the baby face to the left visual field of the mother, which is projected mainly to her right brain hemisphere, this would favor the mother's monitoring of the emotional state of the infant. In parallel, the left-cradled infant looks at the left side of the mother's face, which has been described as being the most expressive [142,200,201]. According to some authors, this direct access to the mother's emotional state would then facilitate creating and reinforcing social bonds within the mother baby dyad [55].

Furthermore, in human mothers, affective symptoms such as stress, depression, and anxiety can alter left cradling, reflecting a reduced ability to be emotionally involved with the infant [54,202–204]. A recent study investigated the link between left-cradling bias and the maternal emotional state in a non-human primate, the olive baboon [49]. The authors found the maternal cradling bias to shift toward a right bias in mothers living in high density groups with higher social pressure, likely involving higher levels of stress for the mothers (e.g., by increased frequency of conflicts and severe aggression). The socially related stress would alter the rightward hemispheric resources allocated to the maternal monitoring and ultimately affect the left-cradling bias [21]. Those results clearly illustrate the phylogenetic continuity between humans and catarrhine monkeys concerning this lateralization and its potential links with hemispheric specialization for emotions, inherited from a common ancestor 25–35 million years ago.

Figure 2. Maternal cradling in olive baboon (*Papio anubis*). A baboon mother cradles her infant on her left side. Photograph credit: Grégoire Boulinguez-Ambroise.

Interestingly, the lateralization of cradling in human mothers is under further investigations to assess the potential of this behavior as a tool to better understand and even early diagnose social disorders in infants, namely autism spectrum disorders (ASD). Several studies have already shown that atypical trajectory in maternal cradling might be one of the early signs of interference in dyadic socio-emotional communication, and thus of potential neurodevelopmental dysfunctions: for instance, right-cradling bias may be associated with a lack of social interactions or degraded interactions within the mother-infant dyad and induce disorders later in life, regarding sociality, namely socio-emotional communication; also, a left-cradling period which lasts too long may reflect the overstimulation in which mothers try to engage ASD infants in response to their lack of responsiveness and social initiative [54,205–209].

Asymmetries in an infant's positioning have been also reported in non-primate species that do not carry their babies. In a wide range of marine and terrestrial mammals, juveniles have a strong preference for keeping their mother on their left side, namely in their left visual field [210]. This has prompted previous authors to propose the idea that the right lateralized "social brain" as described in primates has an ancient evolutionary origin. It would be derived from earlier forms of lateralization in vertebrates, namely lateralization in interactions within the mother-infant dyad that promote bonding and thus maximize the infants' survival.

4. Gestural Laterality and Language Evolutionary Origins

A particular case of socially-directed manual movements are communicative gestures, whose laterality presents specific features compared to manipulative actions and whose characterization in non-human primates provides valuable insights into the evolutive history of human handedness and language.

4.1. A Complex Relationship between Handedness and the Hemispheric Specialization for Language

Humans present a left-hemispheric specialization for language functions, involving in particular the Broca and Wernicke's brain area for the production and processing of speech, respectively in the Inferior Frontal Gyrus and Planum Temporale [211–217]. The strong right-handedness observed in the human species has long been hypothesized to be uniquely related to this brain specialization for language [218,219]. This assumption was based in particular on the mirror neuron system being apparently predominant in the left-hemisphere and driven by neurons of the left Inferior Frontal Gyrus in humans [220,221]. First evidenced in the ventral premotor cortex of rhesus macaques, more specifically in the F5 region which is considered as Broca's area homologue, mirror neurons have the particularity to discharge both during the production of a manual action and during the observation of another individual producing it [222–224]. A large number of studies implying functional neuroimaging in humans have then shown, however, that mirror activity could be evidenced in a wide range of brain regions, both on the left and right hemisphere [225,226]. Moreover, in spite of the above-cited theory, results arising from both neurofunctional and behavioral studies suggest a rather indirect relationship between human language and handedness (see for review [8]). Recent fMRI (functional Magnetic Resonance Imaging) studies revealed independent neuronal circuits for language processing and action observation [227–229]. Häberling et al. [228] notably brought to light three distinct networks within the mirror neuron system which were related to language production and processing, to tool use, and to subjects' handedness, defined in this case as the preferred hand used for writing. Interestingly, among these three networks, only the handedness-linked one was for the most part independent from the Broca's area, and was mainly composed by circuits of the parietal lobe. In addition, it seems that the direction of the laterality for manual actions and the hemispheric specialization for language are relatively disentangled. Indeed, although the incidence of right-hemisphere language dominance is higher among left-handers compared to right-handers, the vast majority of left-handed adults (above 70%) still show a left-hemispheric lateralization for language production [17,230]. The reduced hemispheric lateralization for language production observed amongst left-handed individuals, rather than being due to a reversed asymmetry, might result from a generally weaker lateralization at both the group and individual levels for different cognitive functions [231,232].

By contrast with manipulative actions, the production of communicative manual gestures involves brain regions that are similar to those underlying verbal languages in the left hemisphere [233–238]. Moreover, the tight link between articulated and gestural communication can be observed early in development, the production of pointing gestures playing a key role in the ontogeny of verbal language [239–245]. Population-level right-hand preferences may be observed for the production of communicative gestures in humans, i.e., for co-speech gestures [246,247] (but see [248]), for sign language by deaf adult speakers [249], as well as for deictic and symbolic gestures in preverbal babies, children and adults [3,4,250–254]. Furthermore, even though no significant difference has been found between the direction of manual preference for some communicative gestures and coordinated bimanual actions in adults [251,252], the laterality observed for communicative and non-communicative manual movements seems to be related to different brain region specializations. This is especially underlined by behavioral descriptions in young children, which show that right-hand preference is stronger for gestures (i.e., pointing or signing) than for non-communicative manual actions, suggesting that these two types of manual laterality develop independently [241,242,250,253–256].

Gestural laterality is thus likely to have an evolutionary history inextricably linked to the emergence of intentional communication. In that respect, the great body of research regarding the gestural communication of non-human primates has shed light upon the evolutionary roots of the left-hemispheric specialization for gestures and language.

4.2. Gestural Laterality in Non-Human Primates

As a matter of fact, brachio-manual communicative gestures are found both in human and non-human primates, in which communication relies strongly on the visual sensory channel [257–265]. The question of whether the gestures of humans and other primates (particularly great apes) are homologous has long been a debate, which has been limited by the heterogeneity of studies' focuses depending on the species (see for reviews [266–268]). However, in addition to recent results showing that human infants share the most part of their gestural repertoire with chimpanzees [269], the fact that the functional definitions of primate gestures have been built based on developmental psychology studies [260] allows us to make relative comparisons. The formal gesture definitions used in primate studies may vary from one study to another (see for review [267,268,270,271], yet the communicative nature of these movements is the core elements which functionally differentiate them from other actions. In that respect, the terms "manual gestures" refer to brachio-manual movements which (i) are directed to a recipient; (ii) receive a voluntary response, i.e., induce a change in the recipient's behavior without acting as a direct physical agent, and thus (iii) are mechanically ineffective. Intraspecific manual gestures which fulfill these criteria have long been thought to be unique to humans and great apes [272–274], but in the last years so-defined gestures have been reported in the spontaneous communication of other catarrhine primates (e.g., in olive baboons [275–277]; bonnet macaques, *Macaca radiata* [278]; red-capped mangabeys, *Cercocebus torquatus* [279–281]). To our knowledge, no such forelimb gestures (i.e., apparently intentional) have yet been demonstrated outside the primate lineage, hence the following discussion will focus on this clade.

Interestingly, a right-biased gestural laterality is observed at the population level in great apes, both for gestures directed to humans in experimental contexts, such as pointing or requesting (e.g., in chimpanzees, *Pan troglodytes*, bonobos, *Pan paniscus*, gorillas, *Gorilla gorilla*, and orangutans, *Pongo pygmaeus* [107,282]) and for intraspecific, spontaneous gestures (e.g., in chimpanzees [283–285]; in gorillas [286–289]). Similar findings were reported in primate species that are more phylogenetically distant from humans, especially in olive baboons whose production of threatening "hand-slap" is preferentially produced with the right hand, in intraspecific as well as interspecific contexts [275,276]. Moreover, this gestural laterality is stable through time at the individual level in baboons and chimpanzees [275,290]. As in the case of children, non-human primates' hand preferences for intraspecific gestures are not correlated with manual laterality for non-communicative actions, whether they are manipulative or self-directed (e.g., in chimpanzees [284,285] and in baboons [291]). Experimental studies also evidenced in other Cercopithecidae species that subjects' hand preference for interspecific communicative gestures (i.e., pointing) was uncorrelated to hand preference for manipulative actions (i.e., food grasping), suggesting different cerebral bases for these two types of laterality (in Tonkean macaques, *Macaca tonkeana* [292]; in Campbell's monkeys, *Cercopithecus campbelli*, and red-capped mangabeys [293]). By contrast, Meguerditchian and Vauclair [291] showed that handedness scores computed for different communicative gestures in olive baboons (i.e., "food-beg" and "hand-slap" gestures) were significantly correlated. Additionally, the manual preferences evidenced for pointing gestures in experimental conditions were shown to depend less on the position of the referent object than in the case of grasping actions (in olive baboons [256,294]; in Tonkean macaques [292]; Campbell's monkeys and red-capped mangabeys [293]), similarly to the pattern of manual laterality observed in human children [256]. Moreover, a divergence between gestural laterality patterns was found between platyrrhine monkeys (tufted capuchins) and catarrhine species (human infants, olive baboons, and Tonkean macaques) in a comparative experiment involving pointing gestures [295]. These results suggested that the right-biased gestural laterality observed in catarrhine species may be limited to this clade. However, gesture studies in platyrrhine primates are still rare (but see experimental studies on learnt begging or pointing gestures [296–299]) and do not address the potential laterality of spontaneous brachio-manual gestures in these species, which are phylogenetically more distant from humans than African and Asian monkeys [300]. Thus, supplementary research work might be needed in this area in order to assess when gestural laterality emerged in primate phylogeny.

All the above-cited behavioral data suggest that catarrhine primates all share a left-hemispheric intentional communication system, which support their gesture production. This theory is supported by neuroanatomical and neurofunctional imaging studies revealing a relationship between gestural laterality and brain regions homologous to language-related cortical area in African primates (see for reviews [8,301–303]). One of the first key results in this area has been RMI imaging in great apes showing anatomical asymmetries within cortical regions homologous to Broca's area, which were found to be enlarged in the left hemisphere (in chimpanzees, bonobos, and gorillas [304]). A contralateral association was then evidenced between the direction of gestural laterality and the anatomical asymmetries found in the Inferior Frontal Gyrus and Planum Temporale of adult chimpanzees [290,305,306]. The direct link between the production of communicative gestures and the activation of these cortical regions was then brought to light by functional imaging (PET-MRI: Positron Emission Tomography–MRI) [307]. More recently, Marie et al. [308] showed for the first time a population-level asymmetry of the Planum Temporale in a non-hominoid species, olive baboons. Above the 96 study subjects, 62.5% presented an enlarged Planum Temporale in the left hemisphere, consistently with the population-level asymmetry observed in humans and chimpanzees [305,309–311]. A study conducted in the same baboon population then revealed that this leftward planum temporale asymmetry already existed in the early development of individuals, being observable in newborn baboons and getting stronger in their first year of life [312,313]. Comparably, in humans, the asymmetry of the planum temporale can be observed before birth and continuously develop in favour of the left hemisphere [314]. A longitudinal neuroimaging study evidenced that similar leftward asymmetries of the planum temporale as well as of the Inferior Frontal Gyrus may be observed consistently from 1 to 19 months old in another species of catarrhine monkeys, rhesus macaques [315]. A preprint study authored by Becker et al. [316] reported that olive baboons may also exhibit anatomical asymmetry of markers of Broca's homolog, and that the direction and depth of this asymmetry may be associated with a contralateral gestural lateralization but not with laterality for non-communicative, manipulative actions. At this point, it remains to be investigated whether these anatomical asymmetries in baboon brains are functionally associated with a specialization for the control of gestural communication, similarly to great apes [307], in adulthood as well as across development. By contrast with the trend observed in the human gesture literature, very few studies have explored the development of apes and monkeys' gestural communication [267,317], resulting in a scarcity of data related to the ontogeny of gestural laterality. However, the first promising results cited here pave the way for exciting new research perspectives, exploring whether and how monkeys' gestural laterality develop during their early years of life, potentially in line with the development of cerebral asymmetries.

According to all the commonalities between humans and other catarrhine primates regarding gestural laterality and the associated brain asymmetries, several evolutionary hypotheses proposed that a left-lateralized gestural communication system may have already existed in the brain of the common ancestor of African and Asian primates, over 29 million years ago [8,295,300,302,303,318]. The neural substrates of human intentional communication would then have derived from this hemispheric specialization for gestures, under different evolutionary constraints and at different phylogenetic levels [8,303,319]. Notably, ecological changes might have represented significant pressures shaping catarrhine visual communication, such as a shift from arboreal to terrestrial habitats, associated with an increased visibility and a change of locomotion patterns [320,321]. Moreover, modifications of social systems (and consequently of social complexity) might have affected the extent to which communication relied on brachio-manual gestures, and then on language in the human lineage, depending on the need of sufficiently diverse and flexible communicative signaling to deal with different contexts of cooperation, competition, and cultural transmission [322–326]. Therefore, in line with these theories, the characterization of factors affecting gestural laterality in non-human primates is of great interest for the purpose of depicting the constraints under which humans' gesture and language laterality emerged. We will present, in the following, the main proximate and ultimate causes that have been hypothesized and/or shown to affect the gestural laterality of catarrhine primates.

4.3. Ultimate and Proximal Factors Impacting Primate Gestural Laterality

4.3.1. Effect of Species and Study Population Characteristics

Firstly, the existence of a population-level gestural laterality and its strength appears to depend on the species characteristics, particularly in relation with variation in social systems and ecological characteristics [327]. According to the theory of a social origin of laterality (Section 2.6), it may be predicted that species with high levels of inter-individual cooperation will be more likely to exhibit alignment of individuals' gestural laterality. Moreover, the strength of gestural laterality may depend on social constraints in these species. Observational studies on captive gorillas and chimpanzees brought to light such effects of social dynamics on lateralization of the species' most frequent gestures [288,289,327]. When comparing the production of brachio-manual gestures shared by both species, Prieur et al. [287] found for instance that gorillas were more right-handed than chimpanzees when producing auditory gestures, such as "slap hand". These gestures are more frequent in gorillas probably because of the higher inter-individual distances generally found in this species compared to chimpanzees [328], and are therefore more likely to be socially codified and lateralized. Other species characteristics than sociality might affect gestural laterality, such as the locomotory posture. The theory of a postural origin of manual laterality suggests that the right hand is specialized for complex tasks in terrestrial mammals, the use of one hand or another being less limited by the need for postural stability than in arboreal species [93–95]. In the case of visual communication, it may thus be hypothesized that terrestrial primates are more lateralized when producing brachio-manual gestures than arboreal ones. To our knowledge, there exist no direct comparison of gestural laterality between primate species with different locomotory postures, although comparative studies would be very beneficial to the debate on the origin of primate manual laterality in general [100]. That said, it may be noted that intraspecific gestural laterality has been essentially evidenced at the population-level in terrestrial species, namely chimpanzees, gorillas, and olive baboons [107,275,276,284,285,288,289,327], and spontaneous gestures seem generally more used in primates living in open environments which facilitate the perception of visual signals [329–331]. Finally, the population characteristics, such as the wild or captive environment in which apes and monkeys are studied, may also have an effect on the laterality of manual gestures. Some authors suggested that manual laterality measured in captive primates may be the artefactual results of experimental conditions, particularly those implying the presence of human experimenters ([332–334] but see [188]). Concerning gestural laterality, the effect of captivity is not completely elucidated, particularly because of the lack of direct comparisons between wild and captive populations of primates, and because of the small numbers of studies addressing spontaneous, intraspecific gestures. It is to be noted, however, that some great ape studies show a higher right-hand preference for intraspecific gestures than for human-directed ones, in spontaneous or experimental contexts [107,287].

4.3.2. Effect of Gesture Characteristics

Thus, the preference of one hand to communicate also depends on the characteristics of the gesture itself. The gestural laterality measured in primates differs according to which gesture of the repertoire is studied (e.g., in chimpanzees and gorillas [282,285,289,319,335,336]. As evoked earlier, some authors propose that the most frequent gestures are more likely to be shaped by social pressures, explaining stronger hand preferences when producing them [285]. Moreover, the sensory modality in which the gestural signal is delivered (i.e., visual only, tactile, or audible) impacts its laterality, notably because visual and tactile gestures are more physically directed to the recipient than audible ones, hence are produced preferentially with the ipsilateral hand in relation with the position of the receiver (see in chimpanzees [285]).

4.3.3. Effect of the Interactional Context

The context in which gestures are produced, and particularly the emotional value of certain social situations, has been proven to affect gestural laterality in several primate species. Prieur et al. [285,289] demonstrated that the right hand preference of chimpanzees' and gorillas' gestures was stronger in contexts associated with negative emotional valence than in others. Recently, similar observations were made in red-capped mangabeys [280], in which brachio-manual gestures were more produced with the right hand than the left in aggression and submissive contexts compared to positive or neutral social situations (Figure 3). These results corroborate findings in humans, showing an activation of prefrontal regions of the left brain hemisphere in aggression contexts [164]. This right hand preference for aggressive gestures might be explained by "motivational hypotheses" in line with the "Valence theory" of emotional lateralization (see II.1.), which proposes that the left and right brain hemisphere are respectively specialized in approach- and withdrawal-motivated behaviors [139,156–158,162–164]. Indeed, although aggressive gestures might be underlined by negative emotions in the signaler (for instance, anger), they are also likely to imply a high motivation for approaching the interactant, and thereby to specifically involve left-hemispheric brain regions. In other vertebrate species, non-communicative aggressive behaviors have been found to be lateralized to the right and controlled by the left-hemisphere (e.g., in fishes, *Gambusia holbrooki*, *Xenotoca eiseni*, *Betta splendens* [337]; in green and brown anoles, *Anolis carolinesis* and *A. sagnei* [338]; in mammals such as the European fallow deer, *Dama dama* [339]; but see in domestic and Przewalski horses, *Equus caballus* and *E. przewalskii* [340,341]). Apart from this "motivational hypothesis", the fact that catarrhine primates preferentially gesture with the right hand in aggression situations may be explained by a lesser flexibility of communication in negative contexts, as evidenced for some vocalizations (e.g., alarm calls [342]). Aggressive gestures may be more lateralized than others consequently to a stronger effect of social influences through ontogeny (Section 2.6), and thus be less submitted to the influence of proximate factors. This may be the case of the threatening "hand-slapping" of olive baboons, which has been shown to be highly right-handed at the individual and population-level [275,276].

The emotional value of the interaction is not the only contextual factor which has been shown to affect gestural laterality. The hand used to communicate in primates may also depend on the relative position of the receiver and signaler [280,285,287,327]. One explanation for this is the directionality of gestures, which can result in the use of the ipsilateral hand to efficiently convey visual or tactile signals to a receiver. Moreover, as several primates favor one side or another to approach conspecifics, notably depending on dominance relationships (e.g., in red-capped and grey-cheeked mangabeys, *Cercocebus torquatus* and *Lophocebus albigena* [343]) or depending on the type of interactions (e.g., embracing and grooming in Colombian spider monkeys, *Ateles fusciceps rufiventris* [344]), it may be hypothesized that social laterality affects manual preferences in social interactions, including for gestural communication.

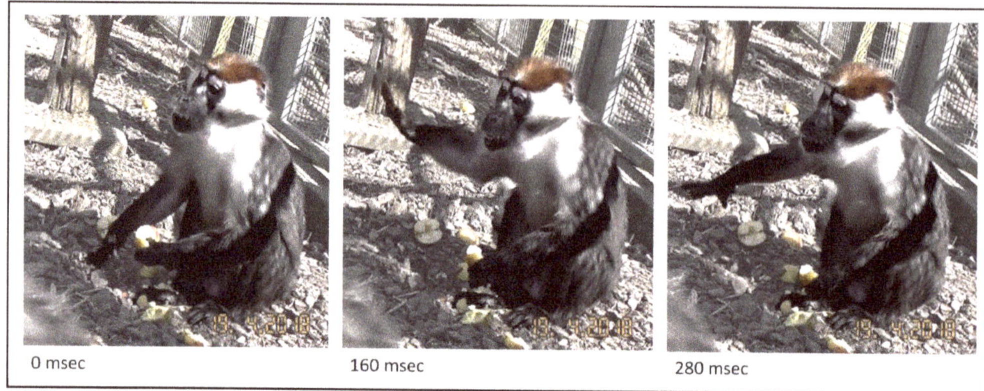

Figure 3. A male red-capped mangabey (*Cercocebus torquatus*) produces a rightward "throw arm" gesture, as part of a threatening display directed to a conspecific. Similarly to chimpanzees and gorillas, red-capped mangabeys gesture preferentially with the right hand in aggression contexts [280]. Photograph credit: Juliette Aychet.

4.3.4. Effect of Signaler and Receiver Characteristics

Primate gestural laterality indeed depends on the relationship between interacting individuals. In captive chimpanzees, gestures are more lateralized to the right hand when directed to dominant conspecifics, and this effect is lessened if the interactants are strongly affiliated [285]. These observations might be explained by the signaler's emotional state varying depending on the identity of the receiver, and particularly the level of psychosocial stress induced by competitive contexts. Such dominance effect has not been evidenced in captive gorillas [327], who exhibit lesser intragroup competition compared to chimpanzees [345–347]. Kinship between signaler and receiver seems to not affect the gestural laterality of gorillas and chimpanzees [285,289,348], however the possible effect of such factors has not been tested in other species.

In addition, demographic factors, i.e., the sex and age of the signaler, has been evidenced to affect gestural laterality in several primate species. Although no effect of sex has been found in chimpanzees and olive baboons' gestural laterality [275,276,291,348], male bonobos have been found to be more right-handed than females for gesture production [186]. Moreover, the converse sex effect has been evidenced in gorillas, in which females are more lateralized in favor of the right hand than are males, and in which males are more right-handed when they gesture toward females than toward male conspecifics [289]. Considering the social structure of these two species, we may hypothesize that these results are related to the dominance relationships of subjects (i.e., female dominance in bonobos and male dominance in gorillas [346,349]), yet the determinants of sex effect on primate gestural laterality is not clear. In humans, more left-handers are found amongst men than women regarding handedness in general, and some cognitive processes are lateralized differentially depending on the individual's sex [6,350–352], though no difference seems to be found for language-related functions and corresponding cerebral asymmetry [352,353]. Different hypotheses have been formulated concerning the effect of sex on human handedness or forelimb asymmetries for non-communicative actions in other mammals (e.g., cats and dogs, *Felis cactus* and *Canis familiaris* [354–356]). Notably, authors suggest a possible effect of sex hormones on cognitive lateralization [350,355], of genetic determinants located on the X chromosome [219,357,358] (but see [31]), and of gender-dependent differences in individual social experience through ontogeny, in the case of humans [359].

Finally, primate gestural laterality is affected by individuals' ages. In chimpanzees, gorillas, and olive baboons, the preference for right-hand gesturing is stronger in adults compared to juveniles [276,282,285,289,335]. This may be due to a maturation of the left hemisphere specialization for intentional communication [276], or it may result from the subjects' individual experience, as adults' gestures are more likely to have been shaped by social experiences [360]. Further studies on the ontogeny of gestural communication would provide a better understanding of this phenomenon. Prieur et al. [285] also observed a senescence effect on chimpanzee gestural laterality, older individuals being less lateralized than young adults possibly because of physical limitations. Nevertheless, other studies have not highlighted any significant effect of age on the gestural laterality of captive chimpanzees and olive baboons [275,361]. On the whole, the extent to which sociodemographic factors affect primate gestural laterality is still poorly or not described in most species.

Multifactorial analyses have been applied to characterize the effect of all these parameters on the gestural laterality of captive chimpanzees and gorillas [285,289,327], and more recently in an exploratory study on captive red-capped mangabeys [280]. Studying primate gesturing with a multifactorial as well as comparative approaches represent promising perspectives for the understanding of the proximate and ultimate factors which shaped human gestural laterality [100,285,289]. In addition, further research efforts are needed with respect to the ontogeny of primate gestural communication and to the potential gestural laterality of non-hominid species, in order to better understand the evolutionary and developmental pathway of this trait.

5. Conclusions–The Way Forward

We aimed to emphasize here the importance of characterizing animal limb preferences to understand the development and evolution of human handedness, by distinguishing laterality for functionally different manual actions (i.e., object manipulation, actions directed to living targets, and non-manipulative, communicative gestures), which might be supported by different hemispheric specializations.

Future research may benefit from recent advances in neuroimaging methods [362], notably functional techniques permitting researchers to link lateralized behaviors to specific brain regions' activity. For instance, the use of functional Near-Infrared Spectroscopy (fNIRS) has been recently validated in non-human primates, allowing non-invasive recording of brain processing lateralization from a functional perspective [363].

Moreover, further developmental studies in different animal species may be needed to unravel the ontogeny of manual lateralities (for instance, regarding the development of gestural communication in non-human primates [267]). Improvements in MRI technology should help to determine whether contralateral hemispheric specialization of the brain is present at early developmental stages, its potential change across ontogeny, and whether it predicts limb preference at later stages [313].

Additionally, one of the major challenges for the understanding of human handedness origins is the improvement of comparative approaches, as still few studies directly involve several species [364]. Research on animal forelimb asymmetries often focuses on mammals (and particularly on the primate lineage), however comparisons of forelimb preferences across a wider range of vertebrate and invertebrate species may provide valuable insights into the evolutionary constraints that have shaped this trait [161]. Behavioral lateralization similarities in species which are phylogenetically distant may result from evolutionary convergence. Their characterization may thus permit us to make hypotheses on the ecological constraints which led to their emergence.

Finally, reliable comparisons of forelimb lateralizations in different species may only be made by homogenizing the task complexity in both experimental and observational studies. Human manual laterality has been argued to be unique because it is observed across different tasks in a large part of the population compared to other species, but one could argue that animals' forelimb asymmetries are rarely assessed for as complex tasks as the ones investigated in humans (e.g., writing or other complex tool uses). Quantitatively, would the right bias observed in humans still be as strong as the ones observed in other species when performing lower demanding tasks? Marchant et al. [365] described humans' manual preferences in diverse spontaneous actions, based on film archives of three traditional societies. They evidenced only a weak overall lateralization for manual actions (barely above 50% of right-hand use in the three study populations), but interestingly found a greater right-hand preference when specifically considering precision tool use (above 84% of right-hand use). The authors thus noted that "the disparity between the ethological and the typical psychological findings on handedness may thus be simply explained: questionnaire and performance testing paradigms focus only on a small and selected proportion of manual activities, those to do with tool use, and especially with skilled, fine-motor tool use. This gives an artefactual, biased picture of extreme lateralization." (p. 256). Task complexity has been hypothesized to affect manual laterality, individuals being more likely to be lateralized for actions with high level of manipulative requirements [111,161]. This has been evidenced in diverse mammal species (e.g., in human [365,366]; non-human primates [367–369]; marsupials [370]; or rodents: [371]), and may be true in other tetrapods, as authors observed stern clawed frog (*Wenopus tropicalis*) changing paw for food manipulation depending on the animacy of the target (Pouydebat et al., unpublished data), due to different levels of manipulative action complexity [372]. To adapt the experimental tasks and protocols in relation to the cognitive and functional capacities of the species (and even their ecology)—in order to propose similar tasks in terms of complexity—represents a real challenge for future studies on limb preference. Moreover, assessing human handedness based on ethological descriptions of spontaneous manual activities may provide more reliable research material to compare with animal observations.

Investigating animal forelimb laterality for diverse (clearly defined) tasks by adopting multi-disciplinary, developmental, and comparative approaches might represent promising perspectives for the understanding of handedness origins.

Author Contributions: Conceptualization, E.P.; writing—original draft preparation, G.B.-A. and J.A.; writing—review and editing, G.B.-A., J.A. and E.P.; supervision, E.P. All authors have read and agreed to the published version of the manuscript.

Funding: The project has received funding from Emergence SU (projet HUMDEXT-Pouydebat).

Institutional Review Board Statement: Not applicable.

Informed Consent Statement: Not applicable.

Acknowledgments: We are very grateful to Antoine Balzeau for inviting us to contribute to this special issue, and to the reviewers for their very helpful comments which permitted us to improve this literature review.

Conflicts of Interest: The authors declare no conflict of interest.

References

1. Marchant, L.F.; McGrew, W.C. Human Handedness: An Ethological Perspective. *Hum. Evol.* **1998**, *13*, 221–228. [CrossRef]
2. Marchant, L.F.; McGrew, W.C. Handedness Is More than Laterality: Lessons from Chimpanzees: Handedness Is Not Laterality. *Ann. N. Y. Acad. Sci.* **2013**, *1288*, 1–8. [CrossRef] [PubMed]
3. Prieur, J.; Barbu, S.; Blois-Heulin, C. Assessment and Analysis of Human Laterality for Manipulation and Communication Using the Rennes Laterality Questionnaire. *R. Soc. Open Sci.* **2017**, *4*, 170035. [CrossRef] [PubMed]
4. Prieur, J.; Barbu, S.; Blois-Heulin, C. Human Laterality for Manipulation and Gestural Communication Related to 60 Everyday Activities: Impact of Multiple Individual-Related Factors. *Cortex* **2018**, *99*, 118–134. [CrossRef]
5. Williams, S.M. Handedness Inventories: Edinburgh versus Annett. *Neuropsychology* **1991**, *5*, 43–48. [CrossRef]
6. Papadatou-Pastou, M.; Ntolka, E.; Schmitz, J.; Martin, M.; Munafò, M.R.; Ocklenburg, S.; Paracchini, S. Human Handedness: A Meta-Analysis. *Psychol. Bull.* **2020**, *146*, 481–524. [CrossRef] [PubMed]
7. Michel, G.F.; Babik, I.; Nelson, E.L.; Campbell, J.M.; Marcinowski, E.C. Evolution and Development of Handedness: An Evo–Devo Approach. In *Progress in Brain Research*; Elsevier: Amsterdam, The Netherlands, 2018; Volume 238, pp. 347–374. ISBN 978-0-12-814671-2.
8. Prieur, J.; Lemasson, A.; Barbu, S.; Blois-Heulin, C. History, Development and Current Advances Concerning the Evolutionary Roots of Human Right-handedness and Language: Brain Lateralisation and Manual Laterality in Non-human Primates. *Ethology* **2019**, *125*, 1–28. [CrossRef]
9. Versace, E.; Vallortigara, G. Forelimb Preferences in Human Beings and Other Species: Multiple Models for Testing Hypotheses on Lateralization. *Front. Psychol.* **2015**, *6*, 233. [CrossRef]

10. Uomini, N.T. Handedness in Neanderthals. In *Neanderthal Lifeways, Subsistence and Technology: One Hundred Fifty Years of Neanderthal Study*; Conard, N.J., Richter, J., Eds.; Vertebrate Paleobiology and Paleoanthropology Series; Springer: Dordrecht, The Netherlands, 2011; pp. 139–154. ISBN 978-94-007-0415-2.

11. Semenov, S.A. *Prehistoric Technology*; Barnes and Nobel: New York, NY, USA, 1964.

12. Frayer, D.W.; Clarke, R.J.; Fiore, I.; Blumenschine, R.J.; Pérez-Pérez, A.; Martinez, L.M.; Estebaranz, F.; Holloway, R.; Bondioli, L. OH-65: The Earliest Evidence for Right-Handedness in the Fossil Record. *J. Hum. Evol.* **2016**, *100*, 65–72. [CrossRef]

13. Corballis, M.C.; Häberling, I.S. The Many Sides of Hemispheric Asymmetry: A Selective Review and Outlook. *J. Int. Neuropsychol. Soc.* **2017**, *23*, 710–718. [CrossRef]

14. Neubauer, S.; Gunz, P.; Scott, N.A.; Hublin, J.-J.; Mitteroecker, P. Evolution of Brain Lateralization: A Shared Hominid Pattern of Endocranial Asymmetry Is Much More Variable in Humans than in Great Apes. *Sci. Adv.* **2020**, *6*, eaax9935. [CrossRef]

15. Toga, A.W.; Thompson, P.M. Mapping Brain Asymmetry. *Nat. Rev. Neurosci.* **2003**, *4*, 37–48. [CrossRef]

16. Gainotti, G. Emotions and the Right Hemisphere: Can New Data Clarify Old Models? *Neuroscientist* **2019**, *25*, 258–270. [CrossRef]

17. Knecht, S.; Dräger, B.; Deppe, M.; Bobe, L.; Lohmann, H.; Flöel, A.; Ringelstein, E.-B.; Henningsen, H. Handedness and Hemispheric Language Dominance in Healthy Humans. *Brain* **2000**, *123*, 2512–2518. [CrossRef] [PubMed]

18. Hellige, J. Unity of Thought and Action: Varieties of Interaction between the Left and Right Cerebral Hemispheres. *Curr. Dir. Psychol. Sci.* **1993**, *2*, 21–26. [CrossRef]

19. Rogers, L.J. Lateralization in Vertebrates: Its Early Evolution, General Pattern, and Development. In *Advances in the Study of Behavior*; Academic Press: Cambridge, MA, USA, 2002; Volume 31, pp. 107–161.

20. Andrew, R.J. The Earliest Origins and Subsequent Evolution of Lateralization. In *Comparative Vertebrate Lateralization*; Rogers, L.J., Andrew, R.J., Eds.; Cambridge University Press: Cambridge, MA, USA, 2002; pp. 70–93. ISBN 978-1-139-43747-9.

21. Vallortigara, G.; Rogers, L.J. Survival with an Asymmetrical Brain: Advantages and Disadvantages of Cerebral Lateralization. *Behav. Brain Sci.* **2005**, *28*, 575–589. [CrossRef] [PubMed]

22. Ströckens, F.; Güntürkün, O.; Ocklenburg, S. Limb Preferences in Non-Human Vertebrates. *Laterality* **2013**, *18*, 536–575. [CrossRef] [PubMed]

23. Vallortigara, G. The Evolutionary Psychology of Left and Right: Costs and Benefits of Lateralization. *Dev. Psychobiol.* **2006**, *48*, 418–427. [CrossRef] [PubMed]

24. Rogers, L.J. Evolution of Hemispheric Specialization: Advantages and Disadvantages. *Brain Lang.* **2000**, *73*, 236–253. [CrossRef] [PubMed]

25. Rogers, L.J.; Zucca, P.; Vallortigara, G. Advantages of Having a Lateralized Brain. *Proc. R. Soc. Lond. Ser. B Biol. Sci.* **2004**, *271*, S420–S422. [CrossRef]

26. Vallortigara, G.; Rogers, L.J. A Function for the Bicameral Mind. *Cortex* **2020**, *124*, 274–285. [CrossRef] [PubMed]

27. Annett, J. Motor Learning: A Review. In *Motor Behavior: Programming, Control, and Acquisition*; Heuer, H., Kleinbeck, U., Schmidt, K.-H., Eds.; Springer: Berlin/Heidelberg, Germany, 1985; pp. 189–212. ISBN 978-3-642-69749-4.

28. Laland, K.N.; Kumm, J.; Van Horn, J.D.; Feldman, M.W. A Gene-Culture Model of Human Handedness. *Behav. Genet.* **1995**, *25*, 433–445. [CrossRef] [PubMed]

29. McManus, I.C.; Bryden, M.P. The Genetics of Handedness, Cerebral Dominance, and Lateralization. In *Handbook of Neuropsychology*; Elsevier Science: New York, NY, USA, 1992; Volume 6, pp. 115–144. ISBN 978-0-444-90492-8.

30. Medland, S.E.; Duffy, D.L.; Wright, M.J.; Geffen, G.M.; Hay, D.A.; Levy, F.; van-Beijsterveldt, C.E.M.; Willemsen, G.; Townsend, G.C.; White, V.; et al. Genetic Influences on Handedness: Data from 25,732 Australian and Dutch Twin Families. *Neuropsychologia* **2009**, *47*, 330–337. [CrossRef] [PubMed]

31. Cuellar-Partida, G.; Tung, J.Y.; Eriksson, N.; Albrecht, E.; Aliev, F.; Andreassen, O.A.; Barroso, I.; Beckmann, J.S.; Boks, M.P.; Boomsma, D.I.; et al. Genome-Wide Association Study Identifies 48 Common Genetic Variants Associated with Handedness. *Nat. Hum. Behav.* **2021**, *5*, 59–70. [CrossRef] [PubMed]

32. De Kovel, C.G.F.; Francks, C. The Molecular Genetics of Hand Preference Revisited. *Sci. Rep.* **2019**, *9*, 5986. [CrossRef]

33. Sha, Z.; Pepe, A.; Schijven, D.; Carrion Castillo, A.; Roe, J.M.; Westerhausen, R.; Joliot, M.; Fisher, S.E.; Crivello, F.; Francks, C. Handedness and Its Genetic Influences Are Associated with Structural Asymmetries of the Cerebral Cortex in 31,864 Individuals. *Proc. Natl. Acad. Sci. USA* **2021**, *118*, e2113095118. [CrossRef]

34. Pfeifer, L.S.; Schmitz, J.; Papadatou-Pastou, M.; Peterburs, J.; Paracchini, S.; Ocklenburg, S. Handedness in Twins: Meta-Analyses. *PsyArXiv* **2021**. [CrossRef]

35. Manns, M.; Basbasse, Y.E.; Freund, N.; Ocklenburg, S. Paw Preferences in Mice and Rats: Meta-Analysis. *Neurosci. Biobehav. Rev.* **2021**, *127*, 593–606. [CrossRef] [PubMed]

36. Ambeskovic, M.; Soltanpour, N.; Falkenberg, E.A.; Zucchi, F.C.R.; Kolb, B.; Metz, G.A.S. Ancestral Exposure to Stress Generates New Behavioral Traits and a Functional Hemispheric Dominance Shift. *Cereb. Cortex* **2017**, *27*, 2126–2138. [CrossRef]

37. Schmitz, J.; Metz, G.A.S.; Güntürkün, O.; Ocklenburg, S. Beyond the Genome—Towards an Epigenetic Understanding of Handedness Ontogenesis. *Prog. Neurobiol.* **2017**, *159*, 69–89. [CrossRef]

38. Damerose, E.; Vauclair, J. Posture and Laterality in Human and Non-Human Primates: Asymmetries in Maternal Handling and the Infant's Early Motor Asymmetries. In *Comparative Vertebrate Lateralization*; Rogers, L.J., Andrew, R., Eds.; Cambridge University Press: Cambridge, MA, USA, 2002; pp. 306–361. ISBN 978-1-139-43747-9.

39. Fong, B.F.; Buis, A.J.E.; Savelsbergh, G.J.P.; de Vries, J.I.P. Influence of Breech Presentation on the Development of Fetal Arm Posture. *Early Hum. Dev.* **2005**, *81*, 519–527. [CrossRef]

40. Hopkins, B.; Rönnqvist, L. Human Handedness: Developmental and Evolutionary Perspectives. In *The Development of Sensory, Motor and Cognitive Capacities in Early Infancy: From Sensation to Cognition*; Sussex, B., Ed.; Psychology Press: Hove, UK, 1998; ISBN 978-1-134-83706-9.

41. Previc, F.H. A General Theory Concerning the Prenatal Origins of Cerebral Lateralization in Humans. *Psychol. Rev.* **1991**, *98*, 299–334. [CrossRef]

42. Ververs, I.A.P.; de Vries, J.I.P.; van Geijn, H.P.; Hopkins, B. Prenatal Head Position from 12–38 Weeks. I. Developmental Aspects. *Early Hum. Dev.* **1994**, *39*, 83–91. [CrossRef]

43. Fagard, J. The Nature and Nurture of Human Infant Hand Preference. *Ann. N. Y. Acad. Sci.* **2013**, *1288*, 114–123. [CrossRef]

44. Sparling, J.W.; Van Tol, J.; Chescheir, N.C. Fetal and Neonatal Hand Movement. *Phys. Ther.* **1999**, *79*, 24–39. [CrossRef]

45. Takeshita, H.; Myowa-Yamakoshi, M.; Hirata, S. A New Comparative Perspective on Prenatal Motor Behaviors: Preliminary Research with Four-Dimensional Ultrasonography. In *Cognitive Development in Chimpanzees*; Matsuzawa, T., Tomonaga, M., Tanaka, M., Eds.; Springer: Tokyo, Japan, 2006.

46. De Vries, J.I.P.; Visser, G.H.A.; Prechtl, H.F.R. The Emergence of Fetal Behaviour. II. Quantitative Aspects. *Early Hum. Dev.* **1985**, *12*, 99–120. [CrossRef]

47. Hepper, P.G.; Mccartney, G.R.; Shannon, E.A. Lateralised Behaviour in First Trimester Human Foetuses. *Neuropsychologia* **1998**, *36*, 531–534. [CrossRef]

48. Hepper, P.G.; Shahidullah, S.; White, R. Handedness in the Human Fetus. *Neuropsychologia* **1991**, *29*, 1107–1111. [CrossRef]

49. Boulinguez-Ambroise, G.; Pouydebat, E.; Disarbois, É.; Meguerditchian, A. Human-like Maternal Left-Cradling Bias in Monkeys Is Altered by Social Pressure. *Sci. Rep.* **2020**, *10*, 11036. [CrossRef] [PubMed]

50. Manning, J.T.; Heaton, R.; Chamberlain, A.T. Left-Side Cradling: Similarities and Differences between Apes and Humans. *J. Hum. Evol.* **1994**, *26*, 77–83. [CrossRef]

51. Hopkins, W.D.; Bard, K.A.; Jones, A.; Bales, S.L. Chimpanzee Hand Preference in Throwing and Infant Cradling: Implications for the Origin of Human Handedness. *Curr. Anthropol.* **1993**, *34*, 786–790. [CrossRef]

52. Boulinguez-Ambroise, G.; Pouydebat, E.; Disarbois, É.; Meguerditchian, A. Maternal Cradling Bias in Baboons: The First Environmental Factor Affecting Early Infant Handedness Development? *Dev. Sci.* **2021**, *25*, e13179. [CrossRef] [PubMed]

53. Forrester, G.S.; Davis, R.; Mareschal, D.; Malatesta, G.; Todd, B.K. The Left Cradling Bias: An Evolutionary Facilitator of Social Cognition? *Cortex* **2019**, *118*, 116–131. [CrossRef] [PubMed]

54. Malatesta, G.; Marzoli, D.; Rapino, M.; Tommasi, L. The Left-Cradling Bias and Its Relationship with Empathy and Depression. *Sci. Rep.* **2019**, *9*, 6141. [CrossRef]

55. Manning, J.T.; Chamberlain, A.T. Left-Side Cradling and Brain Lateralization. *Ethol. Sociobiol.* **1991**, *12*, 237–244. [CrossRef]

56. Regaiolli, B.; Spiezio, C.; Hopkins, W.D. Asymmetries in Mother-Infant Behaviour in Barbary Macaques (Macaca Sylvanus). *PeerJ* **2018**, *6*, e4736. [CrossRef] [PubMed]

57. Fagot, J.; Bard, K.A. Asymmetric Grasping Response in Neonate Chimpanzees (Pan Troglodytes). *Infant Behav. Dev.* **1995**, *18*, 253–255. [CrossRef]

58. Petrie, B.F.; Peters, M. Handedness: Left/Right Differences in Intensity of Grasp Response and Duration of Rattle Holding in Infants. *Infant Behav. Dev.* **1980**, *3*, 215–221. [CrossRef]

59. Adams-Curtis, L.; Fragaszy, D.; England, N. Prehension in Infant Capuchins (Cebus Apella) from Six Weeks to Twenty-Four Weeks: Video Analysis of Form and Symmetry. *Am. J. Primatol.* **2000**, *52*, 55–60. [CrossRef]

60. Westergaard, G.C.; Kuhn, H.E.; Suomi, S.J. Bipedal Posture and Hand Preference in Humans and Other Primates. *J. Comp. Psychol.* **1998**, *112*, 55. [CrossRef]

61. Hook, M.A.; Rogers, L.J. Development of Hand Preferences in Marmosets (Callithrix Jacchus) and Effects of Aging. *J. Comp. Psychol.* **2000**, *114*, 263–271. [CrossRef] [PubMed]

62. Westergaard, G.C.; Champoux, M.; Suomi, S.J. Hand Preference in Infant Rhesus Macaques (Macaca Mulatta). *Child Dev.* **1997**, *68*, 387–393. [CrossRef] [PubMed]

63. Wells, D.L.; Millsopp, S. The Ontogenesis of Lateralized Behavior in the Domestic Cat, Felis Silvestris Catus. *J. Comp. Psychol.* **2012**, *126*, 23–30. [CrossRef] [PubMed]

64. Giljov, A.; Karenina, K.; Ingram, J.; Malashichev, Y. Early Expression of Manual Lateralization in Bipedal Marsupials. *J. Comp. Psychol.* **2017**, *131*, 225–230. [CrossRef]

65. Govind, C.K. Development of Asymmetry in the Neuromuscular System of Lobster Claws. *Biol. Bull.* **1984**, *167*, 94–119. [CrossRef]

66. Govind, C.K.; Pearce, J. Mechanoreceptors and Minimal Reflex Activity Determining Claw Laterality In Developing Lobsters. *J. Exp. Biol.* **1992**, *171*, 149–162. [CrossRef]

67. Hammond, G. Correlates of Human Handedness in Primary Motor Cortex: A Review and Hypothesis. *Neurosci. Biobehav. Rev.* **2002**, *26*, 285–292. [CrossRef]

68. Vallortigara, G.; Bisazza, A. How Ancient Is Brain Lateralization? In *Comparative Vertebrate Lateralization*; Rogers, L.J., Andrew, R., Eds.; Cambridge University Press: Cambridge, MA, USA, 2002; ISBN 978-1-139-43747-9.

69. Hopkins, W.D. Hand Preferences for a Coordinated Bimanual Task in 110 Chimpanzees (Pan Troglodytes): Cross-Sectional Analysis. *J. Comp. Psychol.* **1995**, *109*, 291–297. [CrossRef]

70. Dadda, M.; Bisazza, A. Does Brain Asymmetry Allow Efficient Performance of Simultaneous Tasks? *Anim. Behav.* **2006**, *72*, 523–529. [CrossRef]
71. Hopkins, W.D.; Avants, B.B. Regional and Hemispheric Variation in Cortical Thickness in Chimpanzees (Pan Troglodytes). *J. Neurosci.* **2013**, *33*, 5241–5248. [CrossRef]
72. Hopkins, W.D.; Cantalupo, C. Handedness in Chimpanzees (Pan Troglodytes) Is Associated with Asymmetries of the Primary Motor Cortex but Not with Homologous Language Areas. *Behav. Neurosci.* **2004**, *118*, 1176–1183. [CrossRef] [PubMed]
73. Margiotoudi, K.; Marie, D.; Claidière, N.; Coulon, O.; Roth, M.; Nazarian, B.; Lacoste, R.; Hopkins, W.D.; Molesti, S.; Fresnais, P.; et al. Handedness in Monkeys Reflects Hemispheric Specialization within the Central Sulcus. An In Vivo MRI Study in Right- and Left-Handed Olive Baboons. *Cortex* **2019**, *118*, 203–211. [CrossRef] [PubMed]
74. Phillips, K.A.; Sherwood, C.C. Primary Motor Cortex Asymmetry Is Correlated with Handedness in Capuchin Monkeys (Cebus Apella). *Behav. Neurosci.* **2005**, *119*, 1701–1704. [CrossRef] [PubMed]
75. Sherwood, C.C.; Wahl, E.; Erwin, J.M.; Hof, P.R.; Hopkins, W.D. Histological Asymmetries of Primary Motor Cortex Predict Handedness in Chimpanzees (Pan Troglodytes). *J. Comp. Neurol.* **2007**, *503*, 525–537. [CrossRef] [PubMed]
76. Molesti, S.; Vauclair, J.; Meguerditchian, A. Hand Preferences for Unimanual and Bimanual Coordinated Actions in Olive Baboons (Papio anubis): Consistency over Time and across Populations. *J. Comp. Psychol.* **2016**, *130*, 341–350. [CrossRef]
77. Brown, C.; Magat, M. Cerebral Lateralization Determines Hand Preferences in Australian Parrots. *Biol. Lett.* **2011**, *7*, 496–498. [CrossRef]
78. Kaplan, G.; Rogers, L.J. Brain Size Associated with Foot Preferences in Australian Parrots. *Symmetry* **2021**, *13*, 867. [CrossRef]
79. Magat, M.; Brown, C. Laterality Enhances Cognition in Australian Parrots. *Proc. R. Soc. B Biol. Sci.* **2009**, *276*, 4155–4162. [CrossRef] [PubMed]
80. Brunon, A.; Bovet, D.; Bourgeois, A.; Pouydebat, E. Motivation and Manipulation Capacities of the Blue and Yellow Macaw and the Tufted Capuchin: A Comparative Approach. *Behav. Process.* **2014**, *107*, 1–14. [CrossRef] [PubMed]
81. Byrne, R.A.; Kuba, M.J.; Meisel, D.V.; Griebel, U.; Mather, J.A. Does Octopus Vulgaris Have Preferred Arms? *J. Comp. Psychol.* **2006**, *120*, 198–204. [CrossRef]
82. Chi, J.G.; Dooling, E.C.; Gilles, F.H. Gyral Development of the Human Brain. *Ann. Neurol.* **1977**, *1*, 86–93. [CrossRef]
83. Hering-Hanit, R.; Achiron, R.; Lipitz, S.; Achiron, A. Asymmetry of Fetal Cerebral Hemispheres: In Utero Ultrasound Study. *Arch. Dis. Child. Fetal Neonatal Ed.* **2001**, *85*, F194–F196. [CrossRef]
84. Kasprian, G.; Langs, G.; Brugger, P.C.; Bittner, M.; Weber, M.; Arantes, M.; Prayer, D. The Prenatal Origin of Hemispheric Asymmetry: An In Utero Neuroimaging Study. *Cereb. Cortex* **2011**, *21*, 1076–1083. [CrossRef]
85. Kienast, P.; Schwartz, E.; Diogo, M.C.; Gruber, G.M.; Brugger, P.C.; Kiss, H.; Ulm, B.; Bartha-Doering, L.; Seidl, R.; Weber, M.; et al. The Prenatal Origins of Human Brain Asymmetry: Lessons Learned from a Cohort of Fetuses with Body Lateralization Defects. *Cereb. Cortex* **2021**, *31*, 3713–3722. [CrossRef] [PubMed]
86. Vasung, L.; Rollins, C.K.; Yun, H.J.; Velasco-Annis, C.; Zhang, J.; Wagstyl, K.; Evans, A.; Warfield, S.K.; Feldman, H.A.; Grant, P.E.; et al. Quantitative In Vivo MRI Assessment of Structural Asymmetries and Sexual Dimorphism of Transient Fetal Compartments in the Human Brain. *Cereb. Cortex* **2020**, *30*, 1752–1767. [CrossRef] [PubMed]
87. Bardo, A.; Kivell, T.L.; Town, K.; Donati, G.; Ballieux, H.; Stamate, C.; Edginton, T.; Forrester, G.S. Get a Grip: Variation in Human Hand Grip Strength and Implications for Human Evolution. *Symmetry* **2021**, *13*, 1142. [CrossRef]
88. Govind, C.K. Asymmetry in Lobster Claws. *Am. Sci.* **1989**, *77*, 468–474.
89. Rosenberg, M.S. Fiddler Crab Claw Shape Variation: A Geometric Morphometric Analysis across the Genus Uca (Crustacea: Brachyura: Ocypodidae). *Biol. J. Linn. Soc.* **2002**, *75*, 147–162. [CrossRef]
90. Yosef, R.; Daraby, M.; Semionovikh, A.; Kosicki, J.Z. Individual Laterality in Ghost Crabs (Ocypode Saratan) Influences Burrowing Behavior. *Symmetry* **2021**, *13*, 1512. [CrossRef]
91. Backwell, P.R.Y.; Matsumasa, M.; Double, M.; Roberts, A.; Murai, M.; Keogh, J.S.; Jennions, M.D. What Are the Consequences of Being Left-Clawed in a Predominantly Right-Clawed Fiddler Crab? *Proc. R. Soc. B Biol. Sci.* **2007**, *274*, 2723–2729. [CrossRef] [PubMed]
92. MacNeilage, P.F. Present Status of the Postural Origins Theory. In *Special Topics in Primatology*; Hopkins, W.D., Ed.; The Evolution of Hemispheric Specialization in Primates; Elsevier: Amsterdam, The Netherlands, 2007; Volume 5, pp. 58–91.
93. MacNeilage, P.F.; Studdert-Kennedy, M.G.; Lindblom, B. Primate Handedness Reconsidered. *Behav. Brain Sci.* **1987**, *10*, 247. [CrossRef]
94. MacNeilage, P.F.; Studdert-Kennedy, M.G.; Lindblom, B. Primate Handedness: A Foot in the Door. *Behav. Brain Sci.* **1988**, *11*, 737–746. [CrossRef]
95. MacNeilage, P.F.; Studdert-Kennedy, M.G.; Lindblom, B. Primate Handedness: The Other Theory, the Other Hand and the Other Attitude. *Behav. Brain Sci.* **1991**, *14*, 344–349. [CrossRef]
96. Fleagle, J.G.; Meldrum, D.J. Locomotor Behavior and Skeletal Morphology of Two Sympatric Pitheciine Monkeys, Pithecia Pithecia and Chiropotes Satanas. *Am. J. Primatol.* **1988**, *16*, 227–249. [CrossRef]
97. MacNeilage, P.F. Evolution of Whole-Body Asymmetry Related to Handedness. *Cortex* **2006**, *42*, 94–95. [CrossRef]
98. Harris, L.J. Footedness in Parrots: Three Centuries of Research, Theory, and Mere Surmise. *Can. J. Psychol.* **1989**, *43*, 369–396. [CrossRef] [PubMed]

99. Hopkins, W.D. Posture and Reaching in Chimpanzees (Pan Troglodytes) and Orangutans (Pongo Pygmaeus). *J. Comp. Psychol.* **1993**, *107*, 162–168. [CrossRef] [PubMed]

100. Meguerditchian, A. Pour une large approche comparative entre primates dans les recherches sur les origines de l'Homme: l'exemple de la latéralité manuelle. *BMSAP* **2014**, *26*, 166–171. [CrossRef]

101. Dodson, D.L.; Stafford, D.; Forsythe, C.; Seltzer, C.P.; Ward, J.P. Laterality in Quadrupedal and Bipedal Prosimians: Reach and Whole-Body Turn in the Mouse Lemur (Microcebus Murinus) and the Galago (Galago Moholi). *Am. J. Primatol.* **1992**, *26*, 191–202. [CrossRef]

102. Ward, J.P. Laterality in African and Malagasy Prosimians. In *Creatures of the Dark*; Alterman, L., Doyle, G.A., Izard, M.K., Eds.; Springer: Boston, MA, USA, 1995; pp. 293–309. ISBN 978-1-4419-3250-1.

103. Ottoni, E.B.; Izar, P. Capuchin Monkey Tool Use: Overview and Implications. *Evol. Anthropol. Issues News Rev.* **2008**, *17*, 171–178. [CrossRef]

104. Hook-Costigan, M.A.; Rogers, L.J. Hand Preferences in New World Primates. *Int. J. Comp. Psychol.* **1996**, *9*, 14907.

105. Spinozzi, G.; Castorina, M.G.; Truppa, V. Hand Preferences in Unimanual and Coordinated-Bimanual Tasks by Tufted Capuchin Monkeys (Cebus Apella). *J. Comp. Psychol.* **1998**, *112*, 183–191. [CrossRef]

106. Hopkins, W.D. Chimpanzee Handedness Revisited: 55 Years since Finch (1941). *Psychon. Bull. Rev.* **1996**, *3*, 449–457. [CrossRef]

107. Hopkins, W.D.; Pika, S.; Liebal, K.; Bania, A.; Meguerditchian, A.; Gardner, M.; Schapiro, S.J. Handedness for Manual Gestures in Great Apes. *Dev. Primate Gesture Res.* **2012**, *6*, 93–111.

108. Ward, J.P.; Hopkins, W.D. *Primate Laterality: Current Behavioral Evidence of Primate Asymmetries*, Springer: New York, NY, USA, 1993; ISBN 978-1-4612-4370-0.

109. Giljov, A.; Karenina, K.; Ingram, J.; Malashichev, Y. Parallel Emergence of True Handedness in the Evolution of Marsupials and Placentals. *Curr. Biol.* **2015**, *25*, 1878–1884. [CrossRef] [PubMed]

110. Fagard, J.; Marks, A. Unimanual and Bimanual Tasks and the Assessment of Handedness in Toddlers. *Dev. Sci.* **2000**, *3*, 137–147. [CrossRef]

111. Fagot, J.; Vauclair, J. Manual Laterality in Nonhuman Primates: A Distinction between Handedness and Manual Specialization. *Psychol. Bull.* **1991**, *109*, 76–89. [CrossRef]

112. Meguerditchian, A.; Phillips, K.A.; Chapelain, A.; Mahovetz, L.M.; Milne, S.; Stoinski, T.; Bania, A.; Lonsdorf, E.; Schaeffer, J.; Russell, J.; et al. Handedness for Unimanual Grasping in 564 Great Apes: The Effect on Grip Morphology and a Comparison with Hand Use for a Bimanual Coordinated Task. *Front. Psychol.* **2015**, *6*, 1794. [CrossRef] [PubMed]

113. McGrew, W.C.; Marchant, L.F. Chimpanzees, Tools, and Termites: Hand Preference or Handedness? *Curr. Anthropol.* **1992**, *33*, 114–119. [CrossRef]

114. Vauclair, J.; Meguerditchian, A.; Hopkins, W.D. Hand Preferences for Unimanual and Coordinated Bimanual Tasks in Baboons (Papio anubis). *Cogn. Brain Res.* **2005**, *25*, 210–216. [CrossRef]

115. Giljov, A.; Karenina, K.; Malashichev, Y. Limb Preferences in a Marsupial, Macropus Rufogriseus: Evidence for Postural Effect. *Anim. Behav.* **2012**, *83*, 525–534. [CrossRef]

116. Blois-Heulin, C.; Guitton, J.S.; Nedellec-Bienvenue, D.; Ropars, L.; Vallet, E. Hand Preference in Unimanual and Bimanual Tasks and Postural Effect on Manual Laterality in Captive Red-Capped Mangabeys (Cercocebus Torquatus Torquatus). *Am. J. Primatol.* **2006**, *68*, 429–444. [CrossRef]

117. Blois-Heulin, C.; Bernard, V.; Bec, P. Postural Effect on Manual Laterality in Different Tasks in Captive Grey-Cheeked Mangabey (Lophocebus Albigena). *J. Comp. Psychol.* **2007**, *121*, 205–213. [CrossRef] [PubMed]

118. Westergaard, G.C.; Kuhn, H.E.; Lundquist, A.L.; Suomi, S.J. Posture and Reaching in Tufted Capuchins (Cebus Apella). *Laterality* **1997**, *2*, 65–74. [CrossRef] [PubMed]

119. Joly, M.; Scheumann, M.; Zimmermann, E. Posture Does Not Matter! Paw Usage and Grasping Paw Preference in a Small-Bodied Rooting Quadrupedal Mammal. *PLoS ONE* **2012**, *7*, e38228. [CrossRef]

120. Konerding, W.S.; Hedrich, H.-J.; Bleich, E.; Zimmermann, E. Paw Preference Is Not Affected by Postural Demand in a Nonprimate Mammal (Felis Silvestris Catus). *J. Comp. Psychol.* **2012**, *126*, 15–22. [CrossRef]

121. Ghirlanda, S.; Vallortigara, G. The Evolution of Brain Lateralization: A Game-Theoretical Analysis of Population Structure. *Proc. R. Soc. Lond. B* **2004**, *271*, 853–857. [CrossRef] [PubMed]

122. MacNeilage, P.F.; Rogers, L.J.; Vallortigara, G. Origins of the Left & Right Brain. *Sci. Am.* **2009**, *301*, 60–67. [CrossRef]

123. Rogers, L.J. Laterality in Animals. *Int. J. Comp. Psychol.* **1989**, *3*, 22.

124. Bisazza, A.; Cantalupo, C.; Capocchiano, M.; Vallortigara, G. Population Lateralisation and Social Behaviour: A Study with 16 Species of Fish. *Laterality* **2000**, *5*, 269–284. [CrossRef] [PubMed]

125. Frasnelli, E.; Vallortigara, G.; Rogers, L.J. Left–Right Asymmetries of Behaviour and Nervous System in Invertebrates. *Neurosci. Biobehav. Rev.* **2012**, *36*, 1273–1291. [CrossRef]

126. Perelle, I.B.; Ehrman, L. An International Study of Human Handedness: The Data. *Behav Genet* **1994**, *24*, 217–227. [CrossRef] [PubMed]

127. Raymond, M.; Pontier, D. Is There Geographical Variation in Human Handedness? *Laterality* **2004**, *9*, 35–51. [CrossRef] [PubMed]

128. Fagard, J.; Dahmen, R. Cultural Influences on the Development of Lateral Preferences: A Comparison between French and Tunisian Children. *Laterality* **2004**, *9*, 67–78. [CrossRef] [PubMed]

129. Vuoksimaa, E.; Koskenvuo, M.; Rose, R.J.; Kaprio, J. Origins of Handedness: A Nationwide Study of 30161 Adults. *Neuropsychologia* **2009**, *47*, 1294–1301. [CrossRef]
130. Ghirlanda, S.; Frasnelli, E.; Vallortigara, G. Intraspecific Competition and Coordination in the Evolution of Lateralization. *Philos. Trans. R. Soc. B Biol. Sci.* **2009**, *364*, 861–866. [CrossRef]
131. Loffing, F.; Hagemann, N.; Strauss, B. Automated Processes in Tennis: Do Left-Handed Players Benefit from the Tactical Preferences of Their Opponents? *Null* **2010**, *28*, 435–443. [CrossRef]
132. Richardson, T.; Gilman, R.T. Left-Handedness Is Associated with Greater Fighting Success in Humans. *Sci. Rep.* **2019**, *9*, 15402. [CrossRef]
133. Billiard, S.; Faurie, C.; Raymond, M. Maintenance of Handedness Polymorphism in Humans: A Frequency-Dependent Selection Model. *J. Theor. Biol.* **2005**, *235*, 85–93. [CrossRef]
134. Frasnelli, E.; Vallortigara, G. Individual-Level and Population-Level Lateralization: Two Sides of the Same Coin. *Symmetry* **2018**, *10*, 739. [CrossRef]
135. Abrams, D.M.; Panaggio, M.J. A Model Balancing Cooperation and Competition Can Explain Our Right-Handed World and the Dominance of Left-Handed Athletes. *J. R. Soc. Interface* **2012**, *9*, 2718–2722. [CrossRef] [PubMed]
136. Forrester, G.S.; Leavens, D.A.; Quaresmini, C.; Vallortigara, G. Target Animacy Influences Gorilla Handedness. *Anim. Cogn.* **2011**, *14*, 903–907. [CrossRef] [PubMed]
137. Forrester, G.S.; Quaresmini, C.; Leavens, D.A.; Spiezio, C.; Vallortigara, G. Target Animacy Influences Chimpanzee Handedness. *Anim. Cogn.* **2012**, *15*, 1121–1127. [CrossRef]
138. Baldachini, M.; Regaiolli, B.; Llorente, M.; Riba, D.; Spiezio, C. The Influence of Target Animacy and Social Rank on Hand Preference in Barbary Macaques (Macaca Sylvanus). *Int. J. Primatol.* **2021**, *42*, 155–170. [CrossRef]
139. Demaree, H.A.; Everhart, D.E.; Youngstrom, E.A.; Harrison, D.W. Brain Lateralization of Emotional Processing: Historical Roots and a Future Incorporating "Dominance". *Behav. Cogn. Neurosci. Rev.* **2005**, *4*, 3–20. [CrossRef]
140. Goursot, C.; Düpjan, S.; Puppe, B.; Leliveld, L.M.C. Affective Styles and Emotional Lateralization: A Promising Framework for Animal Welfare Research. *Appl. Anim. Behav. Sci.* **2021**, *237*, 105279. [CrossRef]
141. Borod, J.C.; Koff, E.; Caron, H.S. Right Hemispheric Specialization for the Expression and Appreciation of Emotion: A Focus on the Face Cognitive Processing in the Right Hemisphere. In *Cognitive Processing in the Right Hemisphere*; ScienceOpen: Berlin, Germany, 1983; pp. 83–110.
142. Lindell, A. Chapter 9—Lateralization of the Expression of Facial Emotion in Humans. In *Progress in Brain Research*; Forrester, G.S., Hopkins, W.D., Hudry, K., Lindell, A., Eds.; Cerebral Lateralization and Cognition: Evolutionary and Developmental Investigations of Behavioral Biases; Elsevier: Amsterdam, The Netherlands, 2018; Volume 238, pp. 249–270.
143. Alves, N.T.; Fukusima, S.S.; Aznar-Casanova, J.A. Models of Brain Asymmetry in Emotional Processing. *Psychol. Neurosci.* **2008**, *1*, 63–66. [CrossRef]
144. Fernández-Carriba, S.; Loeches, Á.; Morcillo, A.; Hopkins, W.D. Asymmetry in Facial Expression of Emotions by Chimpanzees. *Neuropsychologia* **2002**, *40*, 1523–1533. [CrossRef]
145. Gruber, T.; Grandjean, D. A Comparative Neurological Approach to Emotional Expressions in Primate Vocalizations. *Neurosci. Biobehav. Rev.* **2017**, *73*, 182–190. [CrossRef]
146. Hauser, M.D. Right Hemisphere Dominance for the Production of Facial Expression in Monkeys. *Science* **1993**, *261*, 475–477. [CrossRef]
147. Hauser, M.D.; Akre, K. Asymmetries in the Timing of Facial and Vocal Expressions by Rhesus Monkeys: Implications for Hemispheric. Specialization. *Anim. Behav.* **2001**, *61*, 391–400. [CrossRef]
148. Leliveld, L.M.C.; Langbein, J.; Puppe, B. The Emergence of Emotional Lateralization: Evidence in Non-Human Vertebrates and Implications for Farm Animals. *Appl. Anim. Behav. Sci.* **2013**, *145*, 1–14. [CrossRef]
149. Best, C.T.; Womer, J.S.; Queen, H.F. Hemispheric Aymmetries in Adults' Perception of Infant Emotional Expressions. *J. Exp. Psychol. Hum. Percept. Perform.* **1994**, *20*, 751–765. [CrossRef] [PubMed]
150. Wallez, C.; Vauclair, J.; Bourjade, M. Baboon (Papio anubis) Chimeric Face Processing by Human (Homo Sapiens) Judges: Influence of Stimuli Complexity on the Perception of Oro-Facial Asymmetries. *J. Comp. Psychol.* **2019**, *133*, 36–45. [CrossRef] [PubMed]
151. Morris, R.D.; Hopkins, W.D. Perception of Human Chimeric Faces by Chimpanzees: Evidence for a Right Hemisphere Advantage. *Brain Cogn.* **1993**, *21*, 111–122. [CrossRef] [PubMed]
152. Wallez, C.; Vauclair, J. Right Hemisphere Dominance for Emotion Processing in Baboons. *Brain Cogn.* **2011**, *75*, 164–169. [CrossRef]
153. Gil-da-Costa, R.; Hauser, M.D. Vervet Monkeys and Humans Show Brain Asymmetries for Processing Conspecific Vocalizations, but with Opposite Patterns of Laterality. *Proc. R. Soc. B Biol. Sci.* **2006**, *273*, 2313–2318. [CrossRef]
154. Siniscalchi, M.; d'Ingeo, S.; Quaranta, A. Lateralized Functions in the Dog Brain. *Symmetry* **2017**, *9*, 71. [CrossRef]
155. Siniscalchi, M.; d'Ingeo, S.; Quaranta, A. Lateralized Emotional Functioning in Domestic Animals. *Appl. Anim. Behav. Sci.* **2021**, *237*, 105282. [CrossRef]
156. Ehrlichman, H. Hemispheric Asymmetry and Positive-Negative Affect. In *Duality and Unity of the Brain: Unified Functioning and Specialisation of the Hemispheres*; Ottoson, D., Ed.; Wenner-Gren Center International Symposium Series; Palgrave Macmillan: London, UK, 1987; pp. 194–206. ISBN 978-1-349-08940-6.

157. Schwartz, G.E.; Ahern, G.L.; Brown, S.-L. Lateralized Facial Muscle Response to Positive and Negative Emotional Stimuli. *Psychophysiology* **1979**, *16*, 561–571. [CrossRef]

158. Silberman, E.K.; Weingartner, H. Hemispheric Lateralization of Functions Related to Emotion. *Brain Cogn.* **1986**, *5*, 322–353. [CrossRef]

159. Hopkins, W.D.; Bennett, A.J. Handedness and Approach-Avoidance Behavior in Chipanzees (Pan). *J. Exp. Psychol. Anim. Behav. Process.* **1994**, *20*, 413–418. [CrossRef] [PubMed]

160. Braccini, S.N.; Caine, N.G. Hand Preference Predicts Reactions to Novel Foods and Predators in Marmosets (Callithrix Geoffroyi). *J. Comp. Psychol.* **2009**, *123*, 18–25. [CrossRef] [PubMed]

161. Rogers, L.J. Hand and Paw Preferences in Relation to the Lateralized Brain. *Philos. Trans. R. Soc. B Biol. Sci.* **2009**, *364*, 943–954. [CrossRef]

162. Harmon-Jones, E.; Harmon-Jones, C.; Price, T.F. What Is Approach Motivation? *Emot. Rev.* **2013**, *5*, 291–295. [CrossRef]

163. Harmon-Jones, E.; Gable, P.A. On the Role of Asymmetric Frontal Cortical Activity in Approach and Withdrawal Motivation: An Updated Review of the Evidence. *Psychophysiology* **2018**, *55*, e12879. [CrossRef] [PubMed]

164. Rohlfs, P.; Ramírez, J.M. Aggression and Brain Asymmetries: A Theoretical Review. *Aggress. Violent Behav.* **2006**, *11*, 283–297. [CrossRef]

165. Gable, P.A.; Dreisbach, G. Approach Motivation and Positive Affect. *Curr. Opin. Behav. Sci.* **2021**, *39*, 203–208. [CrossRef]

166. Killgore, W.D.S.; Yurgelun-Todd, D.A. The Right-Hemisphere and Valence Hypotheses: Could They Both Be Right (and Sometimes Left)? *Soc. Cogn. Affect. Neurosci.* **2007**, *2*, 240–250. [CrossRef]

167. Duboscq, J.; Romano, V.; Sueur, C.; MacIntosh, A.J.J. Scratch That Itch: Revisiting Links between Self-Directed Behaviour and Parasitological, Social and Environmental Factors in a Free-Ranging Primate. *R. Soc. Open Sci.* **2016**, *3*, 160571. [CrossRef]

168. Maestripieri, D.; Schino, G.; Aureli, F.; Troisi, A. A Modest Proposal: Displacement Activities as an Indicator of Emotions in Primates. *Anim. Behav.* **1992**, *44*, 967–979. [CrossRef]

169. Palagi, E.; Norscia, I. Scratching around Stress: Hierarchy and Reconciliation Make the Difference in Wild Brown Lemurs (Eulemur Fulvus). *Stress* **2011**, *14*, 93–97. [CrossRef]

170. Schino, G.; Perretta, G.; Taglioni, A.M.; Monaco, V.; Troisi, A. Primate Displacement Activities as an Ethopharmacological Model of Anxiety. *Anxiety* **1996**, *2*, 186–191. [CrossRef]

171. Sclafani, V.; Norscia, I.; Antonacci, D.; Palagi, E. Scratching around Mating: Factors Affecting Anxiety in Wild Lemur Catta. *Primates* **2012**, *53*, 247–254. [CrossRef] [PubMed]

172. Troisi, A. Displacement Activities as a Behavioral Measure of Stress in Nonhuman Primates and Human Subjects. *Stress* **2002**, *5*, 47–54. [CrossRef] [PubMed]

173. Dimond, S.; Harries, R. Face Touching in Monkeys, Apes and Man Evolutionary Origins and Cerebral Asymmetry. *Neuropsychologia* **1984**, *22*, 227–233. [CrossRef]

174. Rogers, L.J.; Kaplan, G. Hand Preferences and Other Lateral Biases in Rehabilitated Orang-Utans, Pongo Pygmaeus Pygmaeus. *Anim. Behav.* **1996**, *51*, 13–25. [CrossRef]

175. Wagner, K.E.; Hopper, L.M.; Ross, S.R. Asymmetries in the Production of Self-Directed Behavior by Chimpanzees and Gorillas during a Computerized Cognitive Test. *Anim. Cogn.* **2016**, *19*, 343–350. [CrossRef]

176. Forrester, G.S.; Pegler, R.; Thomas, M.S.C.; Mareschal, D. Handedness as a Marker of Cerebral Lateralization in Children with and without Autism. *Behav. Brain Res.* **2014**, *268*, 14–21. [CrossRef] [PubMed]

177. Reissland, N.; Aydin, E.; Francis, B.; Exley, K. Laterality of Foetal Self-Touch in Relation to Maternal Stress. *Laterality* **2015**, *20*, 82–94. [CrossRef] [PubMed]

178. Aruguete, M.S.; Ely, E.A.; King, J.E. Laterality in Spontaneous Motor Activity of Chimpanzees and Squirrel Monkeys. *Am. J. Primatol.* **1992**, *27*, 177–188. [CrossRef] [PubMed]

179. Hopkins, W.D.; Russell, J.L.; Freeman, H.; Reynolds, E.A.M.; Griffis, C.; Leavens, D.A. Lateralized Scratching in Chimpanzees (Pan Troglodytes): Evidence of a Functional Asymmetry during Arousal. *Emotion* **2006**, *6*, 553–559. [CrossRef] [PubMed]

180. Leavens, D.A.; Aureli, F.; Hopkins, W.D.; Hyatt, C.W. Effects of Cognitive Challenge on Self-Directed Behaviors by Chimpanzees (Pan Troglodytes). *Am. J. Primatol.* **2001**, *55*, 1–14. [CrossRef]

181. Bard, K.A.; Hopkins, W.D.; Fort, C.L. Lateral Bias in Infant Chimpanzees (Pan Troglodytes). *J. Comp. Psychol.* **1990**, *104*, 309–321. [CrossRef]

182. Baik, J.-H. Stress and the Dopaminergic Reward System. *Exp. Mol. Med.* **2020**, *52*, 1879–1890. [CrossRef]

183. Sullivan, R.M. Hemispheric Asymmetry in Stress Processing in Rat Prefrontal Cortex and the Role of Mesocortical Dopamine. *Stress* **2004**, *7*, 131–143. [CrossRef] [PubMed]

184. Zhang, X.; Bachmann, P.; Schilling, T.M.; Naumann, E.; Schächinger, H.; Larra, M.F. Emotional Stress Regulation: The Role of Relative Frontal Alpha Asymmetry in Shaping the Stress Response. *Biol. Psychol.* **2018**, *138*, 231–239. [CrossRef]

185. Everson, E.A. *Self-Directed Behaviors, Handedness and Lateralization in the Bonobo (Pan Paniscus)*; UC San Diego: San Diego, CA, USA, 2011.

186. Hopkins, W.D.; de Waal, F.B.M. Behavioral Laterality in Captive Bonobos (Pan Paniscus): Replication and Extension. *Int. J. Primatol.* **1995**, *16*, 261–276. [CrossRef]

187. Suarez, S.D.; Gallup, G.G. Face Touching in Primates: A Closer Look. *Neuropsychologia* **1986**, *24*, 597–600. [CrossRef]

188. Hopkins, W.; Cantalupo, C. Individual and Setting Differences in the Hand Preferences of Chimpanzees (Pan Troglodytes): A Critical Analysis and Some Alternative Explanations. *Laterality* **2005**, *10*, 65–80. [CrossRef]
189. Hopkins, W.D. Comparative and Familial Analysis of Handedness in Great Apes. *Psychol. Bull.* **2006**, *132*, 538–559. [CrossRef]
190. Salva, O.; Regolin, L.; Mascalzoni, E.; Vallortigara, G. Cerebral and Behavioural Asymmetries in Animal Social Recognition. *CCBR* **2012**, *7*, 110–138. [CrossRef]
191. Benelli, G.; Donati, E.; Romano, D.; Stefanini, C.; Messing, R.H.; Canale, A. Lateralisation of Aggressive Displays in a Tephritid Fly. *Sci. Nat.* **2015**, *102*, 1. [CrossRef] [PubMed]
192. Frasnelli, E. Brain and Behavioral Lateralization in Invertebrates. *Front. Psychol.* **2013**, *4*, 939. [CrossRef] [PubMed]
193. Packheiser, J.; Schmitz, J.; Berretz, G.; Papadatou-Pastou, M.; Ocklenburg, S. Handedness and Sex Effects on Lateral Biases in Human Cradling: Three Meta-Analyses. *Neurosci. Biobehav. Rev.* **2019**, *104*, 30–42. [CrossRef] [PubMed]
194. Almerigi, J.B.; Carbary, T.J.; Harris, L.J. Most Adults Show Opposite-Side Biases in the Imagined Holding of Infants and Objects. *Brain Cogn.* **2002**, *48*, 258–263. [PubMed]
195. Salk, L. The Effects of the Normal Heartbeat Sound on the Behaviour of the Newborn Infant: Implications for Mental Health. *World Ment. Health* **1960**, *12*, 168–175.
196. Salk, L. The Role of the Heartbeat in the Relations between Mother and Infant. *Sci. Am.* **1973**, *228*, 24–29. [CrossRef]
197. Todd, B.; Butterworth, G. Her Heart Is in the Right Place: An Investigation of the 'Heartbeat Hypothesis' as an Explanation of the Left Side Cradling Preference in a Mother with Dextrocardia. *Early Dev. Parent.* **1998**, *7*, 229–233. [CrossRef]
198. Richards, J.L.; Finger, S. Mother-Child Holding Patterns: A Cross-Cultural Photographic Survey. *Child Dev.* **1975**, *46*, 1001–1004. [CrossRef]
199. Bryden, M.P.; Ley, R.G. Right Hemispheric Involvement in the Perception and Expression of Emotion in Normal Humans. In *Neuropsychology of Human Emotion*; Heilman, K.M., Satz, P., Eds.; Springer: New York, NY, USA, 1983.
200. Sackeim, H.A.; Gur, R.C.; Saucy, M.C. Emotions Are Expressed More Intensely on the Left Side of the Face. *Science* **1978**, *202*, 434–436. [CrossRef] [PubMed]
201. Vauclair, J.; Donnot, J. Infant Holding Biases and Their Relations to Hemispheric Specializations for Perceiving Facial Emotions. *Neuropsychologia* **2005**, *43*, 564–571. [CrossRef] [PubMed]
202. Reissland, N.; Hopkins, B.; Helms, P.; Williams, B. Maternal Stress and Depression and the Lateralisation of Infant Cradling. *J. Child Psychol. Psychiatry* **2009**, *50*, 263–269. [CrossRef] [PubMed]
203. Suter, S.E.; Huggenberger, H.J.; Schächinger, H. Cold Pressor Stress Reduces Left Cradling Preference in Nulliparous Human Females. *Stress* **2007**, *10*, 45–51. [CrossRef] [PubMed]
204. Weatherill, R.; Almerigi, J.; Levendosky, A.; Bogat, G.A.; von Eye, A.; Harris, L.J. Is Maternal Depression Related to Side of Infant Holding? *Int. J. Behav. Dev.* **2004**, *28*, 421–427. [CrossRef]
205. Fleva, E.; Khan, A. An Examination of the Leftward Cradling Bias among Typically Developing Adults High on Autistic Traits. *Laterality* **2015**, *20*, 711–722. [CrossRef] [PubMed]
206. Malatesta, G.; Marzoli, D.; Apicella, F.; Abiuso, C.; Muratori, F.; Forrester, G.S.; Vallortigara, G.; Scattoni, M.L.; Tommasi, L. Received Cradling Bias During the First Year of Life: A Retrospective Study on Children with Typical and Atypical Development. *Front. Psychiatry* **2020**, *11*, 91. [CrossRef] [PubMed]
207. Malatesta, G.; Marzoli, D.; Tommasi, L. The Association between Received Maternal Cradling and Neurodevelopment: Is Left Better? *Med. Hypotheses* **2020**, *134*, 109442. [CrossRef]
208. Pileggi, L.-A.; Malcolm-Smith, S.; Solms, M. Investigating the Role of Social-Affective Attachment Processes in Cradling Bias: The Absence of Cradling Bias in Children with Autism Spectrum Disorders. *Laterality* **2015**, *20*, 154–170. [CrossRef] [PubMed]
209. Vervloed, M.P.J.; Hendriks, A.W.; van den Eijnde, E. The Effects of Mothers' Past Infant-Holding Preferences on Their Adult Children's Face Processing Lateralisation. *Brain Cogn.* **2011**, *75*, 248–254. [CrossRef]
210. Karenina, K.; Giljov, A.; Ingram, J.; Rowntree, V.J.; Malashichev, Y. Lateralization of Mother–Infant Interactions in a Diverse Range of Mammal Species. *Nat. Ecol. Evol.* **2017**, *1*, 30. [CrossRef] [PubMed]
211. Broca, P. Sur Le Siège de La Faculté Du Langage Articulé. *Proc. Bull. Soc. Anthropol. Paris* **1865**, *6*, 377–393. [CrossRef]
212. Cooper, D.L. Broca's Arrow: Evolution, Prediction, and Language in the Brain. *Anat. Rec.* **2006**, *289B*, 9–24. [CrossRef] [PubMed]
213. Lieberman, P. On the Nature and Evolution of the Neural Bases of Human Language. *Am. J. Phys. Anthr.* **2002**, *119*, 36–62. [CrossRef]
214. Lieberman, P. The Evolution of Human Speech: Its Anatomical and Neural Bases. *Curr. Anthropol.* **2007**, *48*, 39–66. [CrossRef]
215. Vigneau, M.; Beaucousin, V.; Hervé, P.Y.; Duffau, H.; Crivello, F.; Houdé, O.; Mazoyer, B.; Tzourio-Mazoyer, N. Meta-Analyzing Left Hemisphere Language Areas: Phonology, Semantics, and Sentence Processing. *NeuroImage* **2006**, *30*, 1414–1432. [CrossRef]
216. Vigneau, M.; Beaucousin, V.; Hervé, P.-Y.; Jobard, G.; Petit, L.; Crivello, F.; Mellet, E.; Zago, L.; Mazoyer, B.; Tzourio-Mazoyer, N. What Is Right-Hemisphere Contribution to Phonological, Lexico-Semantic, and Sentence Processing?: Insights from a Meta-Analysis. *NeuroImage* **2011**, *54*, 577–593. [CrossRef]
217. Wernicke, C. *Der Aphasische Symptomencomplex: Eine Psychologische Studie Auf Anatomischer Basis*; Cohn & Weigert: Breslau, Poland, 1874.
218. Corballis, M.C. From Mouth to Hand: Gesture, Speech, and the Evolution of Right-Handedness. *Behav. Brain Sci.* **2003**, *26*, 199–208. [CrossRef]

219. Corballis, M.C. Laterality and Human Speciation. In *The Speciation of Modern Homo Sapiens*; Crow, T.J., Ed.; OUP/British Academy: London, UK, 2004; pp. 137–153. ISBN 978-0-19-726311-2.

220. Nishitani, N.; Hari, R. Temporal Dynamics of Cortical Representation for Action. *Proc. Natl. Acad. Sci. USA* **2000**, *97*, 913–918. [CrossRef]

221. Rizzolatti, G.; Fadiga, L.; Gallese, V.; Fogassi, L. Premotor Cortex and the Recognition of Motor Actions. *Cogn. Brain Res.* **1996**, *3*, 131–141. [CrossRef]

222. Fabbri-Destro, M.; Rizzolatti, G. Mirror Neurons and Mirror Systems in Monkeys and Humans. *Physiology* **2008**, *23*, 171–179. [CrossRef] [PubMed]

223. Gallese, V.; Fadiga, L.; Fogassi, L.; Rizzolatti, G. Action Recognition in the Premotor Cortex. *Brain* **1996**, *119*, 593–609. [CrossRef]

224. Rizzolatti, G.; Fadiga, L.; Matelli, M.; Bettinardi, V.; Paulesu, E.; Perani, D.; Fazio, F. Localization of Grasp Representations in Humans by PET: 1. Observation versus Execution. *Exp. Brain Res.* **1996**, *111*, 246–252. [CrossRef] [PubMed]

225. Errante, A.; Fogassi, L. Functional Lateralization of the Mirror Neuron System in Monkey and Humans. *Symmetry* **2021**, *13*, 77. [CrossRef]

226. Molenberghs, P.; Cunnington, R.; Mattingley, J.B. Brain Regions with Mirror Properties: A Meta-Analysis of 125 Human FMRI Studies. *Neurosci. Biobehav. Rev.* **2012**, *36*, 341–349. [CrossRef] [PubMed]

227. Badzakova-Trajkov, G.; Häberling, I.; Roberts, R. Cerebral Asymmetries: Complementary and Independent Processes. *PLoS ONE* **2010**, *5*, e9682. [CrossRef]

228. Häberling, I.S.; Corballis, P.M.; Corballis, M.C. Language, Gesture, and Handedness: Evidence for Independent Lateralized Networks. *Cortex* **2016**, *82*, 72–85. [CrossRef]

229. Pritchett, B.L.; Hoeflin, C.; Koldewyn, K.; Dechter, E.; Fedorenko, E. High-Level Language Processing Regions Are Not Engaged in Action Observation or Imitation. *J. Neurophysiol.* **2018**, *120*, 2555–2570. [CrossRef]

230. Carey, D.P.; Johnstone, L.T. Quantifying Cerebral Asymmetries for Language in Dextrals and Adextrals with Random-Effects Meta Analysis. *Front. Psychol.* **2014**, *5*, 1128. [CrossRef]

231. Johnstone, L.T.; Karlsson, E.M.; Carey, D.P. Left-Handers Are Less Lateralized Than Right-Handers for Both Left and Right Hemispheric Functions. *Cereb. Cortex* **2021**, *31*, 3780–3787. [CrossRef]

232. Woodhead, Z.V.J.; Thompson, P.A.; Karlsson, E.M.; Bishop, D.V.M. An Updated Investigation of the Multidimensional Structure of Language Lateralization in Left- and Right-Handed Adults: A Test–Retest Functional Transcranial Doppler Sonography Study with Six Language Tasks. *R. Soc. Open Sci.* **2021**, *8*, 200696. [CrossRef] [PubMed]

233. Corina, D.P.; Jose-Robertson, L.S.; Guillemin, A.; High, J.; Braun, A.R. Language Lateralization in a Bimanual Language. *J. Cogn. Neurosci.* **2003**, *15*, 718–730. [CrossRef] [PubMed]

234. Emmorey, K.; Mehta, S.; Grabowski, T.J. The Neural Correlates of Sign versus Word Production. *NeuroImage* **2007**, *36*, 202–208. [CrossRef]

235. Gentilucci, M.; Volta, R.D. Spoken Language and Arm Gestures Are Controlled by the Same Motor Control System. *Q. J. Exp. Psychol.* **2008**, *61*, 944–957. [CrossRef] [PubMed]

236. Meguerditchian, A.; Vauclair, J.; Hopkins, W.D. On the Origins of Human Handedness and Language: A Comparative Review of Hand Preferences for Bimanual Coordinated Actions and Gestural Communication in Nonhuman Primates: On the Origins of Human Handedness and Language. *Dev. Psychobiol.* **2013**, *55*, 637–650. [CrossRef] [PubMed]

237. Trettenbrein, P.C.; Papitto, G.; Friederici, A.D.; Zaccarella, E. Functional Neuroanatomy of Language without Speech: An ALE Meta-Analysis of Sign Language. *Hum. Brain Mapp.* **2021**, *42*, 699–712. [CrossRef] [PubMed]

238. Xu, J.; Gannon, P.J.; Emmorey, K.; Smith, J.F.; Braun, A.R. Symbolic Gestures and Spoken Language Are Processed by a Common Neural System. *Proc. Natl. Acad. Sci. USA* **2009**, *106*, 20664–20669. [CrossRef] [PubMed]

239. Bretherton, I.; Bates, E. The Emergence of Intentional Communication. *New Dir. Child Adolesc. Dev.* **1979**, *1979*, 81–100. [CrossRef]

240. Butterworth, G. Pointing Is the Royal Road to Language for Babies. In *Pointing: Where Language, Culture, and Cognition Meet*; Sotaro Kita: Mahwah, NJ, USA, 2003; pp. 9–32.

241. Cochet, H.; Jover, M.; Vauclair, J. Hand Preference for Pointing Gestures and Bimanual Manipulation around the Vocabulary Spurt Period. *J. Exp. Child Psychol.* **2011**, *110*, 393–407. [CrossRef]

242. Esseily, R.; Jacquet, A.-Y.; Fagard, J. Handedness for Grasping Objects and Pointing and the Development of Language in 14-Month-Old Infants. *Laterality* **2011**, *16*, 565–585. [CrossRef]

243. Iverson, J.M.; Goldin-Meadow, S. Gesture Paves the Way for Language Development. *Psychol. Sci.* **2005**, *16*, 367–371. [CrossRef]

244. Nicoladis, E.; Dueck, B.S.; Zarezadehkheibari, S. Hand Preference in Referential Gestures: Relationships to Accessing Words for Speaking in Monolingual and Bilingual Children. *Brain Behav.* **2021**, *11*, e02121. [CrossRef] [PubMed]

245. Vauclair, J.; Cochet, H. La Communication Gestuelle: Une Voie Royale Pour Le Développement Du Langage. *Enfance* **2016**, *2016*, 419–433. [CrossRef]

246. Kimura, D. Manual Activity during Speaking—I. Right-Handers. *Neuropsychologia* **1973**, *11*, 45–50. [CrossRef]

247. Nicoladis, E.; Shirazi, S. Hand Preference in Adults' Referential Gestures during Storytelling: Testing for Effects of Bilingualism, Language Ability, Sex and Age. *Symmetry* **2021**, *13*, 1776. [CrossRef]

248. Lausberg, H.; Kita, S. The Content of the Message Influences the Hand Choice in Co-Speech Gestures and in Gesturing without Speaking. *Brain Lang.* **2003**, *86*, 57–69. [CrossRef]

249. Grossi, G.; Semenza, C.; Corazza, S.; Volterra, V. Hemispheric Specialization for Sign Language. *Neuropsychologia* **1996**, *34*, 737–740. [CrossRef]
250. Cochet, H.; Vauclair, J. Pointing Gestures Produced by Toddlers from 15 to 30 Months: Different Functions, Hand Shapes and Laterality Patterns. *Infant Behav. Dev.* **2010**, *33*, 431–441. [CrossRef] [PubMed]
251. Cochet, H.; Vauclair, J. Hand Preferences in Human Adults: Non-Communicative Actions versus Communicative Gestures. *Cortex* **2012**, *48*, 1017–1026. [CrossRef]
252. Cochet, H.; Vauclair, J. Deictic Gestures and Symbolic Gestures Produced by Adults in an Experimental Context: Hand Shapes and Hand Preferences. *Laterality* **2014**, *19*, 278–301. [CrossRef]
253. Vauclair, J.; Imbault, J. Relationship between Manual Preferences for Object Manipulation and Pointing Gestures in Infants and Toddlers. *Dev. Sci.* **2009**, *12*, 1060–1069. [CrossRef]
254. Bonvillian, J.D.; Richards, H.C.; Dooley, T.T. Early Sign Language Acquisition and the Development of Hand Preference in Young Children. *Brain Lang.* **1997**, *58*, 1–22. [CrossRef]
255. Jacquet, A.-Y.; Esseily, R.; Rider, D.; Fagard, J. Handedness for Grasping Objects and Declarative Pointing: A Longitudinal Study. *Dev. Psychobiol.* **2012**, *54*, 36–46. [CrossRef] [PubMed]
256. Meunier, H.; Vauclair, J.; Fagard, J. Human Infants and Baboons Show the Same Pattern of Handedness for a Communicative Gesture. *PLoS ONE* **2012**, *7*, e33959. [CrossRef]
257. Byrne, R.W.; Cartmill, E.A.; Genty, E.; Graham, K.E.; Hobaiter, C.; Tanner, J.E. Great Ape Gestures: Intentional Communication with a Rich Set of Innate Signals. *Anim. Cogn.* **2017**, *20*, 755–769. [CrossRef]
258. Cartmill, E.A.; Beilock, S.; Goldin-Meadow, S. A Word in the Hand: Action, Gesture and Mental Representation in Humans and Non-Human Primates. *Philos. Trans. R. Soc. Lond. B Biol. Sci.* **2012**, *367*, 129–143. [CrossRef] [PubMed]
259. Gillespie-Lynch, K.; Greenfield, P.M.; Lyn, H.; Savage-Rumbaugh, S. Gestural and Symbolic Development among Apes and Humans: Support for a Multimodal Theory of Language Evolution. *Front. Psychol.* **2014**, *5*, 1228. [CrossRef] [PubMed]
260. Liebal, K.; Waller, B.M.; Burrows, A.M.; Slocombe, K.E. The Morphology of Primate Communication. In *Primate Communication: A Multimodal Approach*; Cambridge University Press: New York, NY, USA, 2014; pp. 31–45. ISBN 978-0-521-19504-1.
261. Liebal, K.; Call, J. The Origins of Non-Human Primates' Manual Gestures. *Philos. Trans. R. Soc. B Biol. Sci.* **2012**, *367*, 118–128. [CrossRef] [PubMed]
262. Liebal, K.; Oña, L. Different Approaches to Meaning in Primate Gestural and Vocal Communication. *Front. Psychol.* **2018**, *9*, 478. [CrossRef] [PubMed]
263. Pika, S.; Liebal, K. *Developments in Primate Gesture Research*; Gesture Studies; John Benjamins Publishing Company: Amsterdam, The Netherlands, 2012; Volume 6, ISBN 978-90-272-2848-2.
264. Redican, W.K. Facial Expressions in Nonhuman Primates. In *Primate Behavior: Developments in Field and Laboratory Research*; Rosenblum, L.A., Ed.; New York Academic Press: New York, NY, USA, 1975; pp. 103–194.
265. Tomasello, M.; Call, J. Thirty Years of Great Ape Gestures. *Anim. Cogn.* **2018**, *22*, 461–469. [CrossRef] [PubMed]
266. Müller, C. Gestures in Human and Nonhuman Primates: Why We Need a Comparative View. *Gesture* **2005**, *5*, 259–283. [CrossRef]
267. Rodrigues, E.D.; Santos, A.J.; Veppo, F.; Pereira, J.; Hobaiter, C. Connecting Primate Gesture to the Evolutionary Roots of Language: A Systematic Review. *Am. J. Primatol.* **2021**, *83*, e23313. [CrossRef] [PubMed]
268. Scott, N.M.; Pika, S. A Call for Conformity. In *Developments in Primate Gesture Research*; Pika, S., Liebal, K., Eds.; John Benjamins Publishing: Amsterdam, The Netherlands, 2012; ISBN 978-90-272-7481-6.
269. Kersken, V.; Gómez, J.-C.; Liszkowski, U.; Soldati, A.; Hobaiter, C. A Gestural Repertoire of 1- to 2-Year-Old Human Children: In Search of the Ape Gestures. *Anim. Cogn.* **2019**, *22*, 577–595. [CrossRef] [PubMed]
270. Bourjade, M.; Cochet, H.; Molesti, S.; Guidetti, M. Is Conceptual Diversity an Advantage for Scientific Inquiry? A Case Study on the Concept of "Gesture" in Comparative Psychology. *Integr. Psychol. Behav. Sci.* **2020**, *54*, 805–832. [CrossRef] [PubMed]
271. Hobaiter, C.; Byrne, R.W. What Is a Gesture? A Meaning-Based Approach to Defining Gestural Repertoires. *Neurosci. Biobehav. Rev.* **2017**, *82*, 3–12. [CrossRef]
272. Pollick, A.S.; Waal, F.B.M.D. Ape Gestures and Language Evolution. *Proc. Natl. Acad. Sci. USA* **2007**, *104*, 8184–8189. [CrossRef]
273. Roberts, A.I.; Vick, S.-J.; Buchanan-Smith, H.M. Usage and Comprehension of Manual Gestures in Wild Chimpanzees. *Anim. Behav.* **2012**, *84*, 459–470. [CrossRef]
274. Roberts, A.I.; Roberts, S.G.B.; Vick, S.-J. The Repertoire and Intentionality of Gestural Communication in Wild Chimpanzees. *Anim. Cogn.* **2014**, *17*, 317–336. [CrossRef]
275. Meguerditchian, A.; Molesti, S.; Vauclair, J. Right-Handedness Predominance in 162 Baboons (Papio anubis) for Gestural Communication: Consistency across Time and Groups. *Behav. Neurosci.* **2011**, *125*, 653–660. [CrossRef]
276. Meguerditchian, A.; Vauclair, J. Baboons Communicate with Their Right Hand. *Behav. Brain Res.* **2006**, *171*, 170–174. [CrossRef]
277. Molesti, S.; Meguerditchian, A.; Bourjade, M. Gestural Communication in Olive Baboons (Papio anubis): Repertoire and Intentionality. *Anim. Cogn.* **2019**, *23*, 19–40. [CrossRef]
278. Gupta, S.; Sinha, A. Gestural Communication of Wild Bonnet Macaques in the Bandipur National Park, Southern India. *Behav. Process.* **2019**, *168*, 103956. [CrossRef] [PubMed]
279. Aychet, J.; Blois-Heulin, C.; Lemasson, A. Sequential and Network Analyses to Describe Multiple Signal Use in Captive Mangabeys. *Anim. Behav.* **2021**, *182*, 203–226. [CrossRef]

280. Aychet, J.; Monchy, N.; Blois-Heulin, C.; Lemasson, A. Context-Dependent Gestural Laterality: A Multifactorial Analysis in Captive Red-Capped Mangabeys. *Animals* **2022**, *12*, 186. [CrossRef]
281. Schel, A.M.; Bono, A.; Aychet, J.; Pika, S.; Lemasson, A. Intentional Gesturing in Red-Capped Mangabeys (Cercocebus Torquatus). Université Paris Nanterre: Nanterre, France, 2022; *under review*.
282. Hopkins, W.D.; Leavens, D.A. Hand Use and Gestural Communication in Chimpanzees (Pan Troglodytes). *J. Comp. Psychol.* **1998**, *112*, 95. [CrossRef] [PubMed]
283. Fletcher, A.W. Clapping in Chimpanzees: Evidence of Exclusive Hand Preference in a Spontaneous, Bimanual Gesture. *Am. J. Primatol.* **2006**, *68*, 1081–1088. [CrossRef] [PubMed]
284. Meguerditchian, A.; Vauclair, J.; Hopkins, W.D. Captive Chimpanzees Use Their Right Hand to Communicate with Each Other: Implications for the Origin of the Cerebral Substrate for Language. *Cortex* **2010**, *46*, 40–48. [CrossRef] [PubMed]
285. Prieur, J.; Pika, S.; Barbu, S.; Blois-Heulin, C. A Multifactorial Investigation of Captive Chimpanzees' Intraspecific Gestural Laterality. *Anim. Behav.* **2016**, *116*, 31–43. [CrossRef]
286. Hopkins, W.D.; Russell, J.L.; Schaeffer, J.A. The Neural and Cognitive Correlates of Aimed Throwing in Chimpanzees: A Magnetic Resonance Image and Behavioural Study on a Unique Form of Social Tool Use. *Phil. Trans. R. Soc. B* **2012**, *367*, 37–47. [CrossRef]
287. Prieur, J.; Barbu, S.; Blois-Heulin, C.; Pika, S. Captive Gorillas' Manual Laterality: The Impact of Gestures, Manipulators and Interaction Specificity. *Brain Lang.* **2017**, *175*, 130–145. [CrossRef]
288. Prieur, J.; Pika, S.; Barbu, S.; Blois-Heulin, C. Gorillas Are Right-Handed for Their Most Frequent Intraspecific Gestures. *Anim. Behav.* **2016**, *118*, 165–170. [CrossRef]
289. Prieur, J.; Pika, S.; Barbu, S.; Blois-Heulin, C. A Multifactorial Investigation of Captive Gorillas' Intraspecific Gestural Laterality. *Laterality Asymmetries Body Brain Cogn.* **2017**, *23*, 538–575. [CrossRef]
290. Meguerditchian, A.; Gardner, M.J.; Schapiro, S.J.; Hopkins, W.D. The Sound of One-Hand Clapping: Handedness and Perisylvian Neural Correlates of a Communicative Gesture in Chimpanzees. *Proc. R. Soc. Lond. B Biol. Sci.* **2012**, *279*, 1959–1966. [CrossRef]
291. Meguerditchian, A.; Vauclair, J. Contrast of Hand Preferences between Communicative Gestures and Non-Communicative Actions in Baboons: Implications for the Origins of Hemispheric Specialization for Language. *Brain Lang.* **2009**, *108*, 167–174. [CrossRef] [PubMed]
292. Meunier, H.; Fizet, J.; Vauclair, J. Tonkean Macaques Communicate with Their Right Hand. *Brain Lang.* **2013**, *126*, 181–187. [CrossRef]
293. Maille, A.; Chapelain, A.; Déruti, L.; Bec, P.; Blois-Heulin, C. Manual Laterality for Pointing Gestures Compared to Grasping Actions in Guenons and Mangabeys. *Anim. Behav.* **2013**, *86*, 705–716. [CrossRef]
294. Bourjade, M.; Meunier, H.; Blois-Heulin, C.; Vauclair, J. Baboons' Hand Preference Resists to Spatial Factors for a Communicative Gesture but Not for a Simple Manipulative Action. *Dev. Psychobiol.* **2013**, *55*, 651–661. [CrossRef] [PubMed]
295. Meunier, H.; Fagard, J.; Maugard, A.; Briseño, M.; Fizet, J.; Canteloup, C.; Defolie, C.; Vauclair, J. Patterns of Hemispheric Specialization for a Communicative Gesture in Different Primate Species: Handedness for Pointing in Primates. *Dev. Psychobiol.* **2013**, *55*, 662–671. [CrossRef] [PubMed]
296. Anderson, J.R.; Kuwahata, H.; Fujita, K. Gaze Alternation during "Pointing" by Squirrel Monkeys (Saimiri Sciureus)? *Anim. Cogn.* **2007**, *10*, 267–271. [CrossRef]
297. Anderson, J.R.; Kuroshima, H.; Hattori, Y.; Fujita, K. Flexibility in the Use of Requesting Gestures in Squirrel Monkeys (Saimiri Sciureus). *Am. J. Primatol.* **2010**, *72*, 707–714. [CrossRef]
298. Defolie, C.; Malassis, R.; Serre, M.; Meunier, H. Tufted Capuchins (Cebus Apella) Adapt Their Communicative Behaviour to Human's Attentional States. *Anim. Cogn.* **2015**, *18*, 747–755. [CrossRef]
299. Hattori, Y.; Kuroshima, H.; Fujita, K. Tufted Capuchin Monkeys (Cebus Apella) Show Understanding of Human Attentional States When Requesting Food Held by a Human. *Anim. Cogn.* **2010**, *13*, 87–92. [CrossRef]
300. Perelman, P.; Johnson, W.E.; Roos, C.; Seuánez, H.N.; Horvath, J.E.; Moreira, M.A.M.; Kessing, B.; Pontius, J.; Roelke, M.; Rumpler, Y.; et al. A Molecular Phylogeny of Living Primates. *PLoS Genet.* **2011**, *7*, e1001342. [CrossRef] [PubMed]
301. Meguerditchian, A. On the Gestural Origins of Language: What Baboons' Gestures and Brain Have Told Us after 15 Years of Research. 2021. hal-03315757. Available online: https://halshs.archives-ouvertes.fr/CERCLE/hal-03315757v1 (accessed on 28 November 2021).
302. Meguerditchian, A.; Vauclair, J. Communicative Signaling, Lateralization and Brain Substrate in Nonhuman Primates: Toward a Gestural or a Multimodal Origin of Language? *Hum. Mente. J. Philos. Stud.* **2014**, *27*, 135–160.
303. Prieur, J.; Lemasson, A.; Barbu, S.; Blois-Heulin, C. Challenges Facing the Study of the Evolutionary Origins of Human Right-Handedness and Language. *Int. J. Primatol.* **2018**, *39*, 183–207. [CrossRef]
304. Cantalupo, C.; Hopkins, W.D. Asymmetric Broca's Area in Great Apes. *Nature* **2001**, *414*, 505. [CrossRef]
305. Hopkins, W.D.; Nir, T.M. Planum Temporale Surface Area and Grey Matter Asymmetries in Chimpanzees (Pan Troglodytes): The Effect of Handedness and Comparison with Findings in Humans. *Behav. Brain Res.* **2010**, *208*, 436–443. [CrossRef]
306. Taglialatela, J.P.; Cantalupo, C.; Hopkins, W.D. Gesture Handedness Predicts Asymmetry in the Chimpanzee Inferior Frontal Gyrus. *Neuroreport* **2006**, *17*, 923–927. [CrossRef]
307. Taglialatela, J.P.; Russell, J.L.; Schaeffer, J.A.; Hopkins, W.D. Communicative Signaling Activates 'Broca's' Homolog in Chimpanzees. *Curr. Biol.* **2008**, *18*, 343–348. [CrossRef]

308. Marie, D.; Roth, M.; Lacoste, R.; Nazarian, B.; Bertello, A.; Anton, J.-L.; Hopkins, W.D.; Margiotoudi, K.; Love, S.A.; Meguerditchian, A. Left Brain Asymmetry of the Planum Temporale in a Nonhominid Primate: Redefining the Origin of Brain Specialization for Language. *Cereb. Cortex* **2018**, *28*, 1808–1815. [CrossRef]
309. Gannon, P.J.; Holloway, R.L.; Broadfield, D.C.; Braun, A.R. Asymmetry of Chimpanzee Planum Temporale: Humanlike Pattern of Wernicke's Brain Language Area Homolog. *Science* **1998**, *279*, 220–222. [CrossRef]
310. Geschwind, N.; Levitsky, W. Human Brain: Left-Right Asymmetries in Temporal Speech Region. *Science* **1968**, *161*, 186–187. [CrossRef]
311. Spocter, M.A.; Sherwood, C.C.; Schapiro, S.J.; Hopkins, W.D. Reproducibility of Leftward Planum Temporale Asymmetries in Two Genetically Isolated Populations of Chimpanzees (Pan Troglodytes). *Proc. R. Soc. B Biol. Sci.* **2020**, *287*, 20201320. [CrossRef]
312. Becker, Y.; Phelipon, R.; Sein, J.; Velly, L.; Renaud, L.; Meguerditchian, A. Planum Temporale Grey Matter Volume Asymmetries in Newborn Monkeys (Papio anubis). *Brain Struct. Funct.* **2021**. [CrossRef]
313. Becker, Y.; Sein, J.; Velly, L.; Giacomino, L.; Renaud, L.; Lacoste, R.; Anton, J.-L.; Nazarian, B.; Berne, C.; Meguerditchian, A. Early Left-Planum Temporale Asymmetry in Newborn Monkeys (Papio anubis): A Longitudinal Structural MRI Study at Two Stages of Development. *NeuroImage* **2021**, *227*, 117575. [CrossRef]
314. Wada, J.A.; Clarke, R.; Hamm, A. Cerebral Hemispheric Asymmetry in Humans: Cortical Speech Zones in 100 Adult and 100 Infant Brains. *Arch. Neurol.* **1975**, *32*, 239–246. [CrossRef]
315. Xia, J.; Wang, F.; Wu, Z.; Wang, L.; Zhang, C.; Shen, D.; Li, G. Mapping Hemispheric Asymmetries of the Macaque Cerebral Cortex during Early Brain Development. *Hum. Brain Mapp.* **2020**, *41*, 95–106. [CrossRef]
316. Becker, Y.; Margiotoudi, K.; Marie, D.; Roth, M.; Nazarian, B.; Lacoste, R.; Anton, J.-L.; Coulon, O.; Claidière, N.; Meguerditchian, A. On the Gestural Origin of Language Lateralisation: Manual Communication Reflects Broca's Asymmetry in Monkeys. *bioRxiv* **2021**. [CrossRef]
317. Liebal, K.; Schneider, C.; Errson-Lembeck, M. How Primates Acquire Their Gestures: Evaluating Current Theories and Evidence. *Anim. Cogn.* **2018**, *22*, 473–486. [CrossRef]
318. Springer, M.S.; Meredith, R.W.; Gatesy, J.; Emerling, C.A.; Park, J.; Rabosky, D.L.; Stadler, T.; Steiner, C.; Ryder, O.A.; Janečka, J.E.; et al. Macroevolutionary Dynamics and Historical Biogeography of Primate Diversification Inferred from a Species Supermatrix. *PLoS ONE* **2012**, *7*, e49521. [CrossRef]
319. Fröhlich, M.; Sievers, C.; Townsend, S.W.; Gruber, T.; van Schaik, C.P. Multimodal Communication and Language Origins: Integrating Gestures and Vocalizations. *Biol. Rev.* **2019**, *94*, 1809–1829. [CrossRef]
320. Gentilucci, M.; Corballis, M.C. From Manual Gesture to Speech: A Gradual Transition. *Neurosci. Biobehav. Rev.* **2006**, *30*, 949–960. [CrossRef]
321. McNeill, D. *How Language Began: Gesture and Speech in Human Evolution*; Cambridge University Press: Cambridge, MA, USA, 2012.
322. Burkart, J.; Martins, E.G.; Miss, F.; Zürcher, Y. From Sharing Food to Sharing Information: Cooperative Breeding and Language Evolution. *Interact. Stud.* **2018**, *19*, 136–150. [CrossRef]
323. Dunbar, R.I.M. The Social Brain: Mind, Language, and Society in Evolutionary Perspective. *Annu. Rev. Anthropol.* **2003**, *32*, 163–181. [CrossRef]
324. Dunbar, R.I.M. The Social Brain Hypothesis and Its Implications for Social Evolution. *Ann. Hum. Biol.* **2009**, *36*, 562–572. [CrossRef]
325. Freeberg, T.M.; Dunbar, R.I.M.; Ord, T.J. Social Complexity as a Proximate and Ultimate Factor in Communicative Complexity. *Phil. Trans. R. Soc. B* **2012**, *367*, 1785–1801. [CrossRef]
326. Kaplan, H.S.; Hooper, P.L.; Gurven, M. The Evolutionary and Ecological Roots of Human Social Organization. *Philos. Trans. R. Soc. B Biol. Sci.* **2009**, *364*, 3289–3299. [CrossRef] [PubMed]
327. Prieur, J.; Pika, S.; Barbu, S.; Blois-Heulin, C. Intraspecific Gestural Laterality in Chimpanzees and Gorillas and the Impact of Social Propensities. *Brain Res.* **2017**, *1670*, 52–67. [CrossRef]
328. Bonnie, K.E.; Bernstein-Kurtycz, L.M.; Shender, M.A.; Ross, S.R.; Hopper, L.M. Foraging in a Social Setting: A Comparative Analysis of Captive Gorillas and Chimpanzees. *Primates* **2019**, *60*, 125–131. [CrossRef] [PubMed]
329. Lemasson, A.; Barbu, S. L'origine Phylogénétique Du Langage: Apports Des Travaux Récents Sur La Communication Vocale Des Cercopithèques. *Faits Lang.* **2011**, *37*, 63–77. [CrossRef]
330. Marler, P. Communication in Monkeys and Apes. In *Primate Behavior—Field Studies of Monkeys and Apes*; DeVore, I., Ed.; Harvard University: Cambridge, MA, USA, 1965; pp. 544–584.
331. Parr, L.; Waller, B.M.; Micheletta, J. Nonverbal Communication in Primates: Observational and Experimental Approaches. In *APA Handbook of Nonverbal Communication*; APA: Washington, DC, USA, 2015; ISBN 978-1-4338-1969-8.
332. McGrew, W.C.; Marchant, L.F. On the Other Hand: Current Issues in and Meta-Analysis of the Behavioral Laterality of Hand Function in Nonhuman Primates. *Am. J. Phys. Anthropol.* **1997**, *104*, 201–232. [CrossRef]
333. McGrew, W.C.; Marchant, L.F. Ethological Study of Manual Laterality in the Chimpanzees of the Mahale Mountains, Tanzania. *Behaviour* **2001**, *138*, 329–358. [CrossRef]
334. Warren, J.M. Handedness and Laterality in Humans and Other Animals. *Physiol. Psychol.* **1980**, *8*, 351–359. [CrossRef]
335. Hobaiter, C.; Byrne, R.W. Laterality in the Gestural Communication of Wild Chimpanzees. *Ann. N. Y. Acad. Sci.* **2013**, *1288*, 9–16. [CrossRef]

336. Hopkins, W.D.; Wesley, M.J. Gestural Communication in Chimpanzees (Pan Troglodytes): The Influence of Experimenter Position on Gesture Type and Hand Preference. *Laterality Asymmetries Body Brain Cogn.* **2002**, *7*, 19–30. [CrossRef]

337. Bisazza, A.; de Santi, A. Lateralization of Aggression in Fish. *Behav. Brain Res.* **2003**, *141*, 131–136. [CrossRef]

338. Deckel, A.W. Laterality of Aggressive Responses in Anolis. *J. Exp. Zool.* **1995**, *272*, 194–200. [CrossRef]

339. Jennings, D.J. Right-Sided Bias in Fallow Deer Terminating Parallel Walks: Evidence for Lateralization during a Lateral Display. *Anim. Behav.* **2012**, *83*, 1427–1432. [CrossRef]

340. Austin, N.P.; Rogers, L.J. Limb Preferences and Lateralization of Aggression, Reactivity and Vigilance in Feral Horses, Equus Caballus. *Anim. Behav.* **2012**, *83*, 239–247. [CrossRef]

341. Austin, N.P.; Rogers, L.J. Lateralization of Agonistic and Vigilance Responses in Przewalski Horses (Equus Przewalskii). *Appl. Anim. Behav. Sci.* **2014**, *151*, 43–50. [CrossRef]

342. Bouchet, H.; Koda, H.; Masataka, N.; Lemasson, A. Vocal Flexibility in Nonhuman Primates and the Origins of Human Language. *Rev. Primatol.* **2016**, *7*. [CrossRef]

343. Baraud, I.; Buytet, B.; Bec, P.; Blois-Heulin, C. Social Laterality and 'Transversality' in Two Species of Mangabeys: Influence of Rank and Implication for Hemispheric Specialization. *Behav. Brain Res.* **2009**, *198*, 449–458. [CrossRef]

344. Boeving, E.R.; Belnap, S.C.; Nelson, E.L. Embraces Are Lateralized in Spider Monkeys (Ateles Fusciceps Rufiventris). *Am. J. Primatol.* **2017**, *79*, e22654. [CrossRef]

345. Forcina, G.; Vallet, D.; Le Gouar, P.J.; Bernardo-Madrid, R.; Illera, G.; Molina-Vacas, G.; Dréano, S.; Revilla, E.; Rodríguez-Teijeiro, J.D.; Ménard, N.; et al. From Groups to Communities in Western Lowland Gorillas. *Proc. R. Soc. B Biol. Sci.* **2019**, *286*, 20182019. [CrossRef]

346. Harcourt, A.H.; Stewart, K.J. Gorilla Society: Conflict, Compromise, and Cooperation between the Sexes. *Mammalia* **2008**, *72*, 139–147.

347. Müller, M.N.; Mitani, J.C. Conflict and Cooperation in Wild Chimpanzees. In *Advances in the Study of Behavior*; Academic Press: Cambridge, MA, USA, 2005; Volume 35, pp. 275–331.

348. Hopkins, W.D.; Russell, J.L.; Cantalupo, C.; Freeman, H.; Schapiro, S.J. Factors Influencing the Prevalence and Handedness for Throwing in Captive Chimpanzees (Pan Troglodytes). *J. Comp. Psychol.* **2005**, *119*, 363–370. [CrossRef]

349. Vervaecke, H.; Vries, H.D.; Elsacker, L.V. An Experimental Evaluation of the Consistency of Competitive Ability and Agonistic Dominance in Different Social Contexts in Captive Bonobos. *Behaviour* **1999**, *136*, 423–442.

350. Hirnstein, M.; Hugdahl, K.; Hausmann, M. Cognitive Sex Differences and Hemispheric Asymmetry: A Critical Review of 40 Years of Research. *Laterality* **2019**, *24*, 204–252. [CrossRef]

351. Papadatou-Pastou, M.; Martin, M.; Munafò, M.R.; Jones, G.V. Sex Differences in Left-Handedness: A Meta-Analysis of 144 Studies. *Psychol. Bull.* **2008**, *134*, 677–699. [CrossRef] [PubMed]

352. Sommer, I.E.; Aleman, A.; Somers, M.; Boks, M.P.; Kahn, R.S. Sex Differences in Handedness, Asymmetry of the Planum Temporale and Functional Language Lateralization. *Brain Res.* **2008**, *1206*, 76–88. [CrossRef]

353. Sommer, I.E.C.; Aleman, A.; Bouma, A.; Kahn, R.S. Do Women Really Have More Bilateral Language Representation than Men? A Meta-Analysis of Functional Imaging Studies. *Brain* **2004**, *127*, 1845–1852. [CrossRef]

354. Laverack, K.; Pike, T.W.; Cooper, J.J.; Frasnelli, E. The Effect of Sex and Age on Paw Use within a Large Sample of Dogs (Canis Familiaris). *Appl. Anim. Behav. Sci.* **2021**, *238*, 105298. [CrossRef]

355. Wells, D.L. Paw Preference as a Tool for Assessing Emotional Functioning and Welfare in Dogs and Cats: A Review. *Appl. Anim. Behav. Sci.* **2021**, *236*, 105148. [CrossRef]

356. Wells, D.L.; Millsopp, S. Lateralized Behaviour in the Domestic Cat, Felis Silvestris Catus. *Anim. Behav.* **2009**, *78*, 537–541. [CrossRef]

357. Jones, G.V.; Martin, M. A Note on Corballis (1997) and the Genetics and Evolution of Handedness: Developing a Unified Distributional Model from the Sex-Chromosomes Gene Hypothesis. *Psychol. Rev.* **2000**, *107*, 213–218. [CrossRef]

358. Priddle, T.H.; Crow, T.J. The Protocadherin 11X/Y (PCDH11X/Y) Gene Pair as Determinant of Cerebral Asymmetry in Modern Homo Sapiens. *Ann. N. Y. Acad. Sci.* **2013**, *1288*, 36–47. [CrossRef] [PubMed]

359. Rogers, L.J. Asymmetry of Brain and Behavior in Animals: Its Development, Function, and Human Relevance. *Genesis* **2014**, *52*, 555–571. [CrossRef] [PubMed]

360. Call, J.; Tomasello, M. (Eds.) *The Gestural Communication of Apes and Monkeys*; Lawrence Erlbaum Associates Publishers: Mahwah, NJ, USA, 2007; ISBN 978-0-8058-6278-2.

361. Hopkins, W.D.; Russell, J.; Freeman, H.; Buehler, N.; Reynolds, E.; Schapiro, S.J. The Distribution and Development of Handedness for Manual Gestures in Captive Chimpanzees (Pan Troglodytes). *Psychol. Sci.* **2005**, *16*, 487–493. [CrossRef]

362. Friedrich, P.; Forkel, S.J.; Amiez, C.; Balsters, J.H.; Coulon, O.; Fan, L.; Goulas, A.; Hadj-Bouziane, F.; Hecht, E.E.; Heuer, K.; et al. Imaging Evolution of the Primate Brain: The next Frontier? *NeuroImage* **2021**, *228*, 117685. [CrossRef]

363. Debracque, C.; Gruber, T.; Lacoste, R.; Grandjean, D.; Meguerditchian, A. Validating the Use of Functional Near-Infrared Spectroscopy in Monkeys: The Case of Brain Activation Lateralization in Papio anubis. *Behav. Brain Res.* **2021**, *403*, 113133. [CrossRef]

364. Ocklenburg, S.; Berretz, G.; Packheiser, J.; Friedrich, P. Laterality 2020: Entering the next Decade. *Laterality* **2021**, *26*, 265–297. [CrossRef]

365. Marchant, L.F.; McGrew, W.C.; Eibl-Eibesfeldt, I. Is Human Handedness Universal? Ethological Analyses from Three Traditional Cultures. *Ethology* **1995**, *101*, 239–258. [CrossRef]
366. Potier, C.; Meguerditchian, A.; Fagard, J. Handedness for Bimanual Coordinated Actions in Infants as a Function of Grip Morphology. *Laterality* **2013**, *18*, 576–593. [CrossRef]
367. Bardo, A.; Pouydebat, E.; Meunier, H. Do Bimanual Coordination, Tool Use, and Body Posture Contribute Equally to Hand Preferences in Bonobos? *J. Hum. Evol.* **2015**, *82*, 159–169. [CrossRef] [PubMed]
368. Laurence, A.; Wallez, C.; Blois-Heulin, C. Task Complexity, Posture, Age, Sex: Which Is the Main Factor Influencing Manual Laterality in Captive Cercocebus Torquatus Torquatus? *Laterality* **2011**, *16*, 586–606. [CrossRef] [PubMed]
369. Pouydebat, E.; Reghem, E.; Gorce, P.; Bels, V. Influence of the Task on Hand Preference: Individual Differences among Gorillas (Gorilla Gorilla Gorilla). *Folia Primatol.* **2010**, *81*, 273–281. [CrossRef] [PubMed]
370. Giljov, A.; Karenina, K.; Malashichev, Y. Forelimb Preferences in Quadrupedal Marsupials and Their Implications for Laterality Evolution in Mammals. *BMC Evol. Biol.* **2013**, *13*, 61. [CrossRef] [PubMed]
371. Stieger, B.; Melotti, L.; Quante, S.M.; Kaiser, S.; Sachser, N.; Richter, S.H. A Step in the Right Direction: The Effect of Context, Strain and Sex on Paw Preference in Mice. *Anim. Behav.* **2021**, *174*, 21–30. [CrossRef]
372. Anzeraey, A.; Aumont, M.; Decamps, T.; Herrel, A.; Pouydebat, E. The Effect of Food Properties on Grasping and Manipulation in the Aquatic Frog Xenopus Laevis. *J. Exp. Biol.* **2017**, *220*, 4486–4491. [CrossRef]

MDPI

St. Alban-Anlage 66

4052 Basel

Switzerland

Tel. +41 61 683 77 34

Fax +41 61 302 89 18

www.mdpi.com

Symmetry Editorial Office

E-mail: symmetry@mdpi.com

www.mdpi.com/journal/symmetry